Studies
in the History of Mathematics and
Physical Sciences

8

Ernst Zermelo as a young man
(courtesy of Mrs. Gertrud Zermelo)

Gregory H. Moore

Zermelo's Axiom of Choice

Its Origins, Development, and Influence

Springer-Verlag
New York Heidelberg Berlin

Gregory H. Moore
Department of Mathematics and
Institute for History and Philosophy
of Science and Technology
University of Toronto
Toronto, Ontario M5S 1A1
Canada

AMS Subject Classifications (1980): 01A55, 01A60, 03-03, 03E25, 04-03, 04A25

Library of Congress Cataloging in Publication Data

Moore, Gregory H.
Zermelo's axiom of choice.
(Studies in the history of mathematics and physical sciences; 8)
Bibliography: p.
Includes index.
1. Axiom of choice. 2. Zermelo, Ernst, 1871–1953.
I. Title. II. Series.
QA248.M59 1982 511.3 82-10429

With 1 Illustration

Typeset by Composition House Ltd., Salisbury, England.
Printed and bound by R. R. Donnelley & Sons, Harrisonburg, VA.
Printed in the United States of America.

9 8 7 6 5 4 3 2 1

ISBN 0-387-90670-3 Springer-Verlag New York Heidelberg Berlin
ISBN 3-540-90670-3 Springer-Verlag Berlin Heidelberg New York

To
Leslie G. Dawson

We shall not cease from exploration
And the end of all our exploring
Will be to arrive where we started
And know the place for the first time.

T. S. Eliot
"Four Quartets"
4.5, 26–29

Preface

This book grew out of my interest in what is common to three disciplines: mathematics, philosophy, and history. The origins of Zermelo's Axiom of Choice, as well as the controversy that it engendered, certainly lie in that intersection. Since the time of Aristotle, mathematics has been concerned alternately with its assumptions and with the objects, such as number and space, about which those assumptions were made. In the historical context of Zermelo's Axiom, I have explored both the vagaries and the fertility of this alternating concern. Though Zermelo's research has provided the focus for this book, much of it is devoted to the problems from which his work originated and to the later developments which, directly or indirectly, he inspired.

A few remarks about format are in order. In this book a publication is indicated by a date after a name; so Hilbert 1926, 178 refers to page 178 of an article written by Hilbert, published in 1926, and listed in the bibliography. A few anonymous articles are cited by journal, according to the list of abbreviations before the bibliography. Sections of chapters have one decimal point, while important propositions have two. Thus 2.5 refers to the fifth section of Chapter 2, and 3.6.4 to the fourth proposition indented in section 6 of Chapter 3. If a theorem or axiom is stated exactly as it was originally given by its author, it occurs in quotation marks. Otherwise, while I have endeavored to remain close to an author's phrasing, some terminology or notation may have been altered for readability. Unless specifically noted, any translation from a foreign language is my own. Finally, certain symbols that occur repeatedly in the text deserve mention: the set $\mathbb{N}$ of all natural numbers $0, 1, 2, \ldots$; the set $\mathbb{R}$ of all real numbers; the set $\mathbb{R}^n$ of all points in

n-dimensional Euclidean space; and the least infinite ordinal ω. Natural numbers are designated by i, j, k, m, n, ordinals by lower-case Greek letters such as α, β, γ, and cardinals by lower-case German letters such as $\mathfrak{m}$, $\mathfrak{n}$, $\mathfrak{p}$, and $\mathfrak{q}$.

Burlingame, California
10 January 1982

GREGORY H. MOORE

Acknowledgments

Gratitude, as the poet William Blake once wrote, is heaven itself. In this spirit it is a pleasure to record the many archival and intellectual debts that I have accumulated while writing this book. I am grateful to the Master and Fellows of Trinity College, Cambridge University, for permission to publish letters from G.H. Hardy to Bertrand Russell. Likewise, I am indebted to the *Bibliothèque* at Chaux-de-Fonds in Switzerland for the right to quote from unpublished correspondence between Louis Couturat and Russell. I also wish to thank the *Universitäts-Bibliotek* of the University of Freiburg im Breisgau for permitting quotations from Zermelo's *Nachlass*. To Kenneth Blackwell and Carl Spadoni of the Russell Archives (McMaster University, Hamilton, Canada) I owe a special debt of gratitude for archival assistance with Russell's letters and manuscripts; the copyright on the letters from Russell to both Couturat and Hardy, quoted with permission, rests with Res.-Lib. Ltd., McMaster University. At the University of Toronto many reference librarians gave unstintingly of their time to aid me in obtaining materials locally or via interlibrary loan. Finally, I am particularly grateful to the *Société Mathématique de France* for permission to publish my translation of the "Cinq lettres" of Baire, Borel, Hadamard, and Lebesgue (see Appendix 1).

My intellectual debts can be acknowledged with even greater pleasure. A philosopher (Bas van Fraassen), a number of mathematicians (Solomon Feferman, Thomas Jech, Azriel Levy, Jan Mycielski, Franklin Tall), and several historians of mathematics (Joseph Dauben, Ivor Grattan-Guinness, Thomas Hawkins, Calvin Jongsma, Barbara Moss) read and commented on various drafts of this book. In particular, Thomas Jech patiently answered innumerable technical questions about the Axiom of Choice and its consequences. The bibliography has benefited from suggestions by J. M. B. Moss, and the translation in Appendix 1 from comments by Jean van Heijenoort. I am indebted to John Addison, Leon Henkin, Robert Solovay, and Alfred Tarski for conversations on particular historical points. After such generous assistance from so many quarters, any remaining errors are mine.

I owe my chief intellectual debt to the late Kenneth O. May, who first inspired me with the potentialities inherent in the history of mathematics. When I came to Toronto to study with him in 1970, he turned my interest in set theory to Zermelo's life and work. Out of the first papers that I wrote for him, this book eventually grew. While I deeply regret that he did not live to see its completion, there is hardly a page that does not bear the imprint of his insight and enthusiasm.

Concerning the actual production of the book, I would like to thank the staff of Springer-Verlag New York Inc. for their unflagging patience and cooperation. Leslie Ovens typed the manuscript, thereby turning innumerable revisions into readable copy. I was generously aided in reading galley proof by Joseph Dauben, Richard Epstein, Raymond Lim, and my father, W. Harvey Moore.

Finally, on a more personal level, I wish to thank my parents, Dr. and Mrs. W. Harvey Moore, and my friends, Eva Silberberg and Marylin Kingston, for their continuing encouragement.

Part of my research was supported by a University Research Fellowship from the Natural Sciences and Engineering Research Council of Canada.

Contents

List of Symbols

ת	The collection of all alephs 52
$^*\omega$	The order-type of ($\mathbb{N}$, $>$) 62
$m(A)$	The measure of the set A 69
A^x	The multiplicative class of the set A 73
$\prod_{x \in A} \overline{\overline{B}}_x$	The product of all cardinals $\overline{\overline{B}}_x$ such that x is in A 74
γ	Zermelo's covering, or choice function 90
L_γ	The set of all Zermelo's γ-sets 91
F_σ	The set of all countable unions of closed sets 103
$F_{\sigma\delta}$	The set of all countable intersections of sets in F_σ 103
$F_{\sigma\delta\sigma}$	The set of all countable unions of sets in $F_{\sigma\delta}$ 103
Δ_4^0	The intersection of $F_{\sigma\delta\sigma}$ and $G_{\delta\sigma\delta}$, which is the family of complements of sets in $F_{\sigma\delta\sigma}$ 103
W	The "set" of all ordinals (*cf.* p. 52) 110
η	The order-type of the rational numbers with $<$ 115
ω_1	The least uncountable ordinal (*cf.* p. 34) 126
θ-chain	Zermelo's generalization of Dedekind's chain 144
$\mathbf{C}_{\mathfrak{n}}^{\mathfrak{m}}$	The Axiom of Choice restricted to disjoint families of $\mathfrak{m}$ sets, each having exactly $\mathfrak{n}$ elements 201
$\mathbf{C}_{\leq\mathfrak{n}}^{\mathfrak{m}}$	The Axiom restricted to disjoint families of $\mathfrak{m}$ sets, each having at most $\mathfrak{n}$ elements 201
$\mathbf{C}^{\mathfrak{m}}$	The Axiom restricted to disjoint families of $\mathfrak{m}$ sets 201
$\mathbf{C}_{\mathfrak{n}}$	The Axiom restricted to disjoint families whose members have exactly $\mathfrak{n}$ elements 201
$\aleph(\mathfrak{m})$	The aleph which is neither greater nor less than $\mathfrak{m}$ 215
$\mathfrak{m} \leq {}^*\mathfrak{n}$	A partial order, based on surjections, on the cardinal numbers 216
CH($\mathfrak{m}$)	The proposition about $\mathfrak{m}$ that there is no $\mathfrak{n}$ such that $\mathfrak{m} < \mathfrak{n} < 2^{\mathfrak{m}}$ 217
$\beta(S)$	The Stone–Čech compactification of the space S 239
Z_2	Cantor's second number-class 246
W_2	The least class of ordinals containing ω and closed under successor and limit of a sequence 247
Ω	The ordinal of W_2 (*cf.* pp. 34, 52, 281) 248
τ	A transfinite choice function 253
ε	A universal choice function, given by the ε-axiom 255
$\mathcal{P}(A)$	The power set of A 263
Ω	The least strongly inaccessible ordinal (*cf.* pp. 34, 52, 248) 281
ω_α	The ordinal of ($\aleph_\alpha$, $<$) 299
$\sum_n^1, \prod_n^1$	The *n*th levels of the arithmetical hierarchy 305

Prologue

David Hilbert once wrote that Zermelo's Axiom of Choice was the axiom "most attacked up to the present in the mathematical literature..." [1926, 178].[1] To this, Abraham Fraenkel later added that "the axiom of choice is probably the most interesting and, in spite of its late appearance, the most discussed axiom of mathematics, second only to Euclid's axiom of parallels which was introduced more than two thousand years ago" [Fraenkel and Bar-Hillel 1958, 56–57]. Rarely have the practitioners of mathematics, a discipline known for the certainty of its conclusions, differed so vehemently over one of its central premises as they have done over the Axiom of Choice. Yet without the Axiom, mathematics today would be quite different.[2] The very nature of modern mathematics would be altered and, if the Axiom's most severe constructivist critics prevailed, mathematics would be reduced to a collection of algorithms. Indeed, the Axiom epitomizes the fundamental changes—mathematical, philosophical, and psychological—that took place when mathematicians seriously began to study infinite collections of sets.

The Axiom of Choice asserts that for every set S there is a function f which associates each non-empty subset A of S with a unique member $f(A)$ of A. From a psychological perspective, one might express the Axiom by saying that an element is "chosen" from each subset A of S. However, if S is infinite, it is difficult to conceive how to make such choices—unless a rule is available to specify an element in each A. The Axiom justifies the existence of a choice

[1] This and all other cited articles by Hilbert can be found in his *Gesammelte Abhandlungen* (Hilbert 1932–1935).

[2] When the term "the Axiom" occurs hereafter, it always refers to the Axiom of Choice.

function f precisely when such a rule is lacking. As a result, the Axiom was vigorously rejected by those mathematicians who identified the existence of a mathematical object with its construction by a rule. Despite this initial widespread distrust, today the vast majority of mathematicians accept the Axiom without hesitation and utilize it in algebra, analysis, logic, set theory, and topology. Furthermore, many propositions have turned out to be equivalent to the Axiom—including Zorn's Lemma, the Löwenheim–Skolem–Tarski Theorem, Tychonoff's Compactness Theorem, and the Trichotomy of Cardinals.

While the Axiom's roots lie in the infinite processes so prominent in nineteenth-century analysis, this assumption was not explicitly stated until a new discipline emerged from analysis: Cantorian set theory. The decisive question, here termed the Well-Ordering Problem, was this: Can every set be well-ordered?[3] Strictly speaking, this Well-Ordering Problem did not originate as a problem at all, but as an assumption. In 1883, Georg Cantor proposed as a valid "law of thought" that every set can be well-ordered.[4] Yet this law of thought, hereafter termed the Well-Ordering Principle, was not accepted by his contemporaries, and a decade later Cantor himself was seeking a proof. It was in order to obtain such a proof that in 1904 Ernst Zermelo first formulated the Axiom of Choice.[5] In this manner, the Axiom passed from unconscious to conscious use and, for many mathematicians at the time, to conscious avoidance. Indeed, Zermelo's solution to the Well-Ordering Problem generated a heated debate, which raged among mathematicians all over Europe, as to whether his Axiom and proof were correct. This controversy soon led him to publish, on the one hand, a spirited defense of his proof and, on the other, his axiomatization for set theory.[6] Within this axiomatization, he believed, both his Axiom and proof were securely embedded. His critics did not agree.

Although it was Zermelo's solution to the Well-Ordering Problem that brought the Axiom of Choice to light, the use of infinitely many arbitrary choices had been growing independently during the last third of the nineteenth century. While some such uses did not need the Axiom, many others did. Yet there remained no awareness that a new mathematical principle was required. In the aftermath of Zermelo's proof, mathematicians actively sought to determine which theorems had unwittingly relied on the Axiom.

---

[3] A set A can be *well-ordered* if there exists a relation S (where aSb may be read as "a is less than b") which orders A and if each non-empty subset B of A contains an element b such that bSc for all other c in B. Such an element b is called the *least element* of B. The relation S *orders* A if for every a, b, c in A, (i) if aSb and bSc, then aSc, and (ii) aSb or $a = b$ or bSa, and (iii) not aSa.

[4] Cantor 1883a, 550. This and all other cited articles by Cantor can be found in his *Gesammelte Abhandlungen* (Cantor 1932).

[5] Zermelo 1904, translated in van Heijenoort 1967.

[6] Zermelo 1908 and 1908a; both are translated in van Heijenoort 1967.

When Zermelo defended the Axiom of Choice by axiomatizing set theory in 1908, a number of his critics attacked that axiomatization. The next decade saw the Axiom explored by various mathematicians in algebra and analysis, but rarely exploited to the full. The situation was profoundly altered when Wacław Sierpiński founded the Warsaw school of mathematics soon after Poland's reunification in 1918. For in Sierpiński, the Axiom gained an advocate who investigated its relationship to many branches of mathematics and who successfully encouraged his students to do likewise. Meanwhile in Germany, Abraham Fraenkel began to study models of Zermelo's axiomatization and to establish the independence of the Axiom, provided that infinitely many urelements were allowed.[7] In 1938 Kurt Gödel obtained deep results about models of set theory by proving the relative consistency of both the Axiom of Choice and the Generalized Continuum Hypothesis. In particular he showed that every model of the usual postulates for set theory, but not necessarily of the Axiom, has a submodel in which both the Axiom of Choice and the Generalized Continuum Hypothesis are true. For all but the most radical constructivists, Gödel's result dispelled any suspicions that the Axiom might lead to a contradiction.

During the quarter-century after Gödel's work, the Axiom was fruitfully applied in diverse fields of mathematics, and many kinds of propositions were shown to be equivalent to it.[8] Moreover, mathematicians studied various propositions which depended on the Axiom but which were suspected of being weaker than it, such as the Prime Ideal Theorem for Boolean algebras. Nevertheless, there was no known way to determine completely the relative strength of these propositions when they involved real numbers, such as the existence of a non-measurable set. This defect in the Fraenkel–Mostowski method of independence proofs was remedied by Paul Cohen's method of forcing, which in 1963 enabled him to demonstrate the independence of the Axiom for systems of set theory lacking urelements. At the same time he established the independence of the Continuum Hypothesis in first-order logic. Set theorists began at once to exploit Cohen's technique and to reformulate it in a general setting. The next two decades produced a cornucopia of independence results involving the Axiom as well as other assumptions. Indeed, the primary focus of research no longer lay within set theory proper, but increasingly centered on *models* of set theory.

However, one consequence of Cohen's method was not altogether welcome. As such semantic investigations of set theory proliferated, the Axiom of Choice and other axioms of set theory (beyond those of Zermelo–Fraenkel) came to be regarded more like the axioms for a group, which has many models, than like the postulates for a categorical system such as the natural numbers.

---

[7] An *urelement* (also called an individual or an atom) was an object which contained no elements, belonged to some set, and yet was not identical with the empty set.

[8] See Appendix 2 as well as Rubin and Rubin 1963.

Since set theory had provided a foundation for mathematics, the unity of mathematics was threatened. In 1967 Andrzej Mostowski commented on this matter:

> Such [post-Cohen independence] results show that axiomatic set theory is hopelessly incomplete . . . . Of course if there are a multitude of set-theories, then none of them can claim the central place in mathematics. Only their common part could claim such a position; but it is debatable whether this common part will contain all the axioms needed for a reduction of mathematics to set theory. [1967, 93–95]

A decade later Jean Dieudonné added:

> The proof by Gödel and P. Cohen that the Axiom of Choice and the Continuum Hypothesis are undecidable, and the numerous metamathematical works which resulted, have greatly changed the views of many mathematicians . . . . Beyond classical analysis (based on the Zermelo–Fraenkel axioms supplemented by the *Denumerable* Axiom of Choice), there is an *infinity* of different possible mathematics, and for the time being no definitive reason compels us to choose one of them rather than another. [1976, 11]

It remains to be seen whether the Axiom of Constructibility, Martin's Axiom, or some large cardinal axiom will eventually restore the unity of set theory and, with it, the foundations of mathematics.

Chapter 1

The Prehistory of the Axiom of Choice

> But as one cannot apply infinitely many times an *arbitrary* rule by which one assigns to a class A an individual of this class, a *determinate* rule is stated here.
>
> Giuseppe Peano [1890, 210]

Throughout its historical development, mathematics has oscillated between studying its assumptions and studying the objects about which those assumptions were made. After the introduction of new mathematical objects, it often happened that the assumptions underlying them remained unspecified for a considerable time; only through extensive use did such assumptions become sufficiently clear to receive an explicit formulation. Usually a body of theorems, consequences of an assumption, were obtained before the assumption itself came to be recognized. At times, indeed, an assumption was specified precisely in order to secure a particular theorem or theorems. Of course, such an assumption ordinarily formed part of a nexus of suppositions with varying degrees of explicitness. What, one may ask, has caused such an assumption to become conscious and explicit? The question grows more complex as soon as we recognize that there was rarely, if ever, a single way of expressing an assumption and that various weakenings or strengthenings of an assumption could serve different mathematical purposes. This preliminary chapter explores how the use of arbitrary choices led, over most of a century, to Zermelo's explicit formulation of the Axiom of Choice.

1.1 Introduction

In 1908 Zermelo proposed a version of the Axiom of Choice that is useful for describing weaker assumptions:

(1.1.1) Given any family T of non-empty sets, there is a function f which assigns to each member A of T an element $f(A)$ of A. [1908a, 274]

Such an f is now called a choice function for T. If (1.1.1) is limited to those families T of a particular cardinality, one obtains a restricted form of the Axiom. Since for any finite T the Axiom is provable, the weakest non-trivial case occurs when T is denumerable.[1] This case is known as the Denumerable Axiom of Choice, and is abbreviated hereafter as the Denumerable Axiom.[2]

At that time Zermelo regarded the Axiom of Choice as codifying an assumption that numerous mathematicians had already made implicitly [1908, 113]. One may then inquire how this assumption developed from earlier mathematics and through what stages it passed on its way to Zermelo's explicit formulation. To answer this question, we need a precise criterion for deciding what constituted an implicit use of the Axiom. The nature of such a use was different during the period before Zermelo's formulation than it was afterward. For this reason the author has termed such an implicit use prior to September 1904 as an "implicit use of the Assumption." Likewise, such an implicit use of the Denumerable Axiom will be called an implicit use of the Denumerable Assumption. After 1904, mathematicians were often conscious that making infinitely many arbitrary choices brought the Axiom into play. Before that date, a mathematician who made such choices seldom recognized that he had done anything unusual or questionable (see 1.2–1.7). Apparently the only exceptions were three Italian mathematicians who began to avoid such choices intentionally during the period 1890–1902 (see 1.8). The earliest of these was Peano, whose views were quoted at the beginning of this chapter.

The hallmark of an implicit use of the Assumption was the occurrence of infinitely many arbitrary choices. Sometimes these choices occurred quite explicitly, sometimes less so. Occasionally a mathematician would employ a proposition which had been proved, by himself or another, using an infinity of arbitrary choices and for which no other proof was then known. Although that mathematician may not have recognized that he had made such arbitrary selections indirectly, we have regarded this case too as an implicit use of the Assumption.

Some uses of the Axiom will be termed avoidable, others unavoidable. Briefly, an avoidable use in a proof was such that the given proof could be modified, with the techniques then available, to specify uniquely whatever the Axiom had been used to select. Sometimes, while the proof could not be modified in this manner, another demonstration of the same theorem was later found which did not rely on such choices. Whenever such is the case, it is noted in the text. Of course, the fact that an implicit use of the Assumption is avoidable does not imply in any way that the mathematician involved should have revised his proof.

An unavoidable use of the Axiom was one such that the proposition in

---

[1] A set is *denumerable* if it can be mapped one-one onto the set $\mathbb{N}$ of natural numbers.

[2] The Denumerable Axiom of Choice has also been called the Countable Axiom of Choice (see Jech 1973, 20). However, we distinguish sharply between a denumerable set and one that is *countable*, *i.e.*, finite or denumerable.

question could not be proved in Zermelo–Fraenkel set theory without urelements (hereafter called **ZF**), or in Zermelo–Fraenkel set theory with urelements (hereafter termed **ZFU**), but could be deduced in **ZF** supplemented by the Axiom of Choice (now known as **ZFC**). At times we refer to an unavoidable use by saying that a proposition P needs or requires the Axiom, though strictly speaking this is an *abus de langage*. When it is now known that a weakened form of the Axiom (such as the Denumerable Axiom) suffices to prove P, this fact appears in the text. To say that P is equivalent to the Axiom means that such an equivalence is provable in **ZF**.

It is possible to define the term "implicit use of the Axiom" much more narrowly than we have done. One might regard the Axiom as used implicitly to prove a proposition P if and only if P is equivalent to the Axiom. For historical purposes, such a definition would be too restrictive, since it would exclude pre-1904 uses which mathematicians described soon afterward as implicit. Moreover, such a definition would distort the historical perspective in a second way. For it would strongly suggest that an implicit use of the Axiom was an implicit use of any other proposition P equivalent to the Axiom (such as Tychonoff's Theorem), even though P was only conceived decades later and did not appear in any direct way in that implicit use of the Axiom.

To make an infinity of arbitrary choices the hallmark of an implicit use of the Assumption, as has been done here, does not eliminate all difficulties. Although the Axiom justifies infinitely many arbitrary choices so long as they are independent of each other, mathematicians sometimes made an infinity of arbitrary selections such that a given choice depended on those previously made. This was the case, for example, in the early attempts to well-order an infinite set M by picking one element after another from M (see 1.5). Shortly after 1904, some mathematicians were satisfied with the Axiom partly because it avoided infinitely many such dependent choices. Later it was recognized that the Axiom could justify dependent choices as well.

Most commonly, dependent choices occurred when a mathematician selected a sequence $a_1, a_2, \ldots$ such that the choice of a_{n+1} depended on a_n. In general, such choices cannot be made by using the Denumerable Axiom, even on the real line.[3] However, the existence of such a sequence can be justified by a proposition which follows from the Axiom of Choice and which Paul Bernays proposed in 1942 as a weakened form of the Axiom useful in analysis. This form is now known as the Principle of Dependent Choices:

(1.1.2) If S is a relation on a set A such that for every x in A there exists some y in A with xSy, then there is a sequence $a_1, a_2, \ldots$ such that for every positive integer n, a_n is in A and a_nSa_{n+1} holds.[4]

---

[3] Jensen 1966, 294.

[4] Bernays 1942, 86. The name of this principle is due to Tarski [1948, 96]. Azriel Levy [1964, 136] generalized it from sequences of type ω to sequences of type α for any infinite ordinal α.

Whenever a mathematician obtained a sequence prior to 1904 by making such arbitrary dependent choices, it will be termed an implicit use of the Dependent Assumption.

To derive propositions which follow from the Axiom but are weaker than it, one might seek an alternative assumption. Such an alternative could be a restricted form of the Axiom, say the Denumerable Axiom or the Principle of Dependent Choices, but might be another type of proposition altogether. The serious investigation of such alternatives, which began in 1918, did not attract much attention at the time except from the Axiom's Italian critics (see 4.7). A much later and more promising alternative was the Axiom of Determinateness (see the Epilogue).

On the other hand, one must not overlook the radical attempt by L. E. J. Brouwer and his followers, beginning in 1907, to reformulate all of mathematics in intuitionistic terms.[5] Roused by a desire to prune Cantor's infinite sets, the intuitionists formulated a cohesive ideology within a more general constructivist framework. Such constructivists were suspicious of the actual infinite (or at least of Cantor's uncountable cardinals), of existence proofs which did not exhibit a uniquely defined object, and of the Principle of the Excluded Middle as applied to infinite sets. The twentieth century has produced a cornucopia of constructivist approaches to mathematics, each of which requires that one abandon a substantial portion of Cantor's results. Although it would take us too far afield to analyze such approaches in general, constructivist opposition to the Axiom will emerge as a central theme (see 2.3–2.4, 2.7–2.9, 3.6, and 4.11).

Within the Cantorian tradition, one can view Zermelo's axiomatization as answering the question: What is a set? This question has served as a theme in the development of set theory, but one not often discussed openly. Nevertheless, it has been implicit in all attempts since 1908 to modify Zermelo's axioms, or to replace them with others (see 4.9), and in the polemics of his French constructivist critics (see 2.3 and 2.4). In 1913 Michele Cipolla inquired whether the Axiom restricted the general concept of set too greatly [1913, 2]. While no one pursued the matter further at the time, this perspective contrasted sharply with other constructivist views that the Axiom was either false or meaningless.

Since the Axiom appears subtly in many proofs and since it is easily overlooked, we now examine some fundamental theorems for which arguments were given before Zermelo that involved its implicit and unavoidable use. Prior to 1904 such implicit uses occurred in real analysis, algebraic number theory, point-set topology, and set theory.[6] Afterwards the controversy

---

[5] See for example Kleene and Vesley 1965, Heyting 1966, Bishop 1967, and Gauthier 1977.

[6] In the mathematical literature the term *point-set topology* is ambiguous. Within this book it refers to the theory, sometimes called *Mannigfaltigkeitslehre* or the theory of point-sets, that Cantor originated and that utilized such notions as limit point, closed set, and derived set in $\mathbb{R}^n$. Largely through the work of Hausdorff, this theory later grew into general topology, which considered more general spaces.

surrounding the Axiom led to research, which continues today, as to whether a theorem requires the Axiom and in how strong a form. Thus the Axiom increased awareness of what one may call, in an informal sense, degrees of non-constructivity. Such research would not have begun, had not many mathematicians regarded the Axiom as a dubious assumption.

The first proposition to be considered, hereafter termed the Countable Union Theorem, was employed in both set theory and analysis, beginning with Cantor's researches in the 1870s:

(1.1.3) The union of a countable family of countable sets is countable.

To understand the Axiom's role in demonstrating the Countable Union Theorem, suppose first that each of the sets $A_1, A_2, \ldots$ is denumerable. It follows that each A_i has as its members $a_{i,1}, a_{i,2}, \ldots$. Thus the union B of all the A_i consists of the elements $a_{i,j}$ where i and j are positive integers. Clearly B is denumerable by Cantor's argument showing the rational numbers to be denumerable. Hence a countable union of countable sets is a subset of a denumerable set, and consequently is countable.

Although the Axiom may not be visible here at first, it entered when we enumerated all the members of all the A_i. There are infinitely many A_i, for each of which there exist many possible bijections onto the set of positive integers.[7] By the Denumerable Axiom we associate with each A_i a unique such bijection a_i. Hence the $a_i(j)$, or $a_{i,j}$, are well-defined. This use of the Denumerable Axiom is unavoidable. Indeed, there exists a model of **ZF** in which the Denumerable Axiom is false and the set $\mathbb{R}$ of all real numbers, though uncountable, is a countable union of countable sets.[8]

Our second example concerns the border between the finite and the infinite, a border that can be very nebulous in the absence of the Axiom:

(1.1.4) Every infinite set A has a denumerable subset.[9]

Cantor in 1895, Borel in 1898, and Russell in 1902 all demonstrated this theorem by using the Denumerable Assumption implicitly.[10] Russell's proof illustrates how the Axiom is involved. Since A is infinite, there exist subsets $A_1, A_2, \ldots$ of A such that, for each n, A_n has exactly n members and is a subset of A_{n+1}. The desired denumerable subset of A is the union of all A_n. Apparently Russell did not notice at the time that to form A_{n+1} once A_n has been obtained, one must select some member of $A - A_n$. Since denumerably many such

[7] A *bijection* from a set A to a set B is a one–one function from A onto B.

[8] Feferman and Levy 1963; Cohen 1966, 143–146.

[9] A set A is *finite* if A is empty or if, for some positive integer n, there is a bijection from A onto $\{1, 2, \ldots, n\}$; otherwise A is *infinite*. For a discussion of the Axiom's role in showing various definitions of finite set to be equivalent, see 1.3 and 4.2.

[10] Cantor 1895, 493; Borel 1898, 12–14; Whitehead 1902, 121–123. As Whitehead acknowledged, Russell wrote the section in which this proof appeared.

choices are needed and since no rule is available in general, the proof requires the Denumerable Axiom. For there is a model of **ZF** in which a certain infinite set of real numbers lacks a denumerable subset.[11]

The third example is a proposition, hereafter called the Partition Principle, which appears all but self-evident:

(1.1.5) If a set M is partitioned into a family S of disjoint non-empty sets, then S is equipollent to a subset of M (*i.e.*, $\overline{\overline{S}} \leq \overline{\overline{M}}$).[12]

In terms of a function f with domain M, (1.1.5) states that $\overline{\overline{f''M}} \leq \overline{\overline{M}}$. During the 1880s Cantor employed a special case of the Partition Principle while investigating the topology of the real line.[13] However, the explicit (though incomplete) formulation of the Partition Principle was due to Cesare Burali-Forti [1896, 46]; see 1.3, 1.4, and 1.8. In general, the proof of this principle depends on selecting an element from each set in S so as to obtain a bijection from S onto a subset of M. Thus the proof relies on the Axiom. Moreover, there exists a model of **ZF** in which the Partition Principle is false.[14] At present it is not known whether the Partition Principle is weaker than the Axiom or equivalent to it.[15]

Lastly, we consider the Trichotomy of Cardinals, a theorem closely related to the proposition that every set can be well-ordered:

(1.1.6) For every cardinal $\mathfrak{m}$ and $\mathfrak{n}$, either $\mathfrak{m} < \mathfrak{n}$ or $\mathfrak{m} = \mathfrak{n}$ or $\mathfrak{m} > \mathfrak{n}$.[16]

When formulated in terms of sets rather than cardinals, (1.1.6) states that any two sets A and B are comparable, *i.e.*, one of them is equipollent to a subset of the other. In 1895 Cantor asserted the Trichotomy of Cardinals without proof. Some four years later he wrote to Dedekind that (1.1.6) followed from the proposition that every set can be well-ordered.[17] Their equivalence remained unproven until Friedrich Hartogs established it in 1915.

---

[11] Cohen 1966, 138. Neither (1.1.3) nor (1.1.4) implies the other in **ZF**. Sageev [1975] has shown that there is a proposition, namely (1.7.10), which yields (1.1.4) but not (1.1.3) in **ZF**. On the other hand, in a personal communication to the author, Jech observed that (1.1.3) but not (1.1.4) holds in what he called the basic Cohen model; *cf.* Jech 1973, 66, 81.

[12] Two sets are *equipollent* if there is a bijection from one of them onto the other.

[13] Cantor 1883b, 413–414; 1884, 464. Now $f''M$ is $\{f(x): x \in M\}$, *i.e.*, the *image* of M under f.

[14] Jech and Sochor 1966, 352.

[15] Certainly the Partition Principle implies many special cases of the Axiom. Sierpiński [1947b, 157] established that the existence of a non-measurable set follows from this principle, while Andrzej Pelc [1978, 587–588] showed—using work by David Pincus—that it yields both the Principle of Dependent Choices and the Axiom restricted to well-ordered families of sets.

[16] $\mathfrak{m} < \mathfrak{n}$ if, for every set A and B of power $\mathfrak{m}$ and $\mathfrak{n}$ respectively, A is equipollent to a subset of B but not to B itself.

[17] Letters of 28 July and 3 August 1899 in Cantor 1932, 443–447.

These four themes—the Countable Union Theorem, the Partition Principle, the border between the finite and the infinite, and finally the intertwined problems of the Well-Ordering Principle and the Trichotomy of Cardinals—form the warp upon which the woof of the Axiom and its early history will be woven. On the other hand, as the next section illustrates, the first two unavoidable implicit uses of the Axiom arose from quite different sources.

1.2 The Origins of the Assumption

After such preliminary remarks, we can indicate the major stages through which the use of arbitrary choices passed on the way to Zermelo's explicit formulation of the Axiom. In particular the outlines of four stages, though not always their precise historical boundaries, are visible. Vestiges of the first stage—choosing an unspecified element from a single set—can be found in Euclid's *Elements*, if not earlier. Such choices formed the basis for the ancient method of proving a generalization by considering an arbitrary but definite object, and then executing the argument for that object. This first stage also included the arbitrary choice of an element from each of finitely many sets. It is important to understand that the Axiom was not needed for an arbitrary choice from a single set, even if the set contained infinitely many elements. For in a formal system a single arbitrary choice can be eliminated through the use of universal generalization or similar rules of inference. By induction on the natural numbers, such a procedure can be extended to any finite family of sets.

The second stage began when a mathematician made an infinite number of choices by stating a rule. Since the second stage presupposed the existence of an infinite family of sets, two promising candidates for its emergence are nineteenth-century analysis and number theory. In the first case there were analysts who arbitrarily chose the terms of an infinite sequence, and, in the second, number-theorists who selected representatives from infinitely many equivalence classes. When some mathematician, perhaps Cauchy, made such an infinity of choices but left the rule unstated, he initiated the third stage.

This oversight—failing to provide a rule for the selection of infinitely many elements—encouraged the fourth stage to emerge. Thus in 1871, as we shall soon describe, Cantor made an infinite sequence of arbitrary choices for which *no* rule was possible, and consequently the Denumerable Axiom was required for the first time. Nevertheless, Cantor did not recognize the impossibility of specifying such a rule, nor did he understand the watershed which he had crossed. After that date, analysts and algebraists increasingly used such arbitrary choices without remarking that an important but hidden assumption was involved. From this fourth stage emerged Zermelo's solution to the

Well-Ordering Problem and his explicit formulation of the Axiom of Choice.

However, during the early years of the nineteenth century mathematics remained very much, as it had been for Euclid, a process of construction. If one wished to prove that a particular type of mathematical object existed, then one had to construct such an object from those previously shown to exist. On the other hand, since the techniques allowed in such a construction were not precisely delimited, the door was open for infinitely many arbitrary choices to enter unnoticed.

These opposite tendencies are visible in the most significant work on number theory to appear during that period, the *Disquisitiones Arithmeticae* which Gauss published in 1801. While discussing binary quadratic forms $ax^2 + 2bxy + cy^2$, Gauss showed that for those forms with a given discriminant $d = b^2 - ac$ there existed a unique integer n such that they could be partitioned into n classes by means of a certain equivalence relation.[1] Since he recognized that there were many ways to select a representative from each equivalence class, he carefully supplied a rule which determined those representatives uniquely. By choosing infinitely many representatives through a rule, though only a finite number of them for each value of d, Gauss paused on the border between the first and second stages.[2] His unflinchingly algorithmic approach to number theory made it unlikely that he ever entered the third stage.

In all probability the third stage, where infinitely many choices were made by an unstated rule, originated in analysis rather than in number theory. This stage was already evident in 1821 when Cauchy demonstrated a version of the Intermediate Value Theorem:

(1.2.1) Any real function f continuous on the closed interval $[a, b]$ has a root there, provided that $f(a)$ and $f(b)$ have opposite signs.

For a given integer m greater than one, Cauchy noted that the finite sequence

$$f(a),\quad f\left(a + \frac{b-a}{m}\right),\quad f\left(a + \frac{2(b-a)}{m}\right),\quad \ldots,\quad f(b)$$

must contain some consecutive pair with opposite signs. He let $f(a_1), f(b_1)$ be one such pair with $a_1 < b_1$ so that $b_1 - a_1 = (b - a)/m$. Next he considered the sequence of points dividing $[a_1, b_1]$ into m equal parts and, as

[1] Gauss 1801, section 223; translated in Gauss 1966. He defined this relation as follows: Two forms A and B are equivalent if they have the same discriminant (he used the term "determinant") and if there is a linear transformation with integer coefficients taking A to B and another such transformation taking B to A.

[2] Gauss [1832, section 42] also stated a rule to determine representatives of equivalence classes when he treated biquadratic residues. Medvedev [1965, 17–18, 23] has erroneously stated that Gauss used the Axiom of Choice implicitly in these works of 1801 and 1832.

before, chose a pair $f(a_2), f(b_2)$ with opposite signs such that $a_2 < b_2$ and $b_2 - a_2 = (b - a)/m^2$. In this fashion Cauchy selected sequences $a_1, a_2, \ldots$ and $b_1, b_2, \ldots$ converging to the same point, say p. Since f was continuous on $[a, b]$ and, for each n, the values $f(a_n)$ and $f(b_n)$ had opposite signs, then $f(p)$ equaled zero as desired.[3]

While Cauchy used infinitely many arbitrary dependent choices to obtain the pair a_n, b_n for every n, he could easily have provided a rule that specified these pairs uniquely. In particular, he could have selected for each n the leftmost pair whose values had opposite signs in the appropriate finite sequence. Consequently, Cauchy's proof remained within the third stage.

The process of repeated interval subdivision which Cauchy employed, traditionally named after Bolzano and Weierstrass, may have originated prior to 1821. Yet Bolzano did not utilize it in his 1817 proof of (1.2.1), despite the claims of various historians.[4] What Bolzano actually did was to give an algorithm for approximating the least upper bound of a bounded subset of $\mathbb{R}$ by summing powers of 2.[5] According to Cantor, the Bolzano–Weierstrass method of interval subdivision had occurred in certain number-theoretic investigations of Lagrange, Legendre, and Dirichlet.[6] It is not known which of their publications he had in mind. For certain cases of that method (to be discussed later in this section), the Axiom is essential. However, it is unlikely that such cases occurred before Weierstrass, if they did so then.

In his paper of 1857 on algebraic number theory, Richard Dedekind employed arbitrary choices in a manner that remained within the first stage but approached both the second and the third. Dedekind considered formal polynomials with integer coefficients, modulo a given prime integer p. After partitioning such polynomials into infinitely many congruence classes modulo p, he remarked that one might select a representative from each such class. Nevertheless, he did not consider more than two such representatives at a time and hence made only a finite number of arbitrary choices in a given proof [1857, 7–8]. Indeed, since the set of these polynomials was denumerable, he could easily have supplied a rule specifying a representative from each class. When he published the second edition of Dirichlet's lectures on number theory in 1871, Dedekind followed the *Disquisitiones Arithmeticae* in partitioning those binary quadratic forms with a given discriminant into congruence classes. On the other hand he refrained from stating the rule, given in the *Disquisitiones*, for selecting representatives from these classes. Instead, he concerned himself with the properties of any arbitrary set of representatives.[7]

---

[3] Cauchy 1821, 460–462; reprinted in Cauchy 1897, 378–380.

[4] See, for example, Boyer [1968, 605] and Kline [1972, 953].

[5] Bolzano 1817, 42–48; translated in Russ 1980, 174–176.

[6] Cantor 1884, 454. He also cited the passage in Cauchy's *Cours d'analyse* where (1.2.1) occurred.

[7] Lejeune Dirichlet and Dedekind 1871, section 59. In the first edition of 1863, Dedekind had not partitioned quadratic forms into equivalence classes.

Here a conceptual shift from the algorithmic toward the non-constructive was under way.

The fourth stage, where a mathematician made infinitely many arbitrary choices for which no rule was possible and for which, consequently, the Axiom of Choice was essential, began by October 1871 when Eduard Heine wrote an article on real analysis. Printed the following year, the article was largely based on unpublished research by Weierstrass. In all probability Heine learned of Weierstrass's work from Georg Cantor, who had studied under Weierstrass at Berlin and who became Heine's colleague at the University of Halle in 1869. Of the theorems in Heine's article, the one involving arbitrary choices was credited to Cantor:

(1.2.2) A real function f is continuous at a point p if and only if f is sequentially continuous at p. [Heine 1872, 183].

In effect, Cantor's theorem stated that two characterizations of continuity are equivalent. The first was the usual definition in real analysis, due to Cauchy and Weierstrass: A real function f is continuous at a point p if for every $\varepsilon > 0$ there is some $\eta > 0$ such that for every x,

$$|x - p| < \eta \quad \text{implies } |f(x) - f(p)| < \varepsilon.$$

The second characterization, by means of sequences rather than intervals, was what will hereafter be termed sequential continuity: A real function f is sequentially continuous at p if, for every sequence $x_1, x_2, \ldots$ converging to p, the sequence $f(x_1), f(x_2), \ldots$ converges to $f(p)$.[8]

Heine's proof, borrowed from Cantor, implicitly used the Assumption to show that sequential continuity at p yielded continuity there: Suppose, Heine began, that f is not continuous at p. Then there is some positive ε such that no matter how small η_0 is, there is always some positive η less than η_0 such that $|f(p + \eta) - f(p)| \geq \varepsilon$:

> So for any one value of η_0, let one such value of η (smaller than this η_0), for which the above difference $[|f(p + \eta) - f(p)|]$ is not smaller than ε be equal to η'. For half as large a value of η_0 the difference for $\eta = \eta''$ cannot be smaller than ε; for an η_0 equal to half the earlier (a quarter of the first) this must occur for $\eta = \eta'''$, and so on. [1872, 183]

Since the sequence $\eta', \eta'', \ldots$ converges to zero, then $p + \eta'$, $p + \eta'', \ldots$ converges to p; but $f(p + \eta')$, $f(p + \eta''), \ldots$ does not converge to $f(p)$, contrary to hypothesis.

Neither Cantor nor Heine gave any indication of suspecting that a new and fundamental assumption was required for this proof. Not until a decade

---

[8] We use the name *sequential continuity* for this property to emphasize how it parallels the definition of continuity while relying on sequences. This notion has also been termed Heine continuity (see Steinhaus 1965, 457).

after the Axiom was formulated did Michele Cipolla in Italy and Wacław Sierpiński in Poland independently recognize that (1.2.2) was unavoidably entangled with the Axiom.[9] Many years later a model of **ZF** was found containing a real function which was sequentially continuous but not continuous.[10]

With the publication of Heine's article, the Denumerable Assumption entered mathematics. Yet at first glance the argument by which Cantor established (1.2.2) did not appear to differ greatly from the argument that Cauchy had given to demonstrate (1.2.1) half a century earlier. When in 1886 Jules Tannery used Cantor's argument to prove (1.2.2) in an introductory textbook on the theory of functions, Tannery did not indicate that anything at all unusual or suspicious had occurred [1886, 108–109]. In fact, Cantor's implicit use of the Assumption while proving (1.2.2) did not lead directly to the formulation of the Denumerable Axiom. Nevertheless, this theorem provided the first of several distinct routes by which the unavoidable use of the Assumption entered mathematics.

Dedekind followed a second such route. In 1871 he published his famous tenth supplement to Dirichlet's lectures on number theory. There he investigated methods for extending the unique factorization into primes, a decomposition known to hold for the integers, to every algebraic number field. Some years earlier Ernst Kummer, who had discovered that such a goal could not be reached directly, introduced his ideal numbers in order to attain it indirectly. Dedekind, on the other hand, proposed quite a different means for arriving at the same goal: the use of *classes* of algebraic numbers. Rather than factoring algebraic integers, he introduced the notion of an ideal as a set of such integers closed under addition and subtraction, and also closed under multiplication by any element in the corresponding algebraic number field.[11] In any such field, every proper ideal could be factored uniquely into prime ideals. Thus, in particular, every proper ideal was included in some prime ideal.[12] This latter theorem, if asserted for every commutative ring, required the Axiom (see 4.5). However, because every algebraic number field was denumerable and so possessed a well-ordering, the Axiom was not needed in Dedekind's proof.

By contrast, Dedekind used arbitrary choices in an unavoidable fashion in 1877 when he published an expository article in French on his theory of ideals

[9] Cipolla 1913, 1–2, and Sierpiński 1916, 689; see 3.6 and 4.1 respectively.

[10] Jaegermann 1965, 699–704.

[11] Lejeune Dirichlet and Dedekind 1871, section 163. An *algebraic number field F* was in effect a subfield of the real algebraic numbers. That is, F was closed under addition, subtraction, multiplication, and division except by zero. In his later work Dedekind used the term *Körper* to mean any subfield of the complex numbers (see Lejeune Dirichlet and Dedekind 1893, section 160).

[12] Lejeune Dirichlet and Dedekind 1871, section 163; Dedekind 1877, 212. A *proper ideal* was an ideal not containing some algebraic integer in the given field F, while a *prime ideal* was a proper ideal not included in any other proper ideal in F. In effect, a prime ideal was maximal.

and modules. In 1871 he had defined a module as any set A of complex numbers closed under addition and subtraction, and had let $a \equiv b \pmod{A}$ mean that $a-b \in A$. The notion of module—which, for mathematics at that time, was extremely general and which included his concept of ideal as a special case—originated from an analogy with the congruence $a \equiv b \pmod{m}$, previously introduced by Gauss for integers a, b, m [Dedekind 1877, 70–71]. After discussing some elementary properties of his modules in his expository article, Dedekind stated the following theorem:

(1.2.3) If A and B are modules, then there exists a subset B_1 of B such that for each b in B there is a unique b_1 in B_1 with $b \equiv b_1 \pmod{A}$.

He designated such a set B_1 as a complete system of representatives for the module B with respect to the module A.[13]

To establish (1.2.3), Dedekind merely partitioned B into congruence classes mod A and selected an element from each class. This procedure did not appear essentially different from the one which he had used previously to obtain a set of representatives for the congruence classes of integers mod m, or of binary forms.[14] In fact, however, there was one essential difference. Whereas the choice of representatives in those earlier cases could be executed by means of a rule, the proof of (1.2.3) could not. Furthermore, if A were the set of all rational numbers and if B were $\mathbb{R}$, then B_1 would constitute a non-measurable set (see 2.3). Thus in (1.2.3) Dedekind was the first to use a form of the Assumption stronger than the denumerable case upon which Cantor had relied to obtain (1.2.2). Yet no more than Cantor or Heine did Dedekind understand that a new axiom was needed to secure his result.

Despite its potential fecundity, Dedekind's argument for (1.2.3) had little direct effect on later mathematicians. This remained true even when he gave a similar argument in his book *Was sind und was sollen die Zahlen?* of 1888. There he went much further by partitioning all sets into equivalence classes, where the equivalence relation was equipollence, and then selecting a representative from every such class.[15] Yet he drew no consequences from this strong result, just as a decade earlier he had not used the full force of (1.2.3). In practice, he had only needed finite sets of representatives for his investigations of algebraic number fields, and so did not require the Axiom.[16] Although

[13] Dedekind 1877, 20–21. The theorem recurred in Lejeune Dirichlet and Dedekind 1879, 483–484, and 1893, 509.

[14] Lejeune Dirichlet and Dedekind 1871, sections 18 and 59.

[15] Dedekind 1888, definition 34. Pincus has given a model of **ZF** in which no such class of cardinal representatives exists, as well as a model of **ZF** in which there is such a class but the Axiom is false [1972, 723; 1974, 322]. Thus the existence of a class of cardinal representatives depends upon the Axiom but is weaker than it.

[16] However, in these investigations Dedekind restricted his discussion of general theories as a concession to other mathematicians; see his letter of 10 June 1876 to Rudolph Lipschitz in Sinaceur 1979, 117–118.

Hilbert, when writing his seminal report of 1897 on such fields, emphasized that Dedekind had founded his investigations of them on the notion of module, Hilbert himself [1897, 245] employed only those modules generated by linear combinations of finitely many basis elements. Likewise, when Heinrich Weber introduced the more general concept of field in 1893, he restricted himself to extension fields obtained by adjoining finitely many new indeterminates, and hence did not require the Assumption [1893, 528]. Only the later development of field theory by Steinitz and of infinite-dimensional vector spaces by Hamel and Hausdorff fully illuminated the role of the Axiom in these matters (see 2.5, 3.5, and 4.5).

By contrast, Cantor's theorem (1.2.2) had a more immediate, if indirect, influence. To illustrate this, we must consider a parallelism that exists between analytic notions defined by neighborhoods and the corresponding notions defined by sequences. Thus (1.2.2) asserted the equivalence of two such notions: continuity based on intervals or neighborhoods, and sequential continuity based on sequences.

The concept of limit point also exemplifies this parallelism. In n-dimensional Euclidean space, on which late nineteenth-century analysts carried out their researches, a point p is a limit point if every neighborhood of p contains some point q in $A-\{p\}$. On the other hand, if there is a sequence $a_1, a_2, \ldots$ of elements of $A-\{p\}$ converging to p, then p is a sequential limit point of A. One can easily demonstrate that every sequential limit point of A is a limit point of A. However, as Sierpiński later remarked, the Denumerable Axiom is needed to deduce the converse [1918, 119–121]. In the absence of the Axiom, there can exist a set A of real numbers for which a certain point p is a limit point, but not a sequential limit point.[17] Essentially, the Axiom prevents this dichotomy from arising by supplying the desired sequence in a way analogous to Cantor's proof that sequential continuity at a point implies continuity there. Although the notions of limit point and of sequential limit point lead to differing definitions of closed set, perfect set, and so forth, these definitions were treated as equivalent in $\mathbb{R}^n$ by various analysts late in the nineteenth century. Thereby they fell into using the Assumption implicitly.

For those definitions based on intervals or neighborhoods, the Axiom turned out to be avoidable in many cases. An important example is the Bolzano–Weierstrass Theorem, actually due to Weierstrass alone: Every infinite bounded subset A of $\mathbb{R}^n$ has a limit point. During 1865 Weierstrass gave a course of lectures, still unpublished, in which he stated his theorem for $n = 2$.[18] A few years later Cantor, a former student of Weierstrass, referred to this theorem in passing for $n = 1$, without supplying a proof [1872, 129]. In his 1874 lectures, Weierstrass demonstrated the theorem in the following form: If a real function f has infinitely many values between two numbers c and

[17] Jech 1973, 141.

[18] Pierre Dugac, lecture at the University of Toronto, 12 April 1979. Kline [1972, 953] also mentioned that this theorem emerged during the 1860s.

d, then the set of values in $[c, d]$ has a limit point b. It was not his theorem, but rather the rudiments of his method of proof, that he owed to Bolzano [1817, 42–48]. For a given integer m greater than one, Weierstrass represented such a limit point b as

$$k + \frac{k_1}{m} + \frac{k_2}{m^2} + \frac{k_3}{m^3} + \cdots,$$

where $k_i \in \{0, 1, \ldots, m - 1\}$. Here k was the least integer, greater than or equal to c, such that the interval $[k, k + 1]$ contained infinitely many values of f. Likewise, k_1 was the least integer such that

$$\left[k + \frac{k_1}{m}, k + \frac{k_1 + 1}{m}\right]$$

contained infinitely many values of f, and so forth for $m^2, m^3, \ldots$. Thus Weierstrass successively approximated his limit point b, and then extended his theorem to $\mathbb{R}^n$, without using arbitrary choices in either case.[19] This proof seems to have circulated informally until Salvatore Pincherle, who had attended Weierstrass's lectures at Berlin in 1878, finally published it [1880, 237–238].

Although the Bolzano–Weierstrass Theorem does not require the Axiom when A is a subset of $\mathbb{R}^n$, it would do so if we replaced the term limit point by sequential limit point:

(1.2.4) Every infinite bounded subset A of $\mathbb{R}^n$ has a sequential limit point.

No one grasped the importance of distinguishing between these two versions of the theorem until Sierpiński investigated the matter decades later [1918, 122]. However, in 1892 Camille Jordan used arbitrary choices to prove (1.2.4) in an article that admirably illustrates how the Assumption crept into theorems about sequential limit points.

Jordan proposed to give a rigorous treatment of multiple Riemann integration, along the lines that Gaston Darboux had introduced for functions of a single variable, by using upper and lower integrals. With this in mind, Jordan introduced what is now known as the Jordan measure of a set in $\mathbb{R}^n$ [1892, 76]. In order to study this notion, he began with a discussion of Cantorian point-set topology. After defining the notion of limit point, he let the distance p_1p_2 between two points $p_1 = (a_1, a_2, \ldots, a_n)$ and $p_2 = (b_1, b_2, \ldots, b_n)$ be

$$\sum_{i=1}^{n} |b_i - a_i|.$$

[19] Unpublished notes, taken in 1874 by G. Hettner, of Weierstrass's lectures on the theory of analytic functions, pp. 305–310, 313–318. These notes are kept in the library of the Mathematical Institute at the University of Göttingen.

A few pages later he demonstrated (1.2.4)—which he termed the Weierstrass Theorem—for the case when $n = 2$.

In the opening lines of the article, Jordan's predisposition to rely on arbitrary choices was already evident. Indeed, he wrote that the definite integral of a function f in $\mathbb{R}^n$ with bounded domain was obtained by decomposing "the domain into infinitesimal elements; one multiplies the measure (*étendue*) of each of these elements by the value of f at a point arbitrarily chosen in the element; and one finds the limit of the sum $\Sigma f d\sigma$ formed in this way" [1892, 69]. This same predisposition was even more apparent when he demonstrated (1.2.4). For, although he stated the Bolzano–Weierstrass Theorem in terms of a limit point, what he proved was the stronger version (1.2.4) yielding a sequential limit point.

Intriguingly, Jordan's demonstration bore a strong resemblance to Cauchy's proof of (1.2.1): Jordan let v and V be two real numbers with $v < V$ such that the set A was bounded by the square having vertices (v, v), (v, V), (V, v), and (V, V). For a given m greater than one, this square could be partitioned into m^2 squares each of length $(V - v)/m$. Since at least one of these smaller squares contained infinitely many points of A, he chose such a square and called it E_1. Repeating this process, he selected a sequence of squares E_n, each of which had length $(V - v)/m^n$, and contained infinitely many points of A. By choosing a point p_n from each $E_n \cap A$, he obtained a sequence $p_1, p_2, \ldots$ in A converging to some point p, as desired [1892, 73–74].

Here Jordan made essential use of the Assumption. Whereas the E_n could be specified by a rule, the p_n could not. In fact, he generated the sequence $p_1, p_2, \ldots$ by infinitely many dependent choices, though his argument could be modified so as to employ only the Denumerable Axiom. Unquestionably his usage of arbitrary choices belonged to the fourth stage, since a model of **ZF** was eventually found that contained an infinite bounded subset A of $\mathbb{R}$ with no sequential limit point.[20]

Jordan's article implicitly used the Denumerable Assumption in two other cases as well. The first of these occurred in his proof that the boundary of a set is closed, for there he presupposed that a limit point is also a sequential limit point. The second instance concerned the notion of the distance $d(A, B)$ between two subsets A and B of $\mathbb{R}^n$. After defining $d(A, B)$ to be the greatest lower bound of $\{pq: p \in A \text{ and } q \in B\}$, he asserted that if A and B are disjoint, bounded, and closed, then there exists some p in A and some q in B such that $pq = d(A, B)$. While this result could be established without the Axiom by using the theorem that a continuous real-valued function on a closed and bounded set attains its minimum, his argument relied instead on the Denumerable Assumption to obtain two sequences $p_1, p_2, \ldots$ and $q_1, q_2, \ldots$ such that the first sequence was in A, the second was in B, and the distances $p_1q_1, p_2q_2, \ldots$ converged to $d(A, B)$ [1892, 73–74].

---

[20] Cohen [1966, 138] provided an infinite set S of real numbers without a denumerable subset. If f is a bijection from $\mathbb{R}$ onto the interval $(0, 1)$, then $f''S$ is the desired set A.

When in 1893 Jordan published the second edition of his *Cours d'analyse*, he revised it to introduce many of the concepts and results found in his article. Yet now he deviated from Cantor by using sequential limit points, rather than limit points, as his fundamental notion.[21] In this vein he defined the sequentially derived set $s(E)$ of E as the set of all sequential limit points of E. Similarly, E was sequentially closed if $s(E)$ was a subset of E, and was sequentially perfect if $E = s(E)$. Also, a point p was a sequentially isolated point of E if p was in E but not in $s(E)$. If we replace the term sequential limit point by limit point in each definition, we obtain the corresponding notions—derived set, closed set, perfect set, and isolated point—that Cantor had introduced by means of neighborhoods.[22]

Jordan considered the concepts of limit point and sequential limit point to be equivalent, as evidenced by his proof of the following result:

(1.2.5) For any subset E of $\mathbb{R}^n$, the sequentially derived set $s(E)$ is sequentially closed.

As given, his argument for (1.2.5) implied only that $s(E)$ was closed, rather than sequentially closed. To establish that $s(E)$ was sequentially closed, he needed a sequence $p_1, p_2, \ldots$ of points in $E - \{p\}$ that converged to p, and thus he relied on the Denumerable Assumption. In a similar fashion he assumed it implicitly while asserting a pair of theorems which he regarded as too elementary to require detailed proof: In $\mathbb{R}^n$, if p is a sequentially isolated point of E, then p is an isolated point of E; a set E of real numbers, bounded above and sequentially closed, contains its least upper bound.[23]

To conclude his treatment of point-set topology, Jordan discussed connected sets. A subset E of $\mathbb{R}^n$ was Jordan-connected (*d'un seul tenant*) if it was bounded, sequentially closed, and could not be decomposed into two sequentially closed sets A and B such that $d(A, B) > 0$. His first result linked the notion with previous ones:

(1.2.6) If a Jordan-connected set E contains at least two points, then E is sequentially perfect.

Since E was sequentially closed, he argued, it remained only to show that every point p was a sequential limit point of E, and in doing so he made infinitely many dependent choices. Both the method of proof and the implicit use of the Assumption were analogous in his final theorem: If p and q are in

---

[21] Jordan 1893, 19. Strictly speaking, he retained the name "limit point" for what is termed here a "sequential limit point."

[22] Jordan [1893, 19] also deviated from Cantor's terminology by designating a sequentially closed set as an *ensemble parfait*, while he used but did not name his version of Cantor's perfect sets.

[23] *Ibid.*, 19–22.

a Jordan-connected subset E of $\mathbb{R}$, then every r between p and q is also in E [1893, 24–25].

While Jordan's implicit uses of the Assumption were directly stimulated by Cantor's researches on point-set topology, this line of argument, involving an infinite sequence of arbitrary dependent choices from subsets of $\mathbb{R}$, had already occurred in France seventy years earlier in the work of Cauchy. What distinguished Jordan's article of 1892 and book of 1893 was that he blended such a sequential approach with Cantorian topological notions in a way that required the Denumerable Axiom but that did not immediately appear to do so. Consequently, it was not evident at the time that a new assumption was involved.

In this context Jordan's work was significant for a second reason: his influence on Henri Lebesgue. This influence, so evident in Lebesgue's early researches on measure and integration, also appears to be present in his use of sequences of arbitrary choices.[24] Since Lebesgue's use of such choices will be treated in detail in 1.7, a single example will suffice here. In an appendix to his lectures on integration, given during 1902–1903 at the *Collège de France*, he discussed a theorem due to Cantor: If all the derived sets $E^{(1)}$, $E^{(2)}, \ldots,$ $E^{(n)}, \ldots$ of a bounded set E of real numbers are non-empty, then their intersection is also non-empty. To prove this theorem, Lebesgue arbitrarily chose a point in each $E^{(n)}$ [1904, 131]. While this particular use of the Assumption was avoidable, many others that he made were not. Despite the fact that he later became a severe critic of the Axiom, he frequently failed to recognize how much he had relied on it in his own research. Surprisingly, neither Lebesgue nor any of his fellow French critics pointed out Jordan's extensive use of arbitrary choices, despite the fact that these critics were well acquainted with his book.

To sum up, by the close of the nineteenth century, analysts were often making an infinite sequence of arbitrary choices in a manner reminiscent of Cauchy seventy years earlier. Such analysts, who accepted the general notion of function (proposed by Dirichlet in 1829) as a correspondence rather than as an analytic expression, did not notice that a number of their results necessitated a fundamentally new assumption: the Axiom of Choice. In fact, however, these results—especially the equivalence in $\mathbb{R}^n$ of continuity and sequential continuity, as well as that of limit point and sequential limit point—required only the Denumerable Axiom. Since this was the weakest non-trivial form of the Axiom of Choice, it is not altogether surprising that they did not recognize their need for it. Such recognition came only when Zermelo's arbitrary choices yielded Cantor's proposition, which many mathematicians found extremely implausible, that every set can be well-ordered. Before turning to that subject, we must consider another pathway along which an infinity of arbitrary choices entered mathematics: the attempts to distinguish precisely between the finite and the infinite.

---

[24] By the turn of the century the casual use of sequences of arbitrary choices was so widespread that Lebesgue may have inherited it from other mathematicians in addition to Jordan.

1.3 The Boundary between the Finite and the Infinite

The actual infinite, which Aristotle had banished in favor of the potential infinite, triumphantly re-entered mathematics through Cantor's theory of sets. Even prior to Cantor, Bernard Bolzano had explored the vagaries of infinite sets in his book *Paradoxes of the Infinite*, published posthumously in 1851. Although Bolzano did not adequately define the notions of finite and infinite set, he knew of two properties that could have served this purpose. One of them supplies the definition adopted here:

(1.3.1) A non-empty set A is finite if for some positive integer n, A is equipollent to $\{1, 2, \ldots, n\}$; otherwise, A is infinite.

The second property is today commonly named after Dedekind:

(1.3.2) A set A is Dedekind-infinite if some proper subset B of A is equipollent to A; otherwise, A is Dedekind-finite.

During the latter half of the nineteenth century a number of mathematicians, particularly Cantor and Dedekind, claimed that a set A is finite if and only if A is Dedekind-finite. By asserting the equivalence of (1.3.1) and (1.3.2), these mathematicians used the Denumerable Assumption implicitly.[1] Thus they created a further route by which the Assumption entered mathematics. As the various possible definitions of finite set were gradually clarified, so too was the explicit use of arbitrary choices that occurred in proving their equivalence.

Aristotle had criticized the concept of the actual infinite because earlier Greek philosophers had used it in ways that he considered illegitimate. Yet he felt unable to dispense altogether with the infinite, since, if he did so, there could not be infinitely many whole numbers. As a compromise, he proposed the notion of the potential infinite, which could always be extended, and then remarked:

> Our account does not rob the mathematicians of their science, by disproving the actual existence of the infinite .... In point of fact they do not need the infinite and do not use it. They postulate only that the finite straight line may be produced as far as they wish.[2]

Such mistrust of the actual infinite remained the orthodox viewpoint for the next two millenia.

---

[1] Strictly speaking, the equivalence of (1.3.1) and (1.3.2)—or what amounts to the same thing in **ZF**, the proposition (1.1.4) that every infinite set has a denumerable subset—is weaker than the Denumerable Axiom. That is, in **ZF** (1.1.4) follows from the Denumerable Axiom, but the converse does not hold; see Levy 1964, 147, and Jech 1973, 124–125.

[2] Aristotle 1941, 268: *Physica*, Book III, Chapter 7. This postulate later appeared as the second in Euclid's *Elements*.

Galileo, who freed himself from many Aristotelian preconceptions, remained somewhat suspicious of the actual infinite. In his book *Two New Sciences*, published in 1638, he discussed a certain property of the whole numbers. On the one hand, he noted, there were fewer such numbers n^2 than numbers n, since there existed numbers not of the form n^2. On the other hand, there were just as many numbers n^2 as numbers n since a one-one correspondence existed between them. In effect, Galileo had given an example of a set of integers that was equipollent to a proper subset and that consequently satisfied definition (1.3.2.). Yet he did not view the matter in this light, and concluded instead that the notions of greater than, equal to, and less than could not apply to infinite quantities.[3]

One misunderstanding found in Galileo's treatment of infinite sets persisted when Bolzano discussed them three centuries later: the conflation of an infinite set of numbers with their sum. This confusion occurred while Bolzano was presenting the set of positive integers as an example of an infinitely great quantity which was not an infinitely great number. (Such a number, he believed, did not exist.) Moreover, although he mentioned a set A of real numbers such that a bijection existed between A and a proper subset B, he still lacked a clear criterion for the relation of "greater than" with respect to equipollence. Likewise, in contrast to the later researches of Cantor, he rejected equipollence as a sufficient criterion for two infinite sets to agree in multitude (*Vielheit*).[4]

In sum, as befits a transitional figure, Bolzano treated the boundary between finite and infinite sets in a manner that blended insight with confusion. To define the notion of finite set, he presupposed the positive integers which he then obtained from that notion. This circularity vitiated his definition, which in other respects resembled (1.3.1). Ambiguity occurred as well in his definition of infinite set: A set A is infinite if every finite set is only a part (*Theil*) of A. For Bolzano, the property (1.3.2) did not constitute a definition of infinite set but merely a relation which might hold between two infinite sets A and B. It is unclear whether he asserted that every infinite set A has an equipollent proper subset B, or only that some infinite sets have such a subset. In any case, he merely stated examples [1851, 6, 28–29]. Despite his occasional lapses, Bolzano labored strenuously to legitimize the notion of infinite set and to distinguish it from the potential infinity of an unbounded variable.

It is uncertain to what extent Bolzano's views on the infinite influenced Cantor, who discussed the *Paradoxes of the Infinite* only in 1883. On that occasion he praised Bolzano's book for asserting that the actual infinite exists but criticized it for failing to provide either a concept of infinite number or the concept of "power" based on equipollence.[5] Although he may have adopted

---

[3] Galilei 1974, 39–42.

[4] Bolzano 1851, 22, 34; translated in Bolzano 1950.

[5] Cantor 1883a, 560–561. Two sets had the same power (*Mächtigkeit*) if they were equipollent. Later he often used the term cardinal number instead of power.

the terms *Menge* (set) and *Vielheit* (multitude or multiplicity) from Bolzano, Cantor's interest in the infinite had originated much earlier in his career.

In 1866 Cantor completed his doctoral dissertation at Berlin on number theory, as developed by Legendre and Gauss, and soon added a lively research interest in analysis as well. After Cantor became a *Privatdozent* at the University of Halle, his colleague Heine led him to study the conditions under which a function is represented by a unique trigonometric series.[6] Around 1870, in the course of these investigations, he began tentatively to extend the concept of number in a way that eventually resulted in his infinite ordinal numbers. Only a decade later did he mention these "symbols of infinity," as he then designated them, in print [1880, 358]. Yet, in this way and others, he increasingly desired to justify and extend the use of the actual infinite in mathematics.

Although Cantor viewed the boundary between the finite and the infinite from a perspective similar to Bolzano's, at first he did not attempt to define that boundary. In his article of 1878, where he established that $\mathbb{R}$ and $\mathbb{R}^n$ have the same power, Cantor described a finite set as one whose power is a positive integer. For such a set, he asserted, every proper subset has a smaller power, whereas an infinite set A has the same power as some proper subset of A. Having clearly stated both properties (1.3.1) and (1.3.2), he could have introduced either of them as a definition for finite and infinite set. He did not do so.[7] By asserting without proof that these two properties are equivalent, he used the Denumerable Assumption implicitly. Yet no trace of explicit arbitrary choices occurred, a circumstance which was to change.

In 1882 Cantor still did not believe that a simple definition of finite set was necessary or even possible. Therefore he was surprised when Dedekind communicated his own definition (1.3.2), the first adequate definition of finite set that did not presuppose the natural numbers.[8] Five years later Cantor finally unveiled his own rather vague alternative definition: "By a *finite* set we understand a set M which arises from *one* original element through the successive addition of new elements in such a way that the original element can be obtained from M by *successively* removing the elements *in reverse order*" [1887, 266]. At no point in his article did he elucidate what he meant by such successive additions and removals. Yet he applied a version of the principle of mathematical induction to show that every set finite in his sense was also Dedekind-finite. Without proof he asserted the converse, that every infinite set was Dedekind-infinite, and hence he implicitly relied on the Assumption once more. This process was characteristic of Cantor's early implicit uses of the Assumption, for he frequently presupposed it in propositions whose proof he considered too obvious to state (see 1.4.)

---

[6] Dauben 1979, 30.

[7] Cantor 1878, 242. Nevertheless, in a letter dated 18 July 1901 to Philip Jourdain, Cantor claimed that the article contained his independent discovery of Dedekind's definition (1.3.2) of infinite set; see Cantor in Grattan-Guinness 1971a, 115.

[8] Letter of 24 January 1888 from Dedekind to H. Weber, in Dedekind 1932, vol. III, 488.

In his *Grundlagen* of 1883, before he had published a definition for finiteness, Cantor claimed a certain difference between finite and infinite sets to be the essential one. While a finite set always retained the same ordinal number no matter how the elements were arranged, an infinite set could be rearranged so as to have more than one ordinal [1883a, 549–550]. Hidden in this characterization was his belief, stated later in the same article, that every set can be well-ordered (see 1.5). If one regarded Cantor's characterization as a definition of finite set and if one made the plausible assumption that this definition was equivalent to the usual one, then from this equivalence would follow his Well-Ordering Principle and hence the Axiom of Choice. However, at this time he was not thinking in terms of any definition for finiteness, much less the equivalence of two such definitions.

Like Cantor, the American logician C. S. Peirce considered the properties (1.3.1) and (1.3.2) to be equivalent. Yet his interest in defining the notion of finite set came not from Cantor, some of whose articles he had read in 1884,[9] but from a desire to justify the syllogism of transposed quantity that the English logician Augustus De Morgan had proposed in 1864:

> Some X is Y;
> for every X something is neither Y nor Z;
> hence, something is neither X nor Z.[10]

In 1885, after remarking that De Morgan's syllogism is valid if and only if the class of all X's is finite, Peirce stated a definition of finite set that closely resembled (1.3.2):

> Now to say that a lot of objects is finite, is the same as to say that if we pass through the class from one to another we shall necessarily come round to one of those individuals already passed; that is, if every one of the lot is in any one-to-one relation to one of the lot, then to every one of the lot some one is in this same relation.[11]

It seems that, in effect, Peirce intended the following definition: A set A is Peirce-finite if every one–one function $f: A \to A$ is onto.[12] Yet in the *Century Dictionary* of 1889, he proposed a definition of finite set that was essentially (1.3.1): "As applied to a class or integer number, capable of being completely counted" [1889, 2225]. Since Peirce presumably regarded his two definitions as equivalent, he too made implicit use of the Denumerable Assumption.

---

[9] Murphey 1961, 240.

[10] Peirce 1885, 201. In effect he also assumed that if b and c are both X's, then they are assigned to different objects each of which is neither Y nor Z. Otherwise the syllogism is not valid.

[11] Peirce 1885, 202. This definition is found in embryo, but not clearly formulated, in Peirce 1881, 95.

[12] Later Peirce considered his definition to be the same as Dedekind's (1.3.2), a conclusion to which Ernst Schröder had come already; see Peirce 1900, 430–431, and Schröder 1898, 305.

However, as he supplied no proof, it is difficult to determine whether he relied on arbitrary choices or some equivalent assumption.

Not until 1888, when Dedekind published his book *Was sind und was sollen die Zahlen?*, was the theory of finite sets developed fully and rigorously. According to its preface, his interest in the subject was of many years standing. In an early draft, written between 1872 and 1878, he clearly stated (1.3.2) as a definition of finite and infinite set.[13] When his definition appeared in print, he remarked that all other attempts known to him to distinguish finite from infinite were so unsuccessful that he would refrain from mentioning them. Using his definition, he demonstrated that every set equipollent to a subset of a Dedekind-finite set is Dedekind-finite. As a consequence of his axioms for the positive integers, he deduced that every Dedekind-infinite set has a denumerable subset. Furthermore, he showed that every finite set is Dedekind-finite [1888, 17, 19–21, 31].

Yet it was his demonstration of the converse that proved to be most significant from the standpoint of the Axiom:

(1.3.3) Every Dedekind-finite set is finite.

Neither Bolzano nor Cantor had supplied a detailed argument for this converse. Within such an argument, denumerably many arbitrary choices or a related assumption had to occur, and so it occurred in Dedekind's. In particular, he deduced the converse from the proposition that if for every n the set $Z_n = \{1, 2, \ldots, n\}$ is equipollent to a subset of a given set S, then S is Dedekind-infinite. The use of arbitrary choices came early in the proof when he asserted that for every n there existed a definite one-one function a_n mapping Z_n into S [1888, 51]. In effect he selected, for every n, one a_n out of the non-empty set of such mappings. Here was the first unequivocal use of arbitrary choices in characterizing the boundary between the finite and the infinite.

Dedekind's proof attains a still greater significance because it prompted another mathematician to question the legitimacy of arbitrary choices. This mathematician was Rodolfo Bettazzi, who, since 1892, had taught at the Military Academy in Turin.[14] During February 1896 he presented two papers [1896, 1896a] before the Turin Academy of Sciences on the subject of Dedekind's *Was sind und was sollen die Zahlen*? In the second of these, Bettazzi objected to selecting all the a_n arbitrarily:

> But since there is more than one such correspondence between *any* Z_n and S, and since Dedekind does not determine a *specific* one among them, then one must take any one of them *arbitrarily*, and do so for each set of correspondences between any Z_n and S; that is, one must choose an object (correspondence) arbitrarily in each

---

[13] This draft is printed in Dugac 1973, 293–309; see especially 294.

[14] Kennedy 1980, 103.

> of the *infinite* sets, which does not seem rigorous; unless one wishes to accept as a postulate that such a choice can be carried out—something, however, which seems ill-advised to us. [1896a, 512]

Thus Bettazzi recognized quite clearly how arbitrary choices were used in Dedekind's proof, as well as the distinction between specifying a particular function and selecting one arbitrarily. However, he was concerned not only that Dedekind made infinitely many arbitrary choices but that he did so from sets with infinitely many members. (Eventually the former infinity was recognized as more fundamental than the latter.)

Yet Bettazzi did more. For he entertained the possibility of postulating arbitrary choices from an infinite number of infinite sets—and then immediately rejected it. This rejection may well bear the mark of Peano's influence. For Peano, one of Bettazzi's colleagues at the Military Academy, had objected earlier to the use of infinitely many arbitrary choices, without however mentioning anyone who had used them.[15] Peano, who frequently attended the meetings of the Academy of Sciences, may have been present when Bettazzi presented his papers.

Bettazzi's second article could have generated a debate over the use of arbitrary choices in set theory. Yet a year after it appeared, he no longer doubted Dedekind's theorem that every Dedekind-finite set is finite. Cesare Burali-Forti, another colleague at the Military Academy, had just presented to the Academy of Sciences a paper on finite sets that persuaded Bettazzi of the theorem's validity.[16] Although Burali-Forti adopted Dedekind's definition of finite set, he differed from its author on a fundamental point. For Burali-Forti doubted that Dedekind's definition sufficed to prove all the desired properties of finite sets—unless one introduced a new postulate: If S is a family of non-empty classes, then S is equipollent to a subclass of the union of S. At present, Burali-Forti added, no one had formulated an adequate list of postulates for set theory; but he insisted that this one belonged on such a list [1896, 34, 36, 46].

By means of his postulate, Burali-Forti developed the theory of Dedekind-finite sets in considerably more depth than even Dedekind had done, and deduced in particular that every Dedekind-finite set is finite. Yet Russell later remarked that this postulate is false in general, unless S is disjoint [1906, 49]. Although on that occasion he did not state a counterexample to the original form of the postulate, it suffices to let S be $\{\{1\}, \{2\}, \{1, 2\}\}$. If one requires S to be disjoint, as Russell suggested, then Burali-Forti's postulate is the Partition Principle. While Cantor had already presupposed a form of this principle in certain proofs (see 1.4), Burali-Forti was the first to give it an explicit, if incomplete, statement. Yet when the postulate is corrected by stipulating S to be

---

[15] Peano 1890, 210; see also 1.8.

[16] Bettazzi 1897, 352.

disjoint, many of Burali-Forti's proofs fail as given—in particular those for the following four propositions:[17]

(1.3.4) If A is Dedekind-finite, then the power set of A is also Dedekind-finite.[18]

(1.3.5) If A is Dedekind-finite and B is Dedekind-infinite, then A is equipollent to a subclass of B.

(1.3.6) The union of a Dedekind-finite family of Dedekind-finite classes is Dedekind-finite.

(1.3.7) If a class V contains every class with exactly one member and if $A \in V$ and $b \notin A$ together imply that $A \cup \{b\}$ is in V, then V contains every non-empty Dedekind-finite class.[19]

In fact, each of these propositions depends on the Denumerable Axiom, and each implies that every Dedekind-finite set is finite.[20]

Thus Burali-Forti's erroneous postulate persuaded Bettazzi that every Dedekind-finite set is finite. Unaware of the error, Bettazzi ceased to be concerned with Dedekind's arbitrary choices. In this way the only mathematician to suspect at an early stage that these two notions of finite set need not be equivalent joined those who employed the two interchangeably. As a

---

[17] However, it can be shown that the Partition Principle does imply these four propositions.

[18] The *power set* of A is the set of all subsets of A.

[19] Burali-Forti 1896, 46, 50–51. Like other logicians influenced by Peano, Burali-Forti used the term class instead of set. In 1906 Mario Pieri, who also belonged to Peano's school of logic and who had been their colleague at the Military Academy until 1900, learned from A. N. Whitehead (via Louis Couturat) that Burali-Forti's postulate was false as originally stated. In reply, Pieri [1906, 204] argued that Burali-Forti's results about Dedekind-finite sets could be secured if, instead, one postulated the first consequence that Burali-Forti had drawn from his original postulate: The union of a Dedekind-infinite class of non-empty classes is Dedekind-infinite. Pieri was correct, since his new postulate implies (1.3.4); but he did not notice that his postulate also depends on the Denumerable Axiom.

[20] To establish this implication for (1.3.4), one uses the result that if the power set of the power set of A is Dedekind-finite, then A is finite. In the case of (1.3.5) one takes B to be the set $\mathbb{N}$ of all natural numbers, and for (1.3.7) one lets V consist of all non-empty finite sets. Finally, we are indebted to Thomas Jech for the following proof that (1.3.6) implies that every Dedekind-finite set is finite:

Assume that A is an infinite, Dedekind-finite set. Let S be the set of all finite, one-one sequences of members of A. Since $\bigcup S$ is $\mathbb{N} \times A$, then $\bigcup S$ is Dedekind-infinite; hence (1.3.6) is false, and the proof is complete, unless S is Dedekind-infinite. In that case, S has a denumerable subset $T = \{t_1, t_2, \ldots\}$. Let $f: \mathbb{N} \to A$ be the infinite sequence obtained by concatenating the members of T in order. Thus $f''\mathbb{N}$ is infinite, since if it were finite then T would be finite as well. So there is a one-one subsequence g of f that is denumerable. But then the range of g is a denumerable subset of A, contradicting the hypothesis that A is Dedekind-finite. Therefore S must have been Dedekind-finite, as desired.

result, neither Bettazzi nor anyone else scrutinized the use of arbitrary choices until Beppo Levi did so, in a different context, five years later (see 1.8).

Among those mathematicians who regarded the notions of finite and Dedekind-finite set as equivalent, the Denumerable Assumption occurred most often while demonstrating (1.1.4): Every infinite set A has a denumerable subset. The early proofs of this proposition all showed Cantor's influence. Although Dedekind had established that every Dedekind-infinite set has a denumerable subset, it was Cantor who in 1895 provided the first demonstration of (1.1.4). In this instance Cantor's use of infinitely many arbitrary choices, so frequently hidden by his refusal to prove "obvious" theorems, was clearly visible. If, he wrote, one has removed distinct elements $t_1, t_2, \ldots, t_n$ from A by a rule, then it is always possible to remove a further element t_{n+1}; thus the set

$$\{t_1, t_2, \ldots, t_n, \ldots\}$$

is a denumerable subset of A [1895, 493].

Yet even here Cantor blurred the distinction between choices made by a rule and those made arbitrarily. Although the first n members of the subset were selected by a rule, he did not explain how such a rule could be extended to the entire denumerable subset. In general, it cannot. What he lacked was an analysis of the term "rule," of what is definable in a language. For Cantor's informal mathematics was motivated, not by such logical concerns about definability, but by a Platonistic vision of the reality of infinite sets—as revealed to him, so he believed, by God.[21]

Over the next decade several mathematicians published demonstrations of (1.1.4) similar to Cantor's. Borel's argument [1898, 12–13] agreed with Cantor's in all essential respects and consisted of successively choosing elements $t_1, t_2, \ldots$ from A. In his proof of 1901 the American mathematician Cassius Keyser did not make his arbitrary choices so visible. He merely asserted that if for every n, $\{1, 2, \ldots, n\}$ can be mapped one-one into A, then A clearly possesses a denumerable subset and so (1.1.4) follows at once [1901, 222–223]. Russell's demonstration of 1902, discussed in 1.1, did not differ substantially from Cantor's and Borel's. As late as 1903, Russell continued to hold that a set is finite if and only if it is Dedekind-finite [1903, 121–123]. In the aftermath of Zermelo's proof, Russell became an agnostic about this equivalence (see 2.7).

Thus by the end of the nineteenth century various mathematicians, influenced by Cantor and Dedekind, were employing arbitrary choices—with a greater or lesser degree of explicitness—to prove that a set is finite if and only if it is Dedekind-finite. Indeed, Cantor's 1895 use of a sequence of dependent arbitrary choices to establish (1.1.4) was analogous to their use in his argument, given two decades earlier, that a real function which is sequentially

---

[21] See Dauben 1979, 238–239.

continuous at a point is also continuous there. While he remained unaware that the use of arbitrary choices involved a new assumption, he did more than anyone else to disseminate theorems relying on such choices in an essential way.

However, what is most intriguing here is what almost happened, but did not. In his book of 1888 Dedekind arbitrarily selected a sequence of functions while giving the first detailed proof that a set is finite if and only if it is Dedekind-finite. (Here Dedekind's proof resembled his selection of representatives from congruence classes in algebraic number theory, discussed in 1.2, as well as the analogous selection used in his book to obtain a representative set of each cardinality.[22]) The explicitness of the arbitrary choices in Dedekind's proof led Bettazzi to mistrust them and to recognize that some new postulate would be necessary if one wished to legitimize them. Nevertheless, probably influenced by Peano, he concluded at once that to postulate such choices would be a mistake. Soon thereafter, Burali-Forti persuaded Bettazzi that every Dedekind-finite set is indeed finite, although in fact the argument was based on an erroneous version of the Partition Principle. Through this error the debate over arbitrary choices was stillborn.

After Zermelo formulated the Axiom of Choice, he cited the equivalence of (1.3.1) and (1.3.2) as an important application of the Axiom. Even so, not all mathematicians agreed. In 1912, Russell and Whitehead considered the possibility that sets existed which were both infinite and Dedekind-finite (see 3.6). The cardinals of such sets were at first called mediate cardinals and, by later mathematicians, Dedekind cardinals. In the absence of the Axiom there turned out to be not two but many definitions of finite set that might not be equivalent. On the other hand, as Tarski later remarked, the equivalence of two definitions of finiteness could have as much deductive strength as the Axiom itself (see 4.2). As it happened, such strength was already potentially present in Cantor's 1883 characterization of finite sets as those having only a single ordinal. But prior to Zermelo, all this lay hidden in darkness. Thanks to Bettazzi, the significance of arbitrary choices in establishing that different definitions of finiteness are equivalent emerged briefly into the light, then vanished.

1.4 Cantor's Legacy of Implicit Uses

While editing Cantor's collected works, Zermelo described the Axiom of Choice as an assumption "which Cantor uses unconsciously and instinctively everywhere, but formulates explicitly nowhere...."[1] Indeed, Cantor did

[22] Dedekind 1888, definition 34.

[1] Zermelo in Cantor 1932, 451.

more than this. His researches on set theory and point-set topology engendered the first substantial body of unavoidable implicit uses of the Assumption. Like other researchers at the time, he remained unaware that a fundamentally new mathematical principle lay hidden in these uses. Yet it was through his discoveries that many mathematicians—including later critics of the Axiom such as Borel and Lebesgue—came to use the Assumption implicitly. Eventually Cantor proposed a principle shown to be intimately related to arbitrary choices (see 1.5), but he did not introduce this Well-Ordering Principle in order to secure the results discussed in the present section. In fact, Zermelo's desire to prove that every set can be well-ordered, a desire which led him to formulate the Axiom of Choice, arose only after Cantor's researches (and to a lesser extent those of Dedekind) had disseminated the direct and indirect use of arbitrary choices quite widely.

Cantor, who in 1871 had initiated the fourth stage in the development of such choices by employing them in real analysis, did not use them in set theory until six years later. Over the next two decades the Assumption occurred repeatedly in certain of his basic results on equipollence, such as the Countable Union Theorem. That these results served principally as lemmas for other theorems helps to explain why Cantor did not detect a new mathematical principle at work: He regarded them as too elementary to demonstrate in detail.

The first of these results originated in 1877. Previously Cantor had shown the set of all real algebraic numbers to be countable, whereas $\mathbb{R}$ itself was uncountable [1874]. Now, while fashioning a proof that $\mathbb{R}^n$ has the same power as $\mathbb{R}$, he corresponded with Dedekind and received much helpful criticism from him. To establish this new theorem, Cantor wished to show that the closed interval [0, 1] retains the same power even if denumerably many elements are removed. With this goal in mind, he stated but did not prove the following lemma in his letter of 29 June 1877:

(1.4.1) If A_k is equipollent to B_k and if $A_k \cap A_m = B_k \cap B_m = 0$ for distinct $k, m = 1, 2, \ldots$, then the union of all A_k is equipollent to the union of all B_k.[2]

There he restricted A_k and B_k to subsets of $\mathbb{R}^n$ rather than allowing them to range over arbitrary sets, which he had not yet considered. The restriction remained in force when he published this lemma in his proof that $\mathbb{R}^n$ and $\mathbb{R}$ are equipollent [1878, 249, 253–254]. While the Assumption was used implicitly in the proof, in 1895 he gave a different demonstration that did not

[2] Cantor in Noether and Cavaillès 1937, 35–37. Here we would like to point out that (1.4.1) easily implies the Countable Union Theorem. For suppose that $A_0, A_1, \ldots$ is a sequence of disjoint denumerable sets. Then each A_n is equipollent to $\mathbb{N} \times \{n\}$. By (1.4.1), the union of all A_n is equipollent to $\bigcup\{\mathbb{N} \times \{n\} : n \in \mathbb{N}\}$, *i.e.*, to $\mathbb{N} \times \mathbb{N}$ and hence to $\mathbb{N}$, as desired.

rely on arbitrary choices. Instead, he noted, since $2^{\aleph_0}$ is the power of the continuum, one could simply calculate

$$2^{\aleph_0} \cdot 2^{\aleph_0} = 2^{\aleph_0 + \aleph_0} = 2^{\aleph_0}.$$[3]

Although (1.4.1) required the Axiom, Cantor never revealed how he employed arbitrary choices in its proof. Like his letter to Dedekind, his articles of 1878 and 1880 provided no details.[4] Nor did he do more than state this elementary result in his later development of transfinite arithmetic [1895, 483, 502]. If one insisted, as some later constructivists were wont to do, that a particular bijection be given from A_k onto B_k in order to consider the two sets equipollent, then the Axiom need not appear in the proof of (1.4.1). However, Cantor, as well as later set theorists, demanded only that some such bijection exist, and so they required the Denumerable Axiom to select one bijection for each k.

Cantor's article of 1878 contained two other propositions, vital to his theory, that depended on the Assumption. The first, a restricted form of the Trichotomy of Cardinals, will be discussed in the next section, while the second was his original statement of the Countable Union Theorem. In that article [1878, 243] and again four years later, Cantor described this theorem as easy to demonstrate—and left the rest to the reader. In the latter article, he applied it to show that the union of all sets A_k is countable, where A_k is the range of

$$f_k(m_1, m_2, \ldots, m_k),$$

f_k is a function, and k as well as each m_i varies over the positive integers [1882, 117].

After completing his 1878 paper on equipollence in $\mathbb{R}^n$, Cantor turned his attention back to derived sets, which had occupied him during his earlier investigations of trigonometric series [1872]. Between 1879 and 1884 he published a series of six articles, all titled "On Infinite Linear Point-Sets." There he studied the topological structure of subsets of $\mathbb{R}^n$, and especially of continua, by iterating his notion of derived set. In this context the Countable Union Theorem performed important services.

Cantor defined the derived set $P^{(1)}$ of P as the set of all limit points of P. For each ordinal α, he let $P^{(\alpha+1)}$ be the derived set of $P^{(\alpha)}$. Finally, if β was a limit ordinal,[5] then $P^{(\beta)}$ was defined to be the intersection of all those $P^{(\alpha)}$ such that $0 < \alpha < \beta$. The Countable Union Theorem, and hence indirectly the Assumption, helped to connect the cardinal number of a set to its derived sets:

(1.4.2) If P is a subset of $\mathbb{R}^n$ and if $P^{(\omega)}$ is denumerable, then P is also denumerable.

---

[3] Cantor 1895, 488. Since for Cantor the real numbers were the continuum *par excellence*, the power of the continuum meant the cardinal number of $\mathbb{R}$.

[4] Cantor 1878, 249; 1880, 356.

[5] A *limit ordinal* is an ordinal other than zero with no immediate predecessor.

Cantor's proof rested on a decomposition of $P^{(1)}$ into $P^{(m)} - P^{(m+1)}$, for every m, and $P^{(\omega)}$. Since $P^{(\omega)}$ was denumerable and since each set $P^{(m)} - P^{(m+1)}$ of isolated points was countable, then $P^{(1)}$ was denumerable by the Countable Union Theorem. Therefore P was also denumerable, as desired. *A fortiori*, the Countable Union Theorem served to obtain a stronger version of (1.4.2), where Cantor replaced ω by an arbitrary denumerable ordinal α, as well as in the following related result:

(1.4.3) If P is a subset of $\mathbb{R}^n$ such that $P^{(\alpha)} = 0$ for some countable α, then both $P^{(1)}$ and P are countable.[6]

Cantor's *Grundlagen*, the fifth article [1883a] in his series "On Infinite Linear Point-Sets," contained his initial treatment of well-ordered sets and infinite ordinal numbers. Although this article will be discussed at length in 1.5, here we must consider how the Countable Union Theorem became entangled with Cantor's number-classes of ordinals. The first number-class consisted of all finite ordinals (or positive integers), while the second was the set of all denumerable ordinals. One important theorem found in his article stated that the power of the second number-class is greater than that of the first. In his aleph-notation, introduced in 1895, this result asserted that $\aleph_1 > \aleph_0$ [1895, 492, 495]. His demonstration of 1883 relied on a proposition vital to the second number-class:

(1.4.4) If $\alpha_1, \alpha_2, \ldots$ is an increasing sequence of ordinals in the second number-class, then the least upper bound of the sequence is also in the second number-class.

In other words, the least upper bound of an increasing sequence of denumerable ordinals is a denumerable ordinal. To obtain this proposition, he applied the Countable Union Theorem and thus indirectly the Denumerable Assumption [1883a, 579–580]. Not until 1915 did Friedrich Hartogs give a proof for the theorem that $\aleph_1 > \aleph_0$ which did not rely on arbitrary choices. By contrast, (1.4.4) required the Axiom. Indeed, some later investigators attempted to determine the structure of the second number-class if one assumed (1.4.4) to be false and so negated the Axiom (see 4.7).

In the course of developing his theory of derived sets, Cantor used an instance of another proposition whose history illustrates how arbitrary choices and related assumptions gradually entered mathematics: the Partition Principle. In particular, he employed two of its special cases, although he did not state either of them explicitly as a theorem. The first of these, here called the Weak Partition Principle, asserted that if a set M is partitioned into

[6] Cantor 1883, 53–54; 1883b, 412–413; 1884, 463. The Countable Union Theorem, and hence the Denumerable Axiom as well, can be eliminated from the proof of (1.4.2) and (1.4.3). It suffices to use the constructive well-ordering of the real intervals with rational endpoints to enumerate the isolated points of P. Such an enumeration was given by Sierpiński [1917].

a family S of disjoint non-empty sets, and if S is equipollent to the second number-class, then M is uncountable (*i.e.*, not $\overline{\overline{M}} < \overline{\overline{S}}$). On the other hand, Cantor's second special case—here called the $\aleph_1$-Partition Principle since $\aleph_1$ is the cardinal of the second number-class—strengthened the conclusion of the first case to $\overline{\overline{S}} \leq \overline{\overline{M}}$ (or, in effect, $\aleph_1 \leq \overline{\overline{M}}$). While he could easily have established the Weak Partition Principle without arbitrary choices, the $\aleph_1$-Partition Principle required them.[7] In fact, the $\aleph_1$-Partition Principle is obtained by applying the Trichotomy of Cardinals to the conclusion of the Weak Partition Principle. These two special cases of the Partition Principle illustrate how, by strengthening in a natural way a proposition which does not depend on the Axiom, one obtains a proposition for which the Axiom is essential.

Cantor employed these special cases while demonstrating two theorems that revealed how, in $\mathbb{R}^n$, the eventual behavior of $P^{(\alpha)}$ depended on the cardinality of $P^{(1)}$:

(1.4.5) If $P^{(1)}$ is countable, then $P^{(\alpha)} = 0$ for some countable α.

(1.4.6) If $P^{(1)}$ is uncountable, then $P^{(\alpha)}$ is perfect for some countable α.

In one proof, published in 1884 but written the previous year, he relied on the Weak Partition Principle to insure that if $P^{(\alpha)} - P^{(\alpha+1)}$ is non-empty for every countable α, then the union of all such sets is uncountable [1884, 464–467]. An earlier demonstration of (1.4.5) had made use of the $\aleph_1$-Partition Principle at the same step [1883b, 413–414]. Cantor attached no particular significance to the difference between the Weak Partition Principle and the $\aleph_1$-Partition Principle, since he accepted the Trichotomy of Cardinals (see 1.5). However, Cantor's student Felix Bernstein implicitly used a stronger case of the Partition Principle in his doctoral dissertation [1901, 44], a fact which led Beppo Levi to state this principle in full generality—and to reject it.[8]

Of the results on derived sets, the best known is the Cantor–Bendixson Theorem. As originally formulated by Cantor in 1883, it contained an error which the young Swedish mathematician Ivar Bendixson soon detected. After this correction the theorem asserted that if $P^{(1)}$ is an uncountable subset of $\mathbb{R}^n$, then $P^{(1)}$ can be partitioned uniquely into a perfect set V and a countable set S, such that $S \cap S^{(\alpha)} = 0$ for some α less than the least uncountable ordinal Ω.[9] Accepting Bendixson's revision, Cantor [1884, 471] restated the theorem

[7] The $\aleph_1$-Partition Principle implies, in particular, that $\mathbb{R}$ has a subset of power $\aleph_1$. Using work by Lebesgue, Sierpiński [1918, 210–211] showed that $\mathbb{R}$ can be partitioned into a family S of $\aleph_1$ non-empty sets without the Axiom. Letting M be $\mathbb{R}$, the $\aleph_1$-Partition Principle then yields that $\aleph_1 \leq \overline{\overline{\mathbb{R}}}$.

[8] Levi 1902, 864; see 1.8. Also see 1.3 for a discussion of Burali-Forti's contribution.

[9] Bendixson 1883, 419; *cf.* Cantor 1883a, 575.

for a closed set, and it is in this form that the Cantor–Bendixson Theorem is most familiar: Every uncountable closed subset of $\mathbb{R}^n$ can be partitioned into a perfect set and a countable set.

While proving this theorem, both Cantor and Bendixson applied the Countable Union Theorem to deduce that if $P^{(1)}$ is uncountable, then $P^{(\Omega)}$ is perfect, and again to show that the set S (defined to be $P^{(1)} - P^{(\Omega)}$) is countable.[10] On the other hand, Sierpiński later provided a demonstration of the Cantor–Bendixson Theorem (as formulated for a closed set) that avoided the Axiom [1917]. Yet he soon noted that if the terms closed and perfect were replaced by sequentially closed and sequentially perfect respectively, then all known proofs of the theorem invoked the Axiom [Sierpiński 1918, 126].

When Cantor developed his final ideas on derived sets, the Assumption appeared in more sophisticated guises. His lengthy letter of 20 October 1884 communicated these ideas to the Swedish mathematician Gösta Mittag-Leffler, editor of the new journal *Acta Mathematica* and Cantor's chief correspondent at the time. To refine his previous decompositions of point-sets, Cantor introduced the coherence Pc of a subset P of $\mathbb{R}^n$ by letting $Pc = P^{(1)} \cap P$. Analogously to the αth derived set, he defined the αth coherence Pc^α to be $(Pc^{\alpha-1})c$, if α was a successor ordinal,[11] and to be the intersection of all Pc^γ with $0 < \gamma < \alpha$ if α was a limit ordinal. The principal theorems stated in his letter all used the Assumption implicitly:

(1.4.7) For each subset P of $\mathbb{R}^n$ there is a countable ordinal α such that Pc^α is empty or dense-in-itself.[12]

(1.4.8) If no subset of P is dense-in-itself, then P is countable.

(1.4.9) Every point-set P that is homogeneous of the $(\alpha + 1)$th order has the $(\alpha + 1)$th power (*i.e.*, $\aleph_\alpha$).[13]

Here P was homogeneous of the $(\alpha + 1)$th order if P was dense-in-itself and if, for each point p in P, every open ball with center p and positive radius had an intersection with P of power $\aleph_\alpha$.[14]

While Cantor did not demonstrate these theorems in his letter, he did so in an article soon published in *Acta Mathematica* [1885, 111–116]. The first two relied indirectly on the Countable Union Theorem. This indirect use of

---

[10] Bendixson 1883, 422–424; Cantor 1884, 466–467.

[11] An ordinal α is a *successor ordinal* if $\alpha = \beta + 1$ for some ordinal β.

[12] A non-empty set P is *dense-in-itself* if P is a subset of $P^{(1)}$.

[13] Cantor in Grattan-Guinness 1970, 76. In 1883 Cantor indexed his infinite cardinals beginning with the ordinal one, but in his later aleph-notation he began with zero.

[14] An *open ball* with center p and radius r is the set of all points whose distance from p is less than r.

the Assumption was mediated by a result which Cantor had obtained previously:

(1.4.10) In $\mathbb{R}^n$, if the intersection of a set P with every bounded set is countable, then P is countable. [1884, 457]

In fact, (1.4.8) can be shown without the Axiom, although (1.4.10) could not.[15] The proposition (1.4.9), which he considered easy to demonstrate and so left to the reader, is of interest here because its proof involves a generalization of the Countable Union Theorem:

(1.4.11) The union of countably many sets, each of power $\aleph_\alpha$, has power $\aleph_\alpha$.

Nevertheless, he did not state (1.4.11) explicitly at this time, and when a related result appeared in his *Beiträge* of 1895, he phrased it in a fashion that did not require the Assumption:

$$\aleph_\alpha \cdot \aleph_\alpha = \aleph_\alpha.$$

Yet the conflation of the two versions, which involved a change in how he defined cardinal multiplication, later led Borel to use the Assumption implicitly (see below and 1.7).

During 1885 Cantor translated many of the concepts that he had previously developed for point-set topology into the corresponding concepts for arbitrary ordered sets. He submitted to *Acta Mathematica* an article which dealt with these new concepts at length but which was never published in its original form during his lifetime. After the article had been partially set in page proofs, Mittag-Leffler urged Cantor not to publish it until he achieved significant new results by its methods, such as a proof or disproof of the Continuum Hypothesis. Otherwise, Mittag-Leffler feared, the abundance of new terminology and the predominantly philosophical tone of the article would tarnish Cantor's reputation among mathematicians. Quickly withdrawing his article, an indignant Cantor ceased submitting his papers to mathematical journals and for several years sent them instead to the philosophers.[16]

The unpublished article contained implicit uses of the Assumption that generalized (1.4.1) in the context of order-types. Given any two order-types α

---

[15] Sierpiński [1917] established without using the Axiom that every subset P of $\mathbb{R}^n$ can be partitioned into a set A, either empty or dense-in-itself, and a countable set B; from this theorem, (1.4.8) follows immediately. In a letter to the author, Thomas Jech showed that (1.4.10) cannot be proved in **ZF**: Consider any model of **ZF**, such as that given by Feferman and Levy [1963], in which there is an uncountable set P of real numbers that is the union of some sequence $P_1, P_2, \ldots$ of disjoint countable sets. For each positive n let f_n be a fixed bijection of $\mathbb{R}$ onto the interval $(n, n + 1)$. Then $\bigcup_{n=1}^{\infty} f''(P_n)$ is uncountable but its intersection with the interval $(-n, n)$, and hence with any bounded set, is countable. Thus in such a model (1.4.10) is false.

[16] For the text of the article and the events surrounding its withdrawal, see Grattan-Guinness 1970 but also May 1971.

and β, Cantor defined their product $\alpha\beta$ as follows: Let B be a set of type β and replace each member b of B by a set A_b of type α, where all A_b are disjoint; the union S of all these A_b has order-type $\alpha\beta$—provided that the order $<$ on S is taken such that (i) for c and d in the same A_b, $c < d$ if and only if this is the case in A_b, and (ii) for c in A_{b_1} and d in A_{b_2}, where b_1 and b_2 are distinct, $c < d$ if and only if b_1 is less than b_2 in B. However, $\alpha\beta$ was well-defined only if every set S' constructed by the same procedure was equipollent to S. In other words, the following generalization of (1.4.1) had to be true:

(1.4.12) If A_x is equipollent to B_x and if $A_x \cap A_y = B_x \cap B_y = 0$ for distinct x, y in a set C, then $\bigcup_{x \in C} A_x$ is equipollent to $\bigcup_{x \in C} B_x$.

Unlike (1.4.1), this proposition allowed A_x and B_x to be arbitrary sets, rather than subsets of $\mathbb{R}^n$, and permitted C to be arbitrarily large rather than denumerable. While (1.4.1) depended on the Denumerable Axiom, here the full force of the Axiom was required. In a similar fashion Cantor assumed (1.4.12) when he introduced order-types and their products for n-fold ordered sets, *i.e.*, sets of n-tuples with an order given on each coordinate.[17]

Although Cantor's article on order-types remained unpublished, a thoroughly revised version appeared in a philosophical journal along with a brief account of its earlier withdrawal [1887, 117]. There he stated (1.4.12) in print for the first time in order to establish that the product of two cardinals is well-defined. Before doing so, he introduced the product $\mathfrak{m}\mathfrak{n}$ of two cardinals in a fashion analogous to the product of two order-types, discussed above: For any sets A and B with $\overline{\overline{A}} = \mathfrak{m}$ and $\overline{\overline{B}} = \mathfrak{n}$, each b in B is replaced by a set A_b equipollent to A, where all A_b are disjoint; then $\mathfrak{m}\mathfrak{n}$ is defined to be $\overline{\overline{S}}$, where S is the union of all A_b with b in B. Since, by (1.4.12), S is equipollent to any other set S' constructed by the same procedure, then $\mathfrak{m}\mathfrak{n}$ is independent of the choice of A and B. Once again Cantor considered the proof of such equipollence to be so elementary that he did not provide it [1887, 121].

In the *Beiträge* of 1895, where Cantor finally unveiled his mature theory of infinite cardinal and ordinal numbers to a mathematical audience, he proposed a different definition for the product of two cardinals. It was this definition which eventually became the standard one: The product $\mathfrak{m}\mathfrak{n}$, with $\mathfrak{m} = \overline{\overline{A}}$ and $\mathfrak{n} = \overline{\overline{B}}$, is the cardinal of the set of all ordered pairs (a, b) with a in A and b in B [1895, 485–486]. By generalizing this approach, he succeeded in defining exponentiation for infinite as well as finite cardinals: $\mathfrak{m}^{\mathfrak{n}}$ is the power of the set of all functions from B to A.[18] Although his new definition for $\mathfrak{m}\mathfrak{n}$ did not involve infinitely many arbitrary choices, Cantor stated that his

[17] Cantor in Grattan-Guinness 1970, 91, 98. As early as [1883, 551], Cantor had defined the special case of the product of two ordinals essentially in the manner stated above for the product of order-types.

[18] Cantor 1895, 486–487; see also Dauben 1979, 173–175.

earlier definition was a consequence of the new one, and thereby he used the Assumption implicitly. In the later work of Whitehead and Russell, discussed in 1.7, the equivalence of these two definitions would become the theorem that the product $\mathfrak{m}\mathfrak{n}$ equals the sum of $\mathfrak{m}$ sets each of power $\mathfrak{n}$, even if $\mathfrak{m}$ is infinite:

$$\overline{\overline{A}} \cdot \overline{\overline{B}} = \sum_{a \in A} \overline{\overline{B}}.$$

When Cantor published the second installment of his *Beiträge* in 1897, the use of arbitrary choices appeared indirectly once more. Here the structure of his second number-class was involved. As he had done more than a decade earlier, he used the Countable Union Theorem to demonstrate the proposition (1.4.4) that the second number-class is closed under the limit of a sequence. This closure enabled him to assert that if α and β are in the second number-class, then so is α^{β}.[19] Likewise he relied on (1.4.4) to deduce the existence of epsilon-numbers in the second number-class, indeed, aleph-one of them.[20]

Thus from 1877 until 1897, the end of his published research, Cantor used the Assumption implicitly again and again to establish fundamental theorems in set theory and point-set topology. However, he rarely employed arbitrary choices in the direct manner found in his proof that every infinite set has a denumerable subset (see 1.3). Instead, the Assumption occurred most frequently in certain basic results, which he considered too elementary to demonstrate, and in those theorems deduced from them. The former consisted of the Countable Union Theorem and the propositions (1.4.1) and (1.4.12) which justified, in effect, the sum of infinitely many cardinals or order-types. A restricted form of the Partition Principle first appeared in his research, but he did not state it explicitly and used it only sporadically. In general, the Denumerable Assumption sufficed for his theorems in point-set topology, because he worked in $\mathbb{R}^n$, and would suffice as well in any other topological space with a countable basis. But when his abstract set theory (embodied in infinite order-types and infinite cardinal numbers) emerged in a mature form, Cantor implicitly relied on the full force of the Assumption.

Despite his repeated implicit uses of the Assumption, Cantor failed to recognize that to make infinitely many arbitrary choices required a new mathematical principle. In part, this was because they occurred primarily in justifying the basic results, mentioned above, which seemed so intuitively obvious. From them the Assumption spread hereditarily to the propositions whose proofs utilized those basic results. It was through such offspring that the Assumption came to be used indirectly by mathematicians such as Borel, Lebesgue, and Russell. Thus, in particular, Burali-Forti [1894, 178] employed it indirectly while expressing Cantor's definition for the product of two order-types within Peano's logical symbolism.

---

[19] Unlike the other implicit uses of the Assumption discussed in this paragraph, this one turned out to be avoidable. See Kuratowski and Mostowski 1968, 276.

[20] Cantor 1897, 222, 231–233, 242–244. An *epsilon-number* is an ordinal α such that $\omega^{\alpha} = \alpha$.

Essentially, the significance of the Assumption in Cantor's work is twofold. First of all, Cantor created a body of important theorems in which the Assumption was unavoidable and through which its implicit use was absorbed by many other mathematicians. Hence, when later researchers examined the underlying presuppositions more closely than Cantor had done, they found that a special axiom of existential character was at work.

Moreover, the second aspect was a transformation in what it meant to prove that a mathematical object "exists." When in 1874 Cantor published a new demonstration of Liouville's Theorem that there exist infinitely many transcendental numbers, he did not define or construct such numbers but showed instead that by assuming their non-existence one was led to a contradiction. Since the time of Euclid, the existence of a mathematical object with a given property had been established by constructing a particular instance of it. On the other hand, Euclid also employed indirect proofs wherein an object was shown to exist in the fashion used by Cantor. The significant difference, not found in Cantor's proof of Liouville's Theorem but inherent in all his theorems requiring the Axiom of Choice, was that here an indirect proof could *not* be replaced by a direct or constructive one. In this way the Axiom epitomized a fundamental shift in the meaning of mathematical existence.

1.5 The Well-Ordering Problem and the Continuum Hypothesis

Although many propositions are logically equivalent to the Axiom of Choice in axiomatic set theory, they have not always been psychologically equivalent. At various periods and to various mathematicians, these propositions have possessed differing degrees of self-evidence. Sometimes one of them seemed true and another false. Thus their logical equivalence was unsuspected or even doubted. Of central importance here are two such propositions, both of which arose before Zermelo formulated the Axiom: Cantor's Well-Ordering Principle (every set can be well-ordered) and the Trichotomy of Cardinals ($\mathfrak{m} < \mathfrak{n}$ or $\mathfrak{m} = \mathfrak{n}$ or $\mathfrak{m} > \mathfrak{n}$). Even though Cantor first mentioned Trichotomy in 1878, it did not assume a pivotal role in his set theory for nearly two decades, when he and other mathematicians seriously attempted to prove it. By contrast, he originally proposed the Well-Ordering Principle in 1883 as a self-evident logical law. Over a decade later, when his principle had gained no supporters, he attempted to deduce it at the same time as Trichotomy. In this way his Well-Ordering Principle was transformed into the Well-Ordering Problem (can every set be well-ordered?) and, with Zermelo's affirmative answer, into the Well-Ordering Theorem. Then the controversy began.

When Cantor originated the Trichotomy of Cardinals in his 1878 article showing that $\mathbb{R}$ and $\mathbb{R}^n$ are equipollent, he did so without fanfare. Rather than presenting Trichotomy either as an axiom or as a proposition in need of proof, he believed it to follow immediately from the definition of equipollence: "If two point-sets [*Mannigfaltigkeiten*] M and N are not of the same power, then either M will have the same power as a proper subset of N, or N will have the same power as a proper subset of M; in the first case we call the power of M smaller, and in the second case larger, than the power of N."[1] Apparently he did not realize at the time that he was not merely stating a definition of the relation $<$ between cardinals. Thus he was unaware that to exclude the existence of sets A and B, neither of which was equipollent to a subset of the other, was not a trivial problem. Cantor, who did not apply Trichotomy in any proof found in his article, clearly had no sense of its considerable deductive strength. At the time such strength was restricted by how he construed the term *Mannigfaltigkeit*. Unquestionably this term included any subset of $\mathbb{R}^n$, but probably not an arbitrary set (*Menge*). For he did not use the term *Menge*, or publish any discussion of arbitrary sets, until four years later [1882, 114]. Only in 1887 did he state Trichotomy for arbitrary sets, and only in 1895 did he discuss it fully. In fact, Trichotomy was one of those fundamental propositions which at first he considered obvious but which, as he refined and generalized his concept of set, required a more critical analysis.

Cantor reached a watershed late in 1882 when he formulated the paramount concept in his theory of sets: well-ordering. A cornucopia of ideas flowed from this notion in the long and enthusiastic letter which he sent to Dedekind on 5 November. What primarily motivated Cantor was the desire to extend the sequence of positive integers in a natural fashion. Previously he had introduced the "symbols of infinity"

$$\infty, \infty + 1, \infty + 2, \ldots$$

only as superscripts α for his derived sets $P^{(\alpha)}$ [1880]. As late as September 1882, when he completed another paper [1883] on derived sets, he emphasized the merely formal significance of these symbols. Yet in his November letter to Dedekind he insisted on their legitimacy as *numbers*. There, in place of ∞, he introduced the ordinal number ω as the limit of the sequence $1, 2, \ldots$. This procedure, which he called the second principle of generation, justified the formation of limit ordinals. The first principle of generation permitted the operation of adding one to any ordinal, an operation used to obtain not only the finite ordinal numbers but also $\omega + 1$ and all other successor ordinals. In contrast, the third principle created the boundaries of his number-classes by stipulating that a number α be in the $(\beta + 1)$th number-class if the set of all ordinal predecessors of α had the power of the βth

[1] Cantor 1878, 242. In other words, the power of a set A is *less than* that of B if A is equipollent to a subset of B but not to B itself.

number-class. Thus the second number-class was the set of all denumerable ordinals. Through this process he conceived of an unlimited sequence of number-classes, which connected the notions of ordinal and cardinal number. In particular, the power of the second number-class was the immediate successor of the power of the first. Here the sequence of cardinals which he later called alephs, the powers of infinite well-ordered sets, came into being.[2] He soon published the strong claim that there exist just as many number-classes (and hence infinite powers) as ordinals.[3]

Of particular importance, Cantor noted in his 1882 letter, was the relationship between his new ordinal numbers and the continuum $\mathbb{R}$. There he gave the name "Two-Class Theorem" to the conjecture which he had first stated four years earlier, in the conclusion to his article showing that $\mathbb{R}$ and $\mathbb{R}^n$ are equipollent:

(1.5.1) Every infinite subset of $\mathbb{R}$ is either denumerable or has the power of the continuum. [1878, 257]

This was the original formulation of the Continuum Hypothesis, a name that Cantor seems never to have used. His letter also contained the new theorem that every infinite subset of the second number-class is either denumerable or has the power of that class. It is quite likely that this theorem, together with (1.5.1), suggested to him the stronger version of the Continuum Hypothesis first given in the letter:

(1.5.2) $\mathbb{R}$ has the power of the second number-class.[4]

In his later aleph-notation of 1895, the Continuum Hypothesis (1.5.2) would state that

$$2^{\aleph_0} = \aleph_1.$$

After 1882 his interest in the Continuum Hypothesis shifted from the earlier Two-Class Theorem to the form (1.5.2), involving well-ordering, that he repeatedly and unsuccessfully attempted to prove.

The difference between these two forms is relevant to the Axiom of Choice. As Cantor understood very well, (1.5.2) implies that the real numbers can be well-ordered, while (1.5.1) does not. Moreover, (1.5.2) is equivalent

[2] Cantor in Noether and Cavaillès 1937, 55–59. However, as Zermelo later noted, Cantor's three principles did not suffice to generate the ωth number-class (see Cantor 1932, 199).

[3] Cantor 1883a, 588. This assertion required that the universe of sets contain exactly κ ordinals, for some κ satisfying the equation $\aleph_\kappa = \kappa$. The least κ for which this is possible is $\lim_{n<\omega} \alpha_n$, where $\alpha_0 = \omega$ and $\alpha_{n+1} = \omega_{\alpha_n}$.

[4] Cantor in Noether and Cavaillès 1937, 57. In particular, the new theorem yielded that there is no $\mathfrak{m}$ such that $\aleph_0 < \mathfrak{m} < \aleph_1$, while (1.5.1) implied that no $\mathfrak{m}$ exists such that $\aleph_0 < \mathfrak{m} < \overline{\overline{\mathbb{R}}}$. From Trichotomy, which Cantor accepted, it then followed that $\mathbb{R}$ is equipollent to the power $\aleph_1$ of the second number-class.

to the conjunction of (1.5.1) and the existence of a well-ordering for $\mathbb{R}$. Once Cantor had made the conceptual leap from his proof that the set of real algebraic numbers can be well-ordered to his belief—motivated by the Continuum Hypothesis (1.5.2)—that $\mathbb{R}$ can be well-ordered, it was a natural next step to assert the Well-Ordering Principle that *every* set can be well-ordered. He took that step, not in his letter to Dedekind, but in the lengthy article (the *Grundlagen*) which developed from it:

> The concept of *well-ordered set* turns out to be essential to the entire theory of point-sets. It is always possible to bring any *well-defined* set into the *form* of a *well-ordered* set. Since this law of thought appears to me to be fundamental, rich in consequences, and particularly marvelous for its general validity, I shall return to it in a later article. [1883a, 550]

With this flourish the Well-Ordering Principle was born. While the promise to return was one that Cantor made for the Continuum Hypothesis as well, he never justified either of them in print to his own satisfaction.

In his 1882 letter to Dedekind, Cantor had claimed that he could prove the Continuum Hypothesis (1.5.2) rigorously, and soon afterward he employed (1.5.2) without special mention in published demonstrations of other results. Twice during 1883 he asserted that a certain subset of $\mathbb{R}^n$ had the power of the second number-class when he had merely shown the subset to be uncountable—in his definition of a semicontinuum and in his first statement of the Cantor–Bendixson Theorem [1883a, 590, 575]. Cantor also extended the Continuum Hypothesis by stating without proof that the set of all real functions has the power of the third number-class [1883a, 590]. Here, well before he established in 1891 that $\mathfrak{m} < 2^{\mathfrak{m}}$ for every cardinal $\mathfrak{m}$, he asserted that $\mathfrak{o} < \mathfrak{o}^{\mathfrak{o}}$, where $\mathfrak{o}$ was the power of the continuum, and that, expressed in his later aleph-notation, $\mathfrak{o}^{\mathfrak{o}} = \aleph_2$, which yields

$$2^{\aleph_1} = \aleph_2.$$

Thus he held that the set of all real functions can be well-ordered and has the next power after $\aleph_1$. Apparently Cantor never generalized his Continuum Hypothesis to any higher powers, as Hausdorff and Tarski were later to do (see 2.5 and 4.3).

The Continuum Hypothesis was also entangled with Cantor's original statement of the Equivalence Theorem in 1882:

> If $M'' \subseteq M' \subseteq M$, and if M has the same power as M'', then M also has the same power as M'.

When he spoke with Dedekind in Harzburg during September, Cantor confessed his inability to demonstrate this result. Yet in his November letter he declared happily: "Now I have found the source of this proposition and can rigorously prove it with the necessary generality, thereby filling an essential gap in the theory of point-sets."[5]

The source of the Equivalence Theorem, which Cantor at first confined to subsets of $\mathbb{R}^n$, was the Continuum Hypothesis (1.5.2). In fact, his restricted form of the Equivalence Theorem followed easily from (1.5.2). When he first published this theorem in 1883, he stated it only under the hypothesis that M has the power of the second number-class [1883a, 574, 582]. On the other hand, his letter to Dedekind ended by posing the theorem, for sets of arbitrary power, as an open problem.[6] During 1884 Cantor affirmed in print that the Equivalence Theorem held for every set M of any power, but he did not offer a proof [1884a, 388]. Perhaps he intended to obtain such a proof by using the Well-Ordering Principle. In any case, the question did not attract the attention of other mathematicians until he raised it anew in his *Beiträge* of 1895 vis-à-vis the Trichotomy of Cardinals.

Cantor's first major advance towards establishing the Continuum Hypothesis occurred in 1884 when he demonstrated a special case of (1.5.1): If an infinite subset A of $\mathbb{R}$ is closed, then A is either denumerable or has the power of the continuum. His proof combined the Cantor–Bendixson Theorem for a closed set with his new result that a perfect subset of $\mathbb{R}$ has the power of the continuum. Once again, Cantor concluded his article with a promise to demonstrate (1.5.1) for *every* infinite subset of $\mathbb{R}$ and to show that $\mathbb{R}$ itself has the power of the second number-class. He hoped to receive substantial assistance from his theorem that no set has a power between those of the first and second number-classes [1884, 488].

On 26 August 1884, believing that at last he had succeeded, Cantor sent his demonstration to Mittag-Leffler: "I am now in possession of an extremely simple proof for the most important theorem of set theory, that the continuum has the power of the second number-class."[7] To begin, Cantor claimed that there exists a closed set M of real numbers with the power of the second number-class. Since, by the Cantor–Bendixson Theorem, M could be decomposed into a perfect set P and a countable set, then P had the power of the second number-class. But any perfect set such as P had the power of the continuum, and consequently $\mathbb{R}$ must also have the power of the second number-class. Thus it only remained, Cantor concluded, to define a single closed set with the power of the second number-class. When all was in order, he would forward the details.

Cantor failed to obtain such a closed subset of $\mathbb{R}$ with power aleph-one, but he soon attacked the problem from a different direction. As he refined his research on derived sets by introducing notions such as the coherence of a set, he discovered another argument for the Continuum Hypothesis. In his lengthy letter of 20 October 1884, discussed in 1.4, he communicated the

[5] Cantor in Noether and Cavaillès 1937, 55; *cf.* Cavaillès 1962, 85n.

[6] Cantor in Noether and Cavaillès 1937, 59.

[7] Cantor in Schoenflies 1927, 16.

underlying ideas to Mittag-Leffler. Central to his new proof was a decomposition theorem: Every subset of $\mathbb{R}$ can be partitioned into three sets, some of which may be empty—a set A with no subset dense-in-itself, a homogeneous set B of first order, and a homogeneous set C of second order.[8] From (1.4.7) and (1.4.8) it followed that A and B were countable and that C, if non-empty, had the power of the second number-class. Thus the Continuum Hypothesis (1.5.2) followed at once.

However, on 14 November he wrote excitedly to Mittag-Leffler that this "proof" was erroneous. His decomposition theorem contained a flaw:

> You know that I have often believed myself to possess a rigorous proof that the linear continuum has the power of the second number-class. Time and again gaps have arisen in my proofs, and each time I exerted myself anew in the same direction. When I believed I had reached the ardently desired goal, suddenly I was startled because in some hidden corner I noticed a false conclusion.
>
> These days, as I again strove toward the goal, what did I find? I found a *rigorous* proof that the continuum does *not* have the power of the second number-class, and, what is more, that it has no power specifiable by an ordinal number.
>
> No matter how dreadful the error, which has been fostered for so long, to eliminate it definitively is that much greater a victory. [Cantor in Schoenflies 1927, 17]

Thus a chastened Cantor came to believe that both the Continuum Hypothesis and the Well-Ordering Principle were false, since the real numbers could not be well-ordered. As a result, there existed no closed subset of $\mathbb{R}$ with power aleph-one. All of this meant that set theory had a structure extremely different from what he had believed hitherto. Uneasily he concluded: "For today I ask you to be content with this brief announcement. I am more astounded over it myself than I can tell you" [*Ibid.*].

Yet the reversals had not run their course. The following day Cantor wrote Mittag-Leffler to reject his own previous reasons for abandoning the Continuum Hypothesis and the Well-Ordering Principle. On the contrary, Cantor now had every reason for believing the Continuum Hypothesis to be true, and he hoped to obtain a definitive proof in the near future. A day later, on 16 November, he explained what had occurred. While composing his recent article [1884a] on point-sets, he had discovered the following theorem: If A and B are dense, denumerable, ordered sets without endpoints (and hence are homogeneous of the first order), then A and B are order-isomorphic. On 14 November he had found, so he thought, a proof for an analogous isomorphism theorem where A and B were homogeneous sets of the second order. However, this last theorem would imply that $\mathbb{R}$ did not have the power of the second number-class. For if $\mathbb{R}$ had such a power and

---

[8] Cantor in Grattan-Guinness 1970, 78. A subset M of $\mathbb{R}$ is *homogeneous of the nth order* if M is dense-in-itself and if for every $p \in M$, the intersection of the interval $(p - r, p + r)$ with M has the power $\aleph_{n-1}$ for every $r > 0$.

if one let A be the open interval (0, 1) and B be the set of all irrationals in this interval, then the order-isomorphism between A and B would yield that the set B of irrationals was perfect—an absurdity. Now, of course, he was convinced that the last "theorem" could not be demonstrated and that his proof of it was in fact false.[9] With this anticlimax, Cantor's most intensive period of research on the Continuum Hypothesis came to an end.

In contrast to the Continuum Hypothesis, the Well-Ordering Principle continued to play a central role in Cantor's published research by connecting cardinals to ordinals. The connection had already appeared in the *Grundlagen*, where he treated each infinite cardinal as the power of the αth number-class for some ordinal α [1883a, 588]. In his aleph-notation of 1895, this proposition would state that every infinite cardinal is an aleph—a result that we shall call the Aleph Theorem.

One form of this theorem, equivalent to the Well-Ordering Principle, occurred in Cantor's letter of 20 October 1884:

(1.5.3) The cardinal numbers are well-ordered by magnitude.[10]

There (1.5.3) served to categorize limit points according to their order. He defined a limit point p (of a point-set P) to be of the αth order if there was some positive r such that, for every open ball B with center p and positive radius less than r, $P \cap B$ had power $\aleph_{\alpha-1}$. Then he asserted:

(1.5.4) In $\mathbb{R}^n$, any limit point of a set P is of the αth order for some α.

A detailed proof of (1.5.4) appeared in his article of 1885 on point-sets: Suppose that p is a limit point of P and that $r_1, r_2, \ldots$ is a sequence of positive real numbers converging to zero. Let α_i be the ordinal corresponding to the power of $P \cap B_i$, where B_i is the open ball with center p and radius r_i. Since $\alpha_1, \alpha_2, \ldots$ is a non-increasing sequence of ordinals, then there exists some n such that $\alpha_m = \alpha_n$ whenever m is greater than n. Hence any ball with center p and sufficiently small positive radius intersects P in the α_nth power, and consequently p is of the α_nth order, as desired [1885, 117–118]. This proof was typical of how (1.5.3) and the Aleph Theorem reduced problems of cardinality to problems concerning ordinals, where the tool of well-ordering was available.

As late as 1891, when Cantor established that

$$\mathfrak{m} < 2^{\mathfrak{m}}$$

for every cardinal $\mathfrak{m}$ and concluded that there is no largest cardinal, he continued to rely uncritically on the Well-Ordering Principle. In particular, he claimed [1891, 77] that in his *Grundlagen* of 1883 he had proved (1.5.3).

[9] Cantor in Schoenflies 1927, 18–19.

[10] Cantor in Grattan-Guinness 1970, 77.

Cantor's purported proof, not clearly stated in the *Grundlagen*, must have used the Well-Ordering Principle. Yet by the time that he published the *Beiträge* in 1895, his attitude toward this principle, the Aleph Theorem, and the Trichotomy of Cardinals had undergone a radical change: All of them, he now believed, required a demonstration.

One of Cantor's principal concerns in the *Beiträge* was the Trichotomy of Cardinals. Previously he had considered this proposition to follow from the definition of $<$ for cardinals—in [1878, 242] for point-sets in $\mathbb{R}^n$ and in [1887, 120] for arbitrary sets. Now in 1895 he presented Trichotomy in the following form:

$$\mathfrak{m} = \mathfrak{n} \quad \text{or} \quad \mathfrak{m} < \mathfrak{n} \quad \text{or} \quad \mathfrak{n} < \mathfrak{m}$$

for any cardinals $\mathfrak{m}$ and $\mathfrak{n}$. All the same, he carefully avoided using Trichotomy in his demonstrations because it "*is by no means immediate and can scarcely be proved at this stage*" [1895, 484; Cantor's emphasis]. On the other hand, he stressed that a demonstration for Trichotomy would immediately yield one for the Equivalence Theorem, which had again become an open problem.[11] Indeed the Equivalence Theorem, which Cantor had originally derived in $\mathbb{R}^n$ from the Continuum Hypothesis, had probably been deduced for arbitrary sets from the Well-Ordering Principle. While he did not refer directly to this principle, he continued to believe that every infinite cardinal is an aleph. In this vein he reaffirmed the truth of (1.5.3) [1895, 495]. Nevertheless, he had not yet arrived at a proof for any of these propositions.

Thus Cantor's *Beiträge*, his last and most comprehensive publication on set theory, left unresolved a number of the most fundamental and interrelated questions: the Trichotomy of Cardinals, the Equivalence Theorem, the Well-Ordering Principle, and the Continuum Hypothesis. It was the first two of these, highlighted in the *Beiträge* as the others were not, that quickly attracted attention.

Although Cantor did not know at the time, the *Beiträge* soon found a perceptive reader in Burali-Forti. In December 1896, shortly after writing his article on Dedekind-finite sets, Burali-Forti presented a paper on the Trichotomy of Cardinals to the Academy of Sciences at Turin. There he decomposed Cantor's new form of Trichotomy into the Equivalence Theorem and the following version of Trichotomy:

(1.5.5) For any classes A and B, either A is equipollent to a subclass of B, or B is equipollent to a subclass of A.

---

[11] Cantor 1895, 484. The Equivalence Theorem followed immediately from the Trichotomy of Cardinals because Cantor had just introduced a new definition for $\overline{\overline{A}} < \overline{\overline{B}}$. In 1895, $\overline{\overline{A}} < \overline{\overline{B}}$ meant that A is equipollent to a proper subset of B and that B is equipollent to no proper subset of A. With Cantor's original definition of 1878, whereby $\overline{\overline{A}} < \overline{\overline{B}}$ meant that A is equipollent to a proper subset of B but not to B itself, the Equivalence Theorem did not follow trivially from Trichotomy.

Burali-Forti demonstrated the Equivalence Theorem in case either A or B was countable, but he did not do so in general. On the other hand, he deduced Trichotomy in the form (1.5.5) from two new postulates—his version of the Partition Principle and the following:

(1.5.6) If A and B are uncountable classes, then there is a function $f: A \rightarrow B$ which is either one–one or onto.

He believed that (1.5.6) was a truth which possessed "the degree of simplicity and clarity appropriate to primitive propositions [postulates]" [1896a, 236].

Burali-Forti divided his proof of (1.5.5) into three cases. The first of these, where either A or B was Dedekind-finite, had appeared in his previous article on finite sets.[12] The second, where either A or B was denumerable, relied on Cantor's 1895 theorem that every infinite set has a denumerable subset. Indirectly both cases used the Denumerable Assumption. However, it was the third case, where A and B were uncountable, that involved his new postulate (1.5.6): Suppose that A is not equipollent to a subclass of B. Then by (1.5.6) there exists a function f mapping A onto B. For each b in B, let $\bar{f}b$ be the inverse image of $\{b\}$, and let $\bar{f}B$ be the class $\{\bar{f}b: b \in B\}$. By the Partition Principle, $\bar{f}B$ is equipollent to a subclass of the union of $\bar{f}B$, that is, a subclass of A. Since $\bar{f}B$ is also equipollent to B, then B is equipollent to a subclass of A, as desired [1896a, 236–237].

What is intriguing about this demonstration, which passed unnoticed both then and later, is its reliance on new postulates to deduce the Trichotomy of Cardinals. We now know that if Burali-Forti's postulate (1.5.6) is assumed, then the Trichotomy of Cardinals holds (and hence the Axiom of Choice as well) even without using the Partition Principle.[13] Regrettably, Burali-Forti did not investigate the Well-Ordering Problem, which Cantor had not stated directly in the *Beiträge*, and thus made no attempt to find an axiomatic solution to it. In fact, the Well-Ordering Problem did not interest mathematicians other than Cantor until Hilbert emphasized its importance in 1900 (see 1.6).

---

[12] Burali-Forti 1896, 50; see 1.3.

[13] We first show that (1.5.6) implies the same proposition with the word uncountable omitted: If A and B are both countable, we are done. So suppose that A is countable but B uncountable. It suffices to deduce that $\aleph_0 \leq \bar{\bar{B}}$ in order to have a one–one mapping from A into B. By (1.5.6) there is some $f: \aleph_1 \rightarrow B$ that is one–one or onto. Since the proof is complete if f is one–one, assume that f is onto. Then we define $g: B \rightarrow \aleph_1$ such that, for each b in B, $g(b)$ is the least α with $f(\alpha) = b$. It follows that g is one–one, giving $\bar{\bar{B}} = \aleph_1$, and so $\aleph_0 \leq \bar{\bar{B}}$, as desired. Finally, to show that (1.5.6), with the word uncountable omitted, implies the Axiom, we suppose that there is no function from A onto B. Then there exists a one-one function $f: A \rightarrow B$. We fix a member u of A and define a function $g: B \rightarrow A$ by letting $g(x) = f^{-1}(x)$ if $x \in f''A$, and $g(x) = u$ if $x \in B - f''A$. Then g maps B onto A. Consequently, either A can be mapped onto B or vice versa, and so Lindenbaum's theorem (4.3.13) holds, from which the Axiom follows.

Whereas Burali-Forti attacked the problems of Trichotomy and the Equivalence Theorem in Peano's symbolic logic, Ernst Schröder approached them from a different, though equally logical, direction. Using the tools of algebraic logic honed by George Boole and C. S. Peirce, Schröder composed an article on those problems in 1896. Published two years later, Schröder's investigations were of uneven value. The result for which he is best known, the first proof of the Equivalence Theorem to appear in print [1898, 336–344], was flawed by an irreparable error. While Cantor, Peano, and Schoenflies all regarded this proof as correct,[14] Alwin Korselt wrote to Schröder about the error on 8 May 1902. In his reply two weeks later Schröder admitted his mistake, which he had acknowledged to Max Dehn the previous year. At the end of May 1902, Korselt submitted his own proof to *Mathematische Annalen* for publication, but for unknown reasons it did not appear there until 1911.[15]

In contrast to the Equivalence Theorem, Schröder's article contained suggestive comments on the Trichotomy of Cardinals. He believed that a proof of Trichotomy could be obtained only with the aid of algebraic logic and that even then the obstacles would be formidable. Nevertheless, he proposed a direction in which to search:

> And with regard to its proof, the essential point will be to find a process of exhaustion, which, by continuing to assign the elements of one of the two sets [A and B] one-one to any elements chosen from the other set, completely exhausts one set or the other. (I let the problem rest here.) [1898, 349]

Perhaps, he concluded, certain theorems concerning order would be helpful in such a proof. Thus, while suggesting the use of arbitrary choices to demonstrate Trichotomy, he did not indicate how to proceed. As we shall see in 1.6, Cantor employed a similar method of attack.

Although Schröder did not investigate the Well-Ordering Problem, he discussed a subsidiary proposition which later became known as the Ordering Principle: Every set can be ordered. After listing various sets for which an order was known, such as any countable set and any set with the power of the continuum, he remained uncertain whether the Ordering Principle was true. In particular, he suspected that it might be impossible to order the domain of everything thinkable, the *absolute Denkbereich*.[16] When he returned to related questions in a paper presented at the First International Congress of Philosophy (Paris, 1900), he stressed that it was still unknown

---

[14] Letter of 30 August 1899 from Cantor to Dedekind, in Cantor 1932, 450; Peano 1906a, 361; Schoenflies 1900, 16.

[15] Korselt 1911, 294–296. This proof resembled Dedekind's unpublished demonstration of 1887 (see footnote 8 in 1.6).

[16] Schröder 1898, 356–358. In 1888 Dedekind had used this domain of everything thinkable to justify the existence of an infinite set (see 3.2).

could be considered as a unity (*Einheit*), a single object, without contradiction. Each of these he termed a consistent multitude (*konsistente Vielheit*) or a set (*Menge*) [1932, 444].

Next, Cantor connected these two kinds of multitudes to his notions of cardinal number and order-type. Each set, or consistent multitude, possessed a cardinal number, and each ordered set had an order-type. On the other hand, he assigned neither a cardinal nor an order-type to any inconsistent multitude. While each well-ordered multitude was called a sequence (*Folge*), he retained the term ordinal number for the order-type of a well-ordered set.

Using these definitions, he established that the multitude Ω of all ordinal numbers in their natural order was a well-ordered, absolutely infinite multitude. In the second installment of the *Beiträge* he had already shown that the ordinals were ordered by magnitude [1897, 216]. It then followed easily from his theorems on well-ordered sets that every non-empty subcollection of Ω had a least element and hence that Ω was well-ordered. On the other hand, if Ω were a consistent multitude, then it would have an ordinal δ greater than every ordinal in Ω—an absurdity since δ was in Ω. Therefore Ω was absolutely infinite, as desired. The system ת of all alephs, he noted, was likewise an absolutely infinite sequence because it was order-isomorphic to Ω.[3]

At this point in his letter Cantor raised the fundamental question: Does the multitude ת contain every infinite cardinal, or, on the contrary, is there some infinite set whose cardinal is not an aleph? He argued that there exists no such infinite set: Suppose that some definite multitude V is not equipollent to an aleph. "Then one easily sees," he continued, "that under the hypothesis the entire system Ω can be mapped one-one into the multitude V, *i.e.*, that there must exist a submultitude V' of V which is equipollent to the system Ω."[4] Hence V' is inconsistent, since Ω is, and consequently V is inconsistent as well. But since cardinal numbers are defined only for consistent classes, then every infinite cardinal is an aleph, as desired. In this way the Aleph Theorem, as well as the Well-Ordering Principle, was demonstrated. Furthermore, since

$$\alpha < \beta \quad \text{or} \quad \alpha = \beta \quad \text{or} \quad \alpha > \beta$$

for any ordinals α and β, he noted that

$$\aleph_\alpha < \aleph_\beta \quad \text{or} \quad \aleph_\alpha = \aleph_\beta \quad \text{or} \quad \aleph_\alpha > \aleph_\beta.$$

By the Aleph Theorem he concluded that Trichotomy held for all cardinals. Thereby Cantor believed that he had resolved the chief problems, other than the Continuum Hypothesis, which remained outstanding in set theory.

While editing Cantor's collected works in 1932, Zermelo commented at length on this attempted proof of the Aleph Theorem. Zermelo found the

[3] Cantor 1932, 444–447. ת, or Tav, is the last letter in the Hebrew alphabet.
[4] Cantor 1932, 447. Implicitly he assumed V to be infinite.

even if one could order the set of all real functions.[17] Like other mathematicians at the time, Schröder remained unaware that the Trichotomy of Cardinals implies the Ordering Principle and, indeed, the Well-Ordering Principle as well.

Independently of Schröder, though also from the standpoint of algebraic logic, C. S. Peirce analyzed both Trichotomy and the question whether every collection can be ordered. In 1893 Peirce concluded that no contradiction would result from assuming that some collection cannot be ordered [1933, 70–71]. On the other hand, during a lecture given in 1897, he mentioned Cantor's relation $<$ between cardinals, and offered an argument that Trichotomy holds. To show that for any collections A and B, either A is equipollent to a subcollection of B or else B is equipollent to a subcollection of A, Peirce began by introducing the collection of all relations with domain A and range B. For each such relation, he considered every possible subrelation S that was a function with domain A. Either there existed a subrelation of S that was a one–one function with domain A (in which case A was equipollent to a subset of B), or else S itself had range B. In the latter case, the relation converse to S was one–one with domain B, and so had a subrelation that was a one–one function $f: B \to A$. Consequently, B was equipollent to a subset of A, as desired [Peirce 1933, 149–152]. This argument, not published at the time, apparently had no influence.

In Peirce's argument there occurred two assumptions, neither of which was noted with care and each of which was later shown to be equivalent to the Axiom: Every relation includes a function with the same domain; every function includes a function (with the same domain) that is either one–one or onto. At the time Peirce was uncertain whether his demonstration was correct. However, writing to Philip Jourdain on 5 December 1908, Peirce could no longer ascertain either the nature or the validity of his previous doubts.[18]

In France another reader of the *Beiträge*, Emile Borel, was also preoccupied with Trichotomy, although more dubious of its truth than either Schröder or Peirce. At the First International Congress of Mathematicians, held in Zurich during 1897, Borel accosted Cantor and questioned him on the subject of cardinals. Thereby Borel learned that Felix Bernstein, who was studying under Cantor at Halle, had recently succeeded in demonstrating the Equivalence Theorem. In his proof Bernstein relied on (1.4.1), and hence indirectly on the Denumerable Assumption, though in an avoidable way.

[17] Schröder 1900, 239; *cf.* Couturat 1900, 400. Writing to Frege on 27 June 1903, Korselt was sceptical of both Cantor's Trichotomy of Cardinals and Schröder's Ordering Principle [Frege 1980, 87–88].
[18] Letter quoted by C. Eisele in Peirce 1976, 879–880. Peirce's concern with proving Trichotomy seems to have arisen from reading Cantor's earlier papers. For in 1900 Peirce wrote to Cantor that he had just obtained a copy of the *Beiträge* (see Peirce 1976, 772).

Soon afterward Borel published the proof in an appendix to his book on set theory and complex functions [1898, 102–104].

At the Congress, Cantor also attempted to persuade Borel that the Trichotomy of Cardinals was true, but without success. Cantor informed Borel that four possibilities could occur vis-à-vis $\overline{\overline{A}}$ and $\overline{\overline{B}}$:

(i) There exists a proper subset of A equipollent to B, but no proper subset of B equipollent to A;
(ii) there exists a proper subset of B equipollent to A, but no proper subset of A equipollent to B;
(iii) there exists a proper subset of A equipollent to B as well as a proper subset of B equipollent to A;
(iv) there exists neither a proper subset of A equipollent to B nor a proper subset of B equipollent to A.

The first case yielded $\overline{\overline{B}} < \overline{\overline{A}}$, the second $\overline{\overline{A}} < \overline{\overline{B}}$, and the third $\overline{\overline{A}} = \overline{\overline{B}}$ by the Equivalence Theorem. If A and B were finite, then the fourth case implied $\overline{\overline{A}} = \overline{\overline{B}}$ as well. However, Borel entertained the possibility that there might exist infinite sets A and B satisfying (iv) and hence that Trichotomy might be false. If this occurred, he insisted, then one could not properly speak of a less-than relation among infinite cardinals; for Trichotomy was a precondition to such a relation. In this vein, Borel inquired whether one could demonstrate that no infinite sets satisfied case (iv) while still preserving the concept of set in full generality. Consequently he refrained from using Cantor's alephs until their foundation was secured.

As the nineteenth century was drawing to a close, Cantor left set theory with a nexus of fundamental, interrelated problems—the Continuum Hypothesis, the Well-Ordering Principle, the Trichotomy of Cardinals, and the Equivalence Theorem—all of which involved arbitrary choices. Central to these problems was the notion of well-ordering, which yielded the hierarchy of alephs. Although Cantor did not resolve any of these problems definitively, he established special cases of each of them. The Equivalence Theorem, the only one of these propositions that did not involve arbitrary choices in an essential way, was the first to be demonstrated completely. The Continuum Hypothesis, which he had striven so hard to prove, remained the most intractable. Influenced by Cantor's *Beiträge*, several mathematicians investigated the Trichotomy of Cardinals. The most intriguing and least noticed of these investigations was Burali-Forti's deduction of Trichotomy from two new set-theoretic postulates. Nevertheless the solution of Trichotomy within an adequate axiomatic framework had to await Zermelo.

Of particular interest is Cantor's changing attitude toward the Trichotomy of Cardinals and the Well-Ordering Principle. Neither of them originated as a problem but as a simple "consequence" of the definition of < for cardinals, and as a "law of thought," respectively. What led Cantor to demand a proof for the Well-Ordering Principle, from which he could then deduce Trichotomy? Perhaps the shift was due to the scepticism which this principle

had engendered among mathematicians. Since the search for a proof of the Well-Ordering Principle, as well as the first statement of Trichotomy as a problem, occurred in the *Beiträge*, it may well be that the desire to systematize set theory prompted Cantor's interest in its foundations, as it later did Bernstein's. Thus rigor became increasingly essential to Cantor, and, along with it, so did a proof for such far-reaching propositions as the Well-Ordering Principle. Cantor's attempt to demonstrate this principle, his last major contribution to mathematics, must now be examined.

1.6 *The Reception of the Well-Ordering Problem*

Around 1895, when Cantor became convinced that the Well-Ordering Principle was a theorem rather than a law of thought, he diligently sought a proof. By 1897 he believed that he had found one. Yet he remained somewhat ill at ease with his proof, since he later refused permission to publish it. Only in 1900, when Hilbert called attention to the need for a well-ordering of $\mathbb{R}$, did the Well-Ordering Problem begin to attract a wider audience.

Hilbert had been the first to learn of Cantor's purported demonstration that every set can be well-ordered. During 1896 or 1897 Cantor sent Hilbert a letter, which has apparently not survived, containing this proof.[1] Fortunately, two similar letters to Dedekind are extant. In the first of these, dated 28 July 1899, Cantor wrote: "The chief question was whether there exist other powers of [infinite] sets besides the alephs. For two years I have possessed a proof that there are no others so that, *e.g.*, the arithmetic linear continuum (the totality of all real numbers) has a definite aleph as its cardinal number."[2]

Cantor's next letter of 3 August 1899 revealed this proof. There he distinguished carefully between two kinds of classes or multitudes (*Vielheiten*). One kind could not be considered as a collection (*Zusammensein*) of all its elements without leading to a contradiction, and hence could not be conceived as finished or completed. Such multitudes he named absolutely infinite (*absolut unendliche*) or inconsistent (*inkonsistente*). The second kind

[1] In all probability, Schoenflies was referring to this letter when he remarked that "it struck those of us at Göttingen as a revelation, and circulated from hand to hand" [1922, 102]. There is some uncertainty concerning the year in which the letter was written. On 4 November 1903, Cantor informed Jourdain that he had sent his proof of the Well-Ordering Principle to Hilbert seven years earlier (presumably in 1896); see Cantor in Grattan-Guinness 1971a, 116. However, in his letter of 28 July 1899 to Dedekind, Cantor stated that he had found the proof two years earlier (presumably in 1897); see Cantor 1932, 443.

[2] Cantor 1932, 443. For a discussion of how either Zermelo or Jean Cavaillès conflated the letters of 28 July and 3 August 1899, see Grattan-Guinness 1974, 126–131, 134–135.

proof to be flawed by the purported existence of a one-one mapping from Ω into V:

> Evidently Cantor conceives of the ordinal numbers in Ω as assigned to successive and arbitrary elements of V in such a way that each element of V is used only *once* . . . . Thus the intuition of time is applied here to a process that surpasses every intuition; a being is simulated who could carry out *successive* arbitrary choices and thereby define an arbitrary subset V' of V, which is *not* definable by the given conditions. Only through using the "Axiom of Choice," which postulates the possibility of a *simultaneous* choice and which Cantor unconsciously and instinctively uses everywhere but explicitly formulates nowhere, could V' be defined as a subset of V. Yet even then, the objection would remain that the proof operates with "inconsistent" multitudes, indeed possibly with self-contradictory concepts, and may on this account be logically inadmissible. Objections of this sort caused the editor [Zermelo] a few years later to base his own proof of the Well-Ordering Theorem . . . purely on the Axiom of Choice *without* using inconsistent multitudes. [Zermelo in Cantor 1932, 451]

In the aftermath of Zermelo's proof, the merits of simultaneous versus successive choices were carefully scrutinized (see 2.3).

The second issue raised by Zermelo, the role of inconsistent classes, deserves a more detailed analysis at this juncture. In 1890 Schröder had already introduced a notion of consistent and inconsistent classes in his *Algebra der Logik*, well before the modern discovery of paradoxes in set theory and logic, and had utilized such inconsistent classes under certain conditions.[5] While Schröder regarded inconsistent classes as a useful device, and not a purely negative one, he did not investigate them in depth. Independently, Cantor did so. Cantor's argument that Ω is an absolutely infinite, or inconsistent, multitude can be viewed as the earliest statement of what eventually became known as Burali-Forti's paradox.

Yet Cantor's letter merits a scrupulous reading on this point, since later opinions have been influenced so strongly by the traumatic view of the paradoxes which Russell set forth in *The Principles of Mathematics* [1903]. One should observe, first of all, that Cantor exhibited no alarm over the state of set theory in his letter—in sharp contrast to Gottlob Frege's dismay upon learning in 1902 of Russell's paradox.[6] What Cantor remarked was merely that certain multitudes are, in effect, too large to be considered as unities (or sets), and so are termed absolutely infinite. Significantly, he retained such absolutely infinite, or inconsistent, multitudes and even employed them in the proof of the Aleph Theorem that he sent to Dedekind. Thus Cantor did not treat these apparent difficulties as paradoxes or contradictions, but as tools with which to fashion new mathematical discoveries. Moreover, he remained only one conceptual leap away from recognizing, as von Neumann eventually did, that each such absolutely infinite class can be treated as a unity so long as one does not permit it to be a member of other

[5] Schröder 1890, 213, 343.

[6] See Frege's letter of 16 June 1902 to Russell in Frege 1980, 130–131.

classes (see 4.9). All the same, events to be discussed later in this section indicate that Cantor was not completely satisfied with his proof for the Well-Ordering Principle.

The question of absolutely infinite multitudes seems to have prompted Cantor to examine the consistency of each of his alephs, for he broached the matter while writing to Dedekind on 28 August 1899. There Cantor declared that one could not even demonstrate the consistency of every finite set. Such consistency was "a simple indemonstrable truth," which he termed the Axiom of Arithmetic [1932, 447–448]. In a similar fashion he regarded the consistency of each aleph as an indemonstrable truth, which he named the Axiom of Extended Transfinite Arithmetic. Nevertheless, over the next decade a number of mathematicians, including Henri Poincaré [1906a, 315] and Arthur Schoenflies [1905, 182], were to question the legitimacy of Cantor's higher alephs.

Evidently Cantor's metamathematical Axiom of Extended Transfinite Arithmetic was the only axiom that he introduced vis-à-vis set theory. He had treated the Well-Ordering Principle as a law of thought (and hence as a principle of logic) rather than as a specifically set-theoretic assumption within mathematics. Cantor, who had a Platonistic conceptual framework, sought to discover truths rather than to ascertain the minimal assumptions necessary for a deductive system. Thus it is of considerable interest that his only axiom for set theory, unpublished at that, was a metamathematical assumption of consistency. On the other hand, his letter of 28 August 1899 contained (as truths rather than as axioms) statements very much like Zermelo's later Axioms of Union and of Separation as well as Fraenkel's Axiom of Replacement. In other words, Cantor asserted that the union of a set of sets is a set, that a submultitude of a set is a set, and that two equipollent multitudes are either both sets or both absolutely infinite [1932, 444]. Influenced by Zermelo and Fraenkel, we now view these statements as axioms for set theory. Nevertheless, it would seriously distort Cantor's perspective to think that he regarded them in this way.[7]

On 30 August 1899, Cantor, who had just received from Dedekind a new and elementary proof for the Equivalence Theorem, wrote to make a request: "It would be *very valuable* if you would also prove the fundamental theorem A [Trichotomy of Cardinals] ··· by the same methods."[8] Thus Cantor hoped that the techniques of Dedekind's *Was sind und was sollen die Zahlen?*, in particular the notion of chain, would permit a direct proof for Trichotomy.[9] In the wake of the Equivalence Theorem, one merely had to

---

[7] For the incorrect assertion that Cantor regarded his statements as axioms, see Bunn 1975, iii, 145–146, 186–187, and 1980, 248.

[8] Cantor 1932, 449–450. Apparently Dedekind had forgotten that on 11 July 1887 he had discovered a similar proof, not published until 1932, of the Equivalence Theorem; see Dedekind 1932, vol. III, 447–448.

[9] If B is a proper subset of A and if $f: A \to A$ is a one–one function which is not onto, then the *f-chain* of B is the intersection of all sets C such that $B \subseteq C$ and that for every $x \in C, f(x) \in C$.

show that if A is equipollent to no proper subset of B and if B is equipollent to no proper subset of A, then A and B are equipollent and hence finite. Dedekind did not generate such a proof, and Cantor remained ambivalent about his own, which he described as "only indirect."

When Hilbert received Cantor's attempted proof of the Well-Ordering Principle in 1896 or 1897, he apparently found it unconvincing. For, as the twentieth century opened, Hilbert drew attention to the problem of well-ordering the real numbers. During his celebrated lecture, delivered in Paris at the Second International Congress of Mathematicians, he treated the Continuum Problem—together with the existence of a well-ordering for the real numbers—as the first of twenty-three problems central to the development of twentieth-century mathematics [1900, 263–264]. Hilbert found it quite plausible that every infinite subset of $\mathbb{R}$ is either denumerable or has the power of the continuum. If true, he added, this proposition (1.5.1) would imply that the power of the continuum is the next after aleph-zero. Probably Hilbert meant that the power of the continuum is aleph-one, since Cantor had previously shown that there exists no cardinal between aleph-zero and aleph-one. Yet for (1.5.1) to imply that the power of the continuum is aleph-one, there must exist a subset of $\mathbb{R}$ with power aleph-one or, equivalently, $\mathbb{R}$ and aleph-one must be comparable.[10] Such had not yet been established.

This gap in Hilbert's argument was intimately connected with the other half of his first problem: "Let us mention another remarkable assertion of Cantor's, which has the closest relationship to the preceding proposition [1.5.1] and which may furnish the key to the proof of that proposition"—the existence of a well-ordering for $\mathbb{R}$. "It seems to me extremely desirable," Hilbert continued, "*to obtain a direct proof for this remarkable assertion of Cantor's*, perhaps by actually describing an order for the real numbers, such that in each subset a least number can be exhibited" [1900, 263]. Clearly Hilbert would have preferred to obtain an effective well-ordering for $\mathbb{R}$.

Although he did not accept Cantor's purported proof that every set can be well-ordered, Hilbert was sympathetic to Cantor's infinite cardinals. In the second problem given in his Paris lecture, namely, to establish the consistency of the real numbers, Hilbert asserted that the consistency of each of Cantor's number-classes and alephs could be established as well. Since, for Hilbert, the consistency of a set of axioms implied the existence of a mathematical object satisfying those axioms, he argued that each aleph existed. On the other hand, he insisted, it could be shown that no consistent axiom system can be given for the collection of all cardinal numbers or of all alephs. Thus he concluded that the collection of all alephs did not exist [1900, 264–265].

---

[10] Two sets A and B are *comparable* if one is equipollent to a subset of the other, *i.e.*, if $\overline{\overline{A}} \le \overline{\overline{B}}$ or $\overline{\overline{B}} \le \overline{\overline{A}}$.

After Hilbert placed the Continuum Hypothesis and the problem of well-ordering the real numbers in the limelight, mathematicians actively pursued these questions in both Germany and England. Yet progress was sporadic at best. The Continuum Hypothesis remained quite intractable, even though Cantor had already proved that every closed subset of $\mathbb{R}$ is either countable or has the power of the continuum. While the problem of finding a subset of $\mathbb{R}$ with power aleph-one was resolved after a fashion, a well-ordering for $\mathbb{R}$ remained elusive. Nevertheless, the search for a demonstration of both the Well-Ordering Principle and the Trichotomy of Cardinals continued. Interwoven with these investigations was the thread that would eventually lead Zermelo to a solution for the Well-Ordering Problem: infinitely many arbitrary choices.

Such choices occurred repeatedly in Schoenflies' lengthy report, which appeared in the *Jahresbericht der Deutschen Mathematiker-Vereinigung* in 1900, on the development of the theory of point-sets. To deduce Cantor's theorem that a sequence $P_1, P_2, \ldots$ of infinite closed and bounded subsets of $\mathbb{R}^n$, with $P_{n+1} \subseteq P_n$ for all n, has a non-empty intersection, Schoenflies selected a point from each P_n. Similarly, Schoenflies made arbitrary choices while showing that the set P of all partitions of a set M into subsets of power $\mathfrak{n}$ has power $\overline{\overline{M}}^{\mathfrak{n}}$. Arbitrary choices were less visible, though necessary, in his proof of the Countable Union Theorem and in his assertion that the product, any finite or infinite number of times, of a given cardinal can be represented by cardinal exponentiation [1900, 58–59, 8–11].

Nevertheless, Schoenflies cast a dubious eye on the Well-Ordering Principle. After noting that in 1883 Cantor had promised to return to it, he added: "Such has not yet occurred, and so for the time being one cannot ascribe any objective validity to the proposition. Indeed, it would be understandable if such an all-embracing general proposition were to meet widespread scepticism" [1900, 49]. Thus for the moment it remained unclear to Schoenflies whether every infinite cardinal, and in particular that of the continuum, was an aleph. Likewise, he added, the Trichotomy of Cardinals continued to lack a proof.[11]

Trichotomy was even more prominent in the doctoral dissertation, influenced strongly by Cantor but supervised by Hilbert, that Felix Bernstein completed at Göttingen in 1901. Bernstein emphasized that two problems presently stood at the center of set-theoretic interest. The first of these was what he named Cantor's Continuum Problem: How many distinct powers $\overline{\overline{A}}$ exist, where A is an infinite subset of $\mathbb{R}$?[12] On the one hand, this problem could be attacked by resolving it for more and more inclusive families of subsets of $\mathbb{R}$. As a first step, Cantor had obtained the answer of two for the

[11] Schoenflies 1900, 49, 15–16. He did not cite Burali-Forti [1896a] on Trichotomy.

[12] Bernstein 1901, 13. This was the first time that the term *Continuum Problem* appeared in print.

family of closed sets. However, Bernstein insisted, Cantor's conjecture of two for all subsets of $\mathbb{R}$ remained uncertain.

Bernstein added that Cantor had also approached the problem from a second direction, via the alephs, and that here the relationship of $\aleph_1$ to the power $2^{\aleph_0}$ of the continuum was fundamental. In this vein Bernstein established that

$$\aleph_1 \leq 2^{\aleph_0}$$

by using a parallelism in the definition of the set $\mathcal{O}$, which consisted of all order-types of denumerable sets, and that of the second number-class, which consisted of all order-types of well-ordered denumerable sets. Since the power of the second number-class was $\aleph_1$, it followed at once that $\aleph_1 \leq \overline{\overline{\mathcal{O}}}$. After giving Cantor's previously unpublished proof that $2^{\aleph_0} \leq \overline{\overline{\mathcal{O}}}$, Bernstein appended his own that $\overline{\overline{\mathcal{O}}} \leq 2^{\aleph_0}$, and thereby used the Assumption implicitly. The Equivalence Theorem yielded that

$$\overline{\overline{\mathcal{O}}} = 2^{\aleph_0},$$

and so Bernstein concluded that $\aleph_1 \leq 2^{\aleph_0}$, as desired [1901, 7–11, 62, 40]. Finally, he disclosed that Cantor had an unpublished proof of the following proposition:

(1.6.1) Every uncountable set has a subset of power $\aleph_1$. [1901, 33]

From (1.6.1) it would result at once that $\aleph_1 \leq 2^{\aleph_0}$. While Cantor's argument for (1.6.1) remains unknown today, it necessarily relied on the Assumption.

The second problem that Bernstein regarded as central to set theory concerned its foundations. Here he did not refer to any paradoxes or contradictions, but rather stated that recently sets had been discovered—such as the set of all cardinals—whose properties did not agree in all essential respects with previously known sets. Thus it behooved mathematicians to determine which laws held for all sets. Likewise, he stressed the need for systematization and for unifying theorems in set theory, without mentioning the possibility of an axiomatic approach [1901, 15].

It was in this context that Bernstein presented his proof of the Equivalence Theorem, together with an investigation of the conditions under which two sets A and B are comparable. Recognizing that the Trichotomy of Cardinals remained a problem of severe difficulty, he offered only a partial solution:

(1.6.2) If $\overline{\overline{A}} + \overline{\overline{B}} = \overline{\overline{A}} \cdot \overline{\overline{B}}$, then the sets A and B are comparable.

(1.6.3) If a family S of sets is closed under finite unions and if $\bar{\bar{A}}^2 = \bar{\bar{A}}$ whenever A is in S, then any two sets in S are comparable.

The proof of (1.6.3) relied on (1.6.2), where Bernstein used arbitrary choices [1901, 29]. Later Tarski demonstrated without using the Axiom that if $\mathfrak{m} + \mathfrak{n} = \mathfrak{m}\mathfrak{n}$ (or if $\mathfrak{m}^2 = \mathfrak{m}$) for all infinite cardinals, then both Trichotomy and the Axiom follow [1924].

Outside Germany, the problems surrounding Trichotomy found an audience primarily among certain English mathematicians (all with connections to Cambridge) who had been impressed by Cantor's researches: Bertrand Russell, G. H. Hardy, and Philip Jourdain. It was in 1896, while reviewing a book by the French Kantian philosopher Arthur Hannequin, that Russell first learned of Cantor.[13] Hannequin's book [1895] was devoted to criticizing the use of indivisibles or atoms, in mathematics and physics, as necessarily self-contradictory. In this context Hannequin argued that Cantor's ordinal number ω must be rejected, since the sequence 1, 2, ... of natural numbers had no limit, and that Cantor's treatment of a continuum as a set of points led to a contradiction: the divisibility of the indivisible [1895, 67–71]. "This," Russell pointed out in his review, "is only our old friend, Kant's second antinomy, but it acquires a new force by the proof of its influence in mathematical method" [1896, 412]. Similarly, Russell accepted Hannequin's objection to infinite ordinals such as ω.

Thus, influenced directly by Hannequin and indirectly by Kant, Russell came to believe that set theory harbored contradictions. In an unpublished essay of 1896, "On Some Difficulties of Continuous Quantity," Russell intended "to show, what mathematicians are in danger of forgetting, that philosophical antinomies find their counterpart in mathematical fallacies. These fallacies seem ... to pervade the Calculus, and even the more elaborate machinery of Cantor's collections (*Mengen*)."[14] When he came under Peano's influence in August 1900, Russell was already predisposed to find paradoxes or contradictions in set theory. What he obtained from Peano was the symbolic machinery to sort what was "correct" in Cantor from what was not. The result, less than a year later, was Russell's paradox.

On the other hand, probably through Peano's influence, Russell came to accept Cantor's theory of infinite ordinal and cardinal numbers in its broad outlines. As a result, Russell composed an article on well-ordered relations wherein he extended Peano's symbolic logic in order to express Cantor's theorems on well-ordering. Nevertheless, Russell concluded that the class of all ordinals was not well-ordered by magnitude and that "there is no reason, so far as I know, for believing that every class can be well-ordered" [1902, 33, 43].

---

[13] Letter of 11 September 1917 from Russell to Jourdain, in Grattan-Guinness 1977, 143.

[14] Quoted in Moore and Garciadiego 1981, 324.

Russell's first conclusion, and probably his second, resulted from reading the article of 1897 in which Burali-Forti argued that neither $\alpha \leq \beta$ nor $\beta \leq \alpha$ need hold for order-types α and β of perfectly ordered sets.[15] Nevertheless, Cantor demonstrated later the same year that either $\alpha \leq \beta$ or $\beta \leq \alpha$ must hold if α and β are order-types of well-ordered sets [1897, 216]. Shortly thereafter, Burali-Forti noted carefully that perfectly ordered classes were not necessarily well-ordered [1897a, 260]. Here he was correct. Yet his argument could be transferred to Cantor's well-ordered sets, as Russell recognized in *The Principles of Mathematics* [1903, 323]. In this fashion Russell created, in effect, what is now known as Burali-Forti's paradox. As late as 1906, Burali-Forti insisted that his 1897 article contained no paradox.[16] While no one disagreed before 1903, after that date mathematicians and historians increasingly viewed the matter from Russell's perspective.

To escape from this paradox of the largest ordinal, Russell argued in the *Principles* that the class of all ordinals was not well-ordered by magnitude [1903, 323]. In all probability, it was this conclusion that led him to differ with Cantor on both the Well-Ordering Principle and a number of related matters:

> Cantor assumes as an axiom that every class is the field of some well-ordered series, and deduces that all cardinals can be correlated with ordinals [1.5.3] . . . . This assumption seems to me unwarranted, especially in view of the fact that no one has yet succeeded in arranging a class of 2^{α_0} terms in a well-ordered series. We do not know that of any two different cardinal numbers one must be the greater, and it may be that 2^{α_0} is neither greater nor less than α_1 and α_2 and their successors . . . .[17]

Thus Russell saw no reason to believe that the Well-Ordering Principle was true, and remained equally sceptical of the Aleph Theorem, the Trichotomy of Cardinals, and the proposition that $\aleph_1 \leq 2^{\aleph_0}$.

G. H. Hardy, Russell's friend and colleague, did not concur in rejecting the proposition that $\aleph_1 \leq 2^{\aleph_0}$. During 1903 Hardy set out to demonstrate both this proposition and the following generalization:

(1.6.4) $\aleph_{\alpha+1} \leq 2^{\aleph_\alpha}$ for every ordinal α.

---

[15] Burali-Forti 1897. He termed a class M *perfectly ordered* by a relation S if M is ordered by S and if (i) M has a first element, (ii) every element of M which has a successor has an immediate successor, and (iii) every element x of M with an immediate predecessor has a predecessor y with no immediate predecessor such that the class of all z in M which are both successors of y and predecessors of x is finite.

[16] Burali-Forti in Couturat 1906, 228–229. Concerning the origins of Burali-Forti's paradox, see Moore and Garciadiego 1981.

[17] Russell 1903, 322–323. He used α_n to stand for $\aleph_n$ and series to mean sequence.

The proposition (1.6.4) was a corollary of Hardy's argument for a stronger result:

(1.6.5) Every infinite cardinal is either an aleph or is greater than all alephs.

Teeming with infinities of successive arbitrary choices, his argument reveals how far from obvious it was to mathematicians in 1903 that such choices implied the Aleph Theorem and the Well-Ordering Principle. In fact, Hardy remained uncertain that the Aleph Theorem was true. His argument allowed for the possibility that a cardinal might be greater than every aleph:

> For, given an aggregate whose cardinal $>\alpha_0$, we can choose from it successively individuals
>
> $$u_1, u_2, \ldots, u_\omega, \ldots, u_\alpha, \ldots,$$
>
> corresponding to all the numbers of the first and second [number] classes; if the process came to an end, the cardinal of the aggregate would be α_0. Its cardinal therefore $\geq\alpha_1$; and if $>\alpha_1$, $\geq\alpha_2$, and so on. And if $>\alpha_n$, for all finite values of n, it must be $\geq\alpha_\omega$; for we can choose individuals from the aggregate corresponding to all the numbers of the first, second, third, ..., nth, ... classes. And by a repetition of these two arguments, we can show that if there is no α_β equal to the cardinal of the aggregate, it must be at least equal to the cardinal of the aggregate of all α_β's, and so greater than any α_β.[18]

Hardy obtained this argument by generalizing Cantor's proof of the theorem (1.1.4) that every infinite set has a denumerable subset. On the other hand, Hardy's argument also resembles closely the one that Cantor had sent to Dedekind in 1899 for the Aleph Theorem. This suggests that Cantor's proof for the Aleph Theorem, which originated in 1896 or 1897, may also have grown out of his 1895 proof of (1.1.4). Indeed, Cantor's two proofs used arbitrary choices in a similar way, and one was a natural extension of the other.

In conclusion, Hardy supplied what he termed "the actual construction of a set of points of cardinal $\aleph_1$" [1903, 88]. Here his use of arbitrary choices reveals how vague the criteria for such a construction had become. Hardy assigned each non-zero countable ordinal to an increasing sequence of natural numbers. Then each such sequence could be correlated easily with a unique binary decimal. The ordinal 1 corresponded to the sequence 1, 2, 3, ... If β corresponded to $b_1, b_2, \ldots$, then $\beta + 1$ corresponded to $b_2, b_3, \ldots$. To define the sequence corresponding to a limit ordinal γ, he diagonalized the sequences corresponding to a sequence of ordinals whose limit was γ. "We have an infinite freedom of choice," he noted, "whenever we wish to define the sequence corresponding to any [ordinal] number which has no immediate predecessor" [1903, 90]. Such choices insured that

---

[18] Hardy 1903, 88. He followed Russell in using α_β for $\aleph_\beta$.

if $\beta < \gamma$, then the sequence corresponding to γ was eventually greater than the one corresponding to β. We know now that arbitrary choices, or an equivalent assumption, were essential to demonstrate that $\aleph_1 \leq 2^{\aleph_0}$.[19]

Influenced by Hardy's article, Philip Jourdain attempted to modify the proof of (1.6.5) to yield one for both the Aleph Theorem and the Well-Ordering Principle. Jourdain apparently sent a draft of his argument to Hardy, since on 14 October 1903 Hardy wrote to Russell on this matter. In his letter Hardy rejected Jourdain's argument on the grounds that it denied the existence of the class of all ordinals. Hardy also revealed that, while writing his 1903 article, he had attempted unsuccessfully to show that the class of all ordinals is equipollent to the class of all entities.[20]

On 29 October 1903 Jourdain sent Cantor, with whom he had been corresponding for two years, his new argument that every infinite cardinal is an aleph: "*In other words*: *every consistent Manifold* [*set*] *can be put in the form of a well-ordered Manifold*."[21] Cantor's reply of 4 November must have startled Jourdain. It revealed that Cantor had sent essentially the same proof to Hilbert seven years earlier and to Dedekind in 1899. After stating this proof, Cantor encouraged Jourdain to develop his own for publication. Yet four months later, when Jourdain requested permission to publish the letter containing Cantor's proof, he was rebuffed.[22]

In all probability Cantor realized that his proof contained ambiguities and did not wish them exposed to public censure. The principal ambiguity lay in his notion of consistent and inconsistent multitudes (*Vielheiten*), of which only the consistent ones were defined to be sets (*Mengen*). How could one determine whether a given multitude was consistent? In his November letter to Jourdain, Cantor framed his reply to this question essentially as he had done four years earlier to Dedekind: "A consistent multitude is one such that the assumption of *collecting all its elements* into *one object* does not lead to a contradiction ... ."[23] Thus the multitude of all ordinals and that of all alephs were inconsistent, as was Dedekind's multitude of everything thinkable. Yet, short of discovering a contradiction, how was one to decide whether a particular multitude was inconsistent? Although Cantor gave no such independent criterion, Jourdain soon provided one.

On Cantor's recommendation, Jourdain prepared his demonstration of the Well-Ordering Principle, via the Aleph Theorem, for publication. The resulting article appeared in the *Philosophical Magazine* for January 1904. To distinguish consistent classes from inconsistent ones, Jourdain

---

[19] This inequality is equivalent to a special case of the Axiom. Sierpiński [1954, 53] showed that $\aleph_1 \leq 2^{\aleph_0}$ if and only if there exists a function that assigns to each denumerable set D of real numbers a real number not in D.

[20] The letter is printed in Moore and Garciadiego 1981, 332.

[21] Jourdain in Grattan-Guinness 1971a, 115. Jourdain's emphasis.

[22] Cantor in Grattan-Guinness 1971a, 116–118.

[23] Cantor in Grattan-Guinness 1971a, 116.

introduced a formal criterion: A class is inconsistent if it has a subclass equipollent to the class W of all ordinals [1904, 67]. In order to circumvent Burali-Forti's paradox based on the class W, Jourdain only allowed consistent classes to have an order-type or a cardinal number—a restriction similar to Cantor's. Even though Jourdain's criterion for inconsistent classes was more cogent than Cantor's, it gained few adherents. What it lacked was the further condition that every inconsistent class be equipollent to the class of all sets, a condition which von Neumann introduced two decades later.[24] Yet in 1904 Jourdain came closer than anyone else had previously to understanding that such very large classes are not dangerous so long as they are handled with sufficient care.

Jourdain began his argument for the Aleph Theorem by accepting Hardy's proposition that every infinite cardinal is either an aleph or is greater than all alephs. Then, despite Russell's claim to the contrary in the *Principles*, Jourdain asserted that the class W is well-ordered. His assertion rested on the following theorem:

(1.6.6) If an ordered class M has no subset of type $^*\omega$, then M is well-ordered.[25]

To establish (1.6.6), Jourdain arbitrarily chose a descending sequence of elements from a subset M' of M with no least element, thereby relying necessarily on the Dependent Assumption. With (1.6.6) in hand, the Aleph Theorem followed easily: If there is a cardinal $\overline{\overline{M}}$ greater than every aleph, then the class M has a subclass equipollent to W; consequently, M is inconsistent and so lacks a cardinal. Therefore, by Hardy's result (1.6.5), every infinite cardinal is an aleph [Jourdain 1904, 67].

Actually Jourdain's argument showed only that, under the given hypothesis, there exists for each ordinal β a subclass of M equipollent to β. What he needed to obtain was a subclass of M equipollent to the class W of *all* ordinals. To do so, he made as many successive arbitrary choices as there were ordinals. This assumption was even stronger than the Well-Ordering Principle (or the Axiom of Choice) and, in fact, was equivalent to asserting that the class of all sets can be well-ordered.

[24] While it is uncertain in exactly which formal systems for set theory Jourdain's and von Neumann's criteria are equivalent, historically they were quite distinct. When Jourdain used his criterion to prove the Well-Ordering Theorem, he relied as well on as many successive arbitrary choices as there were ordinals. On the other hand, von Neumann later deduced from his criterion that the class of all sets is well-ordered, without any use of arbitrary choices or need for the Axiom of Choice. In **ZF**, certainly, von Neumann's criterion implies Jourdain's. Yet Jourdain's might be true and von Neumann's false in a model where the class of all ordinals was not equipollent to that of all sets.

[25] Jourdain 1904, 64–65. An ordered set M has type $^*\omega$ if M is order-isomorphic to the set of all negative integers with the usual less-than relation.

At the end of his article Jourdain mentioned several important consequences of his Aleph Theorem in cardinal arithmetic. First of all, the power of the continuum was an aleph. Moreover, the Trichotomy of Cardinals held—a corollary which, mistakenly, he believed insufficient to yield the Aleph Theorem. In addition, there resulted a positive solution to a problem that A. N. Whitehead had left open in 1902 (see 1.7):

(1.6.7) If $\mathfrak{m}$ is infinite and $\mathfrak{n} \leq \mathfrak{m}$, then $\mathfrak{m} + \mathfrak{n} = \mathfrak{m}$.[26]

A few months later Jourdain contributed another article containing several further corollaries of the Aleph Theorem. The first of these had also been left as an open problem by Whitehead in 1902:

(1.6.8) If $\mathfrak{m}$ is infinite and $\mathfrak{n} \leq \mathfrak{m}$, then $\mathfrak{m} \cdot \mathfrak{n} = \mathfrak{m}$.

In particular, $\mathfrak{m}^2 = \mathfrak{m}$ for every infinite $\mathfrak{m}$. From this Jourdain deduced the following proposition, which Schoenflies had stated without proof in 1900:

(1.6.9) If $\mathfrak{m}$ is infinite, then $\mathfrak{m}^{\mathfrak{m}} = 2^{\mathfrak{m}}$.

Expressing the hope that he would soon be able to investigate the Continuum Hypothesis in depth, Jourdain concluded his article [1904a, 301–303].

Jourdain's triumph was soon darkened by a cloud of insight. In a paper dated 6 September 1904, he retracted his claim that the real numbers can be well-ordered. His proof of the Aleph Theorem, he admitted, would justify such a conclusion only if one had also shown that no subset of $\mathbb{R}$ is equipollent to the class W of all ordinals: "It is very important to prove that $2^{\aleph_0}$ is equal to some Aleph in order to be certain that the number-continuum is not what I have called an 'inconsistent aggregate'" [1905, 42]. Thus, despite all his efforts, despite his conviction that $2^{\aleph_0} = \aleph_1$ was probably true, Jourdain remained unable to establish even that there exists a well-ordering of $\mathbb{R}$. Yet during that same September, Zermelo was crystallizing his own demonstration that the real numbers, and indeed every set, can be well-ordered.

To sum up, as the twentieth century opened, no consensus existed among mathematicians vis-à-vis the Well-Ordering Problem. In his Paris address Hilbert stressed the need to obtain a particular well-ordering for $\mathbb{R}$, while Russell and Schoenflies remained dubious about the truth of the Well-Ordering Principle. On the other hand, Hardy believed himself to have shown the ostensibly weaker proposition that every infinite cardinal is either an aleph or is greater than all alephs, and in particular that $\aleph_1 \leq 2^{\aleph_0}$. Influenced by Hardy, Jourdain discovered an argument for the Aleph

[26] Jourdain 1904, 67–68, 73.

Theorem and the Well-Ordering Principle, only to learn that Cantor had obtained a similar argument seven years earlier. However, Cantor had reservations about this argument, probably because it relied on the inconsistent class of all ordinals, and refused to permit Jourdain to publish it. When Jourdain's version appeared, it met with indifference and disbelief.[27]

Hidden in Cantor's and Jourdain's arguments was a potent new assumption based on infinitely many successive, dependent, arbitrary choices. Yet neither of them realized that a new assumption was involved. Apropos of his proof, Jourdain later acknowledged:

> The validity of the process of making an infinite series of arbitrary selections was simply assumed by me in consequence of Hardy's work [1903]; but, in common with most other mathematicians, I was quite unconscious at that time of the fact that any unproved assumption was made by the admission of the principle of selection [Axiom of Choice].[28]

It remained for Zermelo to recognize that the use of infinitely many arbitrary choices had to be postulated and that such choices could be made independently of each other in order to avoid contaminating them with intuitions of time.

1.7 *Implicit Uses by Future Critics*

It is a historic irony that many of the mathematicians who later opposed the Axiom of Choice had used it implicitly in their own researches. This occurred chiefly to those who were pursuing Cantorian set theory or applying it in real analysis. At the turn of the century such analysts included René Baire, Emile Borel, and Henri Lebesgue in France, as well as W. H. Young in Germany. In England, on the other hand, Bertrand Russell and Alfred North Whitehead were investigating transfinite ordinal and cardinal numbers as part of their researches on the foundations of mathematics. At times these various mathematicians used the Assumption in the guise of infinitely many arbitrary choices, and at times it appeared indirectly via the Countable Union Theorem and similar results. Certainly the indirect uses did not indicate any conscious adoption of non-constructive premises. Yet when infinitely many arbitrary choices occurred in the work of a mathe-

---

[27] See 2.7. Only A. E. Harward [1905], an amateur mathematician in the Indian Civil Service, accepted Jourdain's proof in print. It was rejected by Schoenflies [1908, 36], Zermelo [1908, 120–121], and others.

[28] Jourdain 1921, 244.

matician such as Borel, they revealed a certain ambivalence toward the methods permissible in mathematics. This methodological ambivalence—vis-à-vis Cantorian set theory, the notion of an arbitrary real function, and non-constructive assumptions—typified the researches of Baire, Borel, and Lebesgue. Zermelo's 1904 proof of the Well-Ordering Theorem led them to develop their constructivist philosophy of mathematics and to become increasingly intolerant of non-constructive methods such as the Axiom (see 2.3).

In his doctoral thesis, published in 1895, Borel was already relying on non-constructive methods. Influenced by Poincaré, Borel's thesis dealt with the analytic continuation of a complex function across a curve containing a dense set of singularities. In the course of his investigations, Borel first formulated the denumerable case of the Heine–Borel Theorem (later misnamed by Schoenflies) for the real numbers: If a closed interval $[a, b]$ is covered by denumerably many open intervals, then a finite subcollection of those intervals covers $[a, b]$. To establish this theorem, Borel made aleph-one successive arbitrary selections. He chose (a_0, b_0) to be one of the intervals containing the point a, (a_1, b_1) to be an interval containing b_0, (a_2, b_2) an interval containing b_1, and so forth. If the point b was not in (a_n, b_n) for any natural number n, then the sequence $b_0, b_1, \ldots$ converged to a point b_ω in $[a, b]$. Hence there was an interval $(a_{\omega+1}, b_{\omega+1})$ containing b_ω. As $a_{\omega+1}$ was between b_m and b_{m+1} for some m, one could discard the intervals (a_{m+k}, b_{m+k}) for all positive k. If, continuing in this fashion, one did not attain b, then for every countable ordinal α one would have a point b_α, and so there would be aleph-one intervals—a conclusion which contradicted the original hypothesis. Therefore a finite subcollection of intervals covering $[a, b]$ could be obtained by this process [1895, 51–52].

In Borel's proof the Assumption served to justify selecting (a_α, b_α) for each countable ordinal α. However, he could easily have avoided such an implicit use. Since the original cover was denumerable and hence well-ordered, at each stage he could have picked the appropriate interval whose rank was least in that well-ordering. Three years later, when he provided a second demonstration of this theorem in his book on complex functions, Borel employed the Bolzano–Weierstrass method of interval subdivision [1898, 42–43]. While he used the Assumption implicitly to choose subintervals, again it would not have been difficult for him to specify a rule for his choices, if he had wished to do so.

On the other hand, Borel's book contained several theorems, all concerning cardinality, for which the Assumption was unavoidable. The first of these, discussed in 1.3, was his proof of Cantor's result that every infinite set has a denumerable subset [1898, 12–13]. The second provided a modest generalization of the Countable Union Theorem:

(1.7.1) The union of denumerably many sets A_1, $A_2, \ldots,$ each of the power of the continuum, also has the power of the continuum.

To show (1.7.1), Borel remarked that for each n, A_n had the power of the open interval $(n - 1, n)$ in $\mathbb{R}$, and hence the union of all A_n had the power of the continuum [1898, 16]. In effect he arbitrarily chose a bijection f_n from A_n onto $(n - 1, n)$ for every n in order to obtain a one-one function from $\bigcup_{n=1}^{\infty} A_n$ into $\mathbb{R}$, as desired. These two theorems depended only on the Denumerable Axiom, whereas a third required $2^{\aleph_0}$ arbitrary choices. Indeed, the third proposition generalized the Countable Union Theorem still further:

(1.7.2) The union of a family A of sets, such that A as well as every member of A has the power of the continuum, also has the power of the continuum. [1898, 16–17]

While Borel asserted (1.7.2) without proof, it is likely that he would have proceeded to use arbitrary choices as he had done in (1.7.1). Finally, on the real line he introduced his Borel-measurable sets, subsequently known as Borel sets, by starting with intervals and then closing under the operations of complement and of countable union. Even though he did not use the Assumption implicitly here, he seemed to suggest that some subsets of $\mathbb{R}$ were not Borel-measurable—an assertion which depended on the Axiom [1898, 46–49]. Later the Axiom turned out to be entangled with many of the fundamental properties of Borel sets.

Like his compatriot Borel, René Baire obtained his doctorate in mathematics at the *École Normale Supérieure*. Baire's thesis of 1899 contained deep results in the theory of real functions, a subject which was not being pursued vigorously in France. In particular, he classified discontinuous functions by treating them as limits of sequences of certain functions. As the framework for this classification, he accepted Dirichlet's general definition of a real function as any correspondence $x \to f(x)$ between real numbers. "In this definition," Baire remarked, "it is beside the point to inquire how the correspondence can be established effectively, or even whether it is possible to establish it effectively" [1899, 1]. At that time he took a much more tolerant view of non-effective or non-constructive methods than he was to do after Zermelo's proof.

This early tolerance was equally apparent when Baire generalized Cantor's results (1.4.5) and (1.4.6) on derived sets to the following theorem:

(1.7.3) For every countable ordinal α, let P_α be a closed, bounded subset of $\mathbb{R}$. Suppose that (i) if $\beta < \alpha$, then $P_\alpha \subseteq P_\beta$ and that (ii) if $\beta < \alpha$ and P_β is finite, then P_α is empty. It follows that for some countable γ, $P_\gamma = P_{\gamma+1}$.

Baire formulated (1.7.3) in order to relativize the hierarchy of derived sets in a fashion suitable for studying certain sets of discontinuities. The Denumerable Assumption entered his demonstration via Cantor's theorem

(1.4.4) that the limit of a sequence of countable ordinals is a countable ordinal. Moreover, Baire presupposed that every sequentially closed subset of $\mathbb{R}$ is closed [1899, 51–52].

Baire's thesis focused on his classification of real functions and, to a lesser extent, on his notions of first and second category. The rudiments of his classification, as well as a number of his results, had already appeared in a paper presented to the Paris Academy of Sciences on 6 June 1898. There he defined the Baire class zero as the set of all continuous real functions. For any other countable α, the Baire class α was the set of all real functions f, not in any previous class, such that

$$f(x) = \lim_{n\to\infty} f_n(x)$$

for every real x, where $f_1, f_2, \ldots$ was some sequence of real functions in previous classes. His classification did not exhaust all real functions, he noted, since the union of all his classes only had the power of the continuum, whereas the set of all discontinuous functions had a higher power [1898, 1623]. Here he used the Assumption indirectly by relying on the theorem, closely related to (1.7.2), that the union of aleph-one sets, each having the power of the continuum, also has the power of the continuum.

Within his thesis Baire employed (1.7.3), and hence the Denumerable Assumption, to characterize those functions $f: \mathbb{R} \to \{0, 1\}$ in class one [1899, 53]. More generally, he asserted but did not prove the following characterization of his classes with finite index:

(1.7.4) For any given positive integer n, every real function f of class n, and every real x,

$$f(x) = \sum_{a_1=1}^{\infty} \sum_{a_2=1}^{\infty} \cdots \sum_{a_n=1}^{\infty} P_{a_1,a_2,\ldots,a_n}(x)$$

where each $P_{a_1,a_2,\ldots,a_n}(x)$ is some polynomial in x with real coefficients. [1899, 69]

As Sierpiński later remarked [1918, 135], this theorem requires the Axiom whenever $n > 1$. For $n = 1$, (1.7.4) was merely the Weierstrass Approximation Theorem that every continuous real function is the limit of a sequence of polynomials. Knowing that every function f of class two was a limit of some sequence of functions of class one, Baire chose such a sequence arbitrarily for each f, and thereby obtained a double sequence of polynomials. Then he seems to have generalized this procedure to class n by induction. Finally, he demonstrated that no new functions would be gained by extending his classification to uncountable ordinals. For the limit of a sequence of functions, each of countable Baire class, was itself a function in a countable Baire class—a result which again relied on (1.4.4) and the Denumerable Assumption [1899, 70].

Questions surrounding Baire's classification became a subject of dispute in the aftermath of Zermelo's proof. In particular, it was later recognized that Baire's classification of real functions was essentially equivalent to the Borel sets for $\mathbb{R}$, and that the existence of a function escaping Baire's hierarchy was intimately connected with the existence of a measurable non-Borel set. To establish the existence of such a set required the Axiom (see 2.3 and 4.1).

The second major contribution of Baire's thesis was the notion, later pursued in general topology, of sets of first and second category. A subset A of $\mathbb{R}$ was of first category if it was the union of countably many nowhere dense sets; otherwise A was of second category.[1] Chief among his results was the Baire Category Theorem ($\mathbb{R}$ is of second category), which later mathematicians extended to more general topological spaces. While it is now known that Baire's theorem can be shown without the Axiom, his proof relied on selecting a point with a certain property from each of infinitely many nested intervals and thus used the Dependent Assumption. By contrast, the Axiom could not be avoided in his next theorem:[2]

(1.7.5) In $\mathbb{R}$ the union of countably many sets, each of first category, is also of first category.

Though he considered this proposition to follow immediately from his definition of first category, he seems to have based this conclusion on the Countable Union Theorem and thus indirectly on the Assumption [1899, 66].

It is intriguing that Baire had a premonition of how non-constructive methods were connected to his research. In a letter of 25 October 1898 to Vito Volterra, who was keenly interested in his work, Baire expressed uncertainty about the existence of Baire class α for $\alpha \geq 3$. Moreover, he wondered whether it was even possible to define a real function not in his classification:

> One would know how to obtain a function *beyond* all the [Baire] classes if one knew how to *partition the continuum* into two sets, each of second category on every interval. One would let $f = 0$ for the first set and $f = 1$ for the second. I have not succeeded in *defining* such a partition. Clearly it is a delicate matter not just to answer, but even to pose, such a question. The meaning of the phrase "to define a set" would have to be made more precise. Yet I am convinced that the act of imposing on a set the condition of being definable must considerably restrict the notion of set. Above all, the difficulties result from the way in which we know an *arbitrary irrational number* . . . . May we not hope to learn how far we are permitted to use the notion of *arbitrary* [real] *function*?[3]

---

[1] Baire 1899, 65; *cf.* 1898, 1623. A subset A of $\mathbb{R}$ is *nowhere dense* if no open interval is a subset of the closure of A. The *closure* of A is the union of A and the derived set of A.

[2] In the model of **ZF** given by Feferman and Levy [1963], $\mathbb{R}$ is a countable union of countable sets, and so (1.7.5) is false there by the Baire Category Theorem.

[3] Baire in Dugac 1976a, 347–348.

In a sense Baire's qualms were justified. When he hypothesized that $\mathbb{R}$ could be partitioned into two sets, each of second category on every open interval, he obtained sets lacking what was later called the Baire property.[4] However, the existence of a subset of $\mathbb{R}$ lacking the Baire property depends on the Axiom, since there would then exist a real function escaping from Baire's classification. The related questions of definability that he raised in his letter later became significant in determining the relationship of the Axiom to constructive mathematical methods (see 2.3).

Influenced by Baire and Borel among others, Henri Lebesgue undertook his first researches on the theory of real functions. This work led to profound discoveries, first developed fully within Lebesgue's thesis of 1902, concerning measure and integration. There he posed his well known Measure Problem, which asked for a function m such that for every bounded subset A of $\mathbb{R}^n$, $m(A)$ is a non-negative real number satisfying three conditions:

(i) $m(A)$ is positive for some set A.
(ii) Congruent sets have equal measure.
(iii) The measure of countably many disjoint sets is the sum of the measures of these sets (countable additivity).

In order to resolve this problem, he introduced the measurable sets,[5] which extended the family of Borel sets and which permitted him to define the Lebesgue integral. At the same time he acknowledged that the measurable sets might not contain every bounded subset of $\mathbb{R}^n$, and so might not solve the Measure Problem [1902, 236]. Not long afterward, the discovery of a bounded non-measurable set proved to be one of the more controversial consequences of the Axiom (see 2.3).

In view of Lebesgue's later opposition to the Axiom, it is worthwhile examining how arbitrary choices occurred in an essential way at the very heart of his research: his proof that the measurable subsets of $\mathbb{R}$ are countably additive. To begin, he needed to establish that the union of countably many measurable sets is measurable. First he supposed that $A_1, A_2, \ldots$ was a sequence of disjoint measurable subsets of $\mathbb{R}$ and that $[a, b]$ was an interval including all these A_i. Then he enclosed each A_i in the union of some countable family $\mathscr{F}_i$ of open intervals, and each $[a, b]-A_i$ in the union of another

[4] A subset A of $\mathbb{R}$ has the *Baire property* if for every interval P, relative to which A is of second category, there is some open subinterval E such that $(\mathbb{R}-A) \cap E$ is of first category relative to $\mathbb{R}$. Later Kuratowski [1933, 51, 55] referred to this property as the Baire property in the large sense, to distinguish it from what he called the Baire property in the restricted sense, *i.e.*, the above property with the term "interval P" replaced by "perfect set P." In contrast to the usage adopted here, the Baire property in the restricted sense was sometimes termed the Baire property, *e.g.*, by Sierpiński [1923] and Luzin [1927].

[5] Let A be a bounded subset of $\mathbb{R}$ and suppose that k is the length of some fixed interval I which includes A. The outer measure of A is defined to be the least upper bound of the sum of the lengths of intervals covering A. On the other hand, the inner measure of A is defined as $k - b$, where b is the least upper bound of the sum of the lengths of the intervals covering $I-A$. Then S is said to be *measurable* if its outer measure and its inner measure are equal.

countable family $\mathscr{G}_i$ of open intervals, such that each set $(\bigcup \mathscr{F}_i) \cap (\bigcup \mathscr{G}_i)$ had a total length d_i. Here the sum of all the d_i was taken to be an arbitrarily small positive number [1902, 237–239].

Yet Lebesgue provided no rule for choosing the families $\mathscr{F}_i$ and $\mathscr{G}_i$. Nor could he have done so, thereby avoiding arbitrary choices, since the Denumerable Axiom is necessary to ensure the countable additivity of Lebesgue measure.[6] This fact was first recognized, though not proved, by Sierpiński [1916, 690]. Thus Lebesgue was eventually faced with the dilemma of either accepting the Denumerable Axiom or else restricting his theory of integration in an essential way.

At Göttingen the English mathematician W. H. Young, another future opponent of the Axiom, was pursuing real analysis in a fashion similar to Baire, Borel, and Lebesgue. Like them, Young employed infinitely many arbitrary choices. During 1902 he used the Assumption implicitly to demonstrate Cantor's topological theorem that every family of disjoint open intervals on $\mathbb{R}$ is countable, later known as the Countable Chain Condition. This theorem did not require the Assumption since each such interval contained a uniquely specifiable rational number, the least in Cantor's well-ordering of the rationals. However, Young's proof used the Assumption indirectly when he relied on the following proposition:

(1.7.6) The union of countably many finite sets is countable. [1902, 248]

In the absence of the Axiom, as Russell noted in 1906, even this weak form of the Countable Union Theorem could fail.[7]

On the other hand, Young could not have avoided the Denumerable Assumption in his proof, published in 1903, of a theorem which he formulated independently but which is often credited to Ernst Lindelöf alone:[8]

---

[6] As was mentioned in footnote 2, there is a model of **ZF** in which $\mathbb{R}$ is a countable union of countable sets. Hence Lebesgue measure is not countably additive in this model. For otherwise $\mathbb{R}$ would be a countable union of sets with measure zero, and would therefore have measure zero itself.

[7] Russell 1906, 47–48. However, one can prove (1.7.6) without the Axiom whenever all the sets are subsets of some ordered set, such as $\mathbb{R}$. In particular, one can define a function f from the union of the disjoint finite sets $A_1, A_2, \ldots$ to the denumerable set $\mathbb{N} \times \mathbb{N}$ by letting $f(x) = (i, j)$, where x is the jth member of A_i in the given order on A_i. Since f is one–one, then (1.7.6) follows from the stated hypothesis.

[8] In fact (1.7.7), even restricted to the case when M is $\mathbb{R}$, implies that every Dedekind-finite subset of $\mathbb{R}$ is finite. The following argument grew out of conversations with Thomas Jech: Suppose A is an infinite, Dedekind-finite subset of $\mathbb{R}$, and let B be the image of A under an order-preserving bijection of $\mathbb{R}$ onto the interval $(0, 1)$. Since B is an infinite bounded set, it has a limit point p. Either $B \cap (0, p)$ or $B \cap (p, 1)$ is infinite and Dedekind-finite; suppose the former. If D is the image of $B \cap (0, p)$ under an order-preserving bijection from $(0, p)$ onto $\mathbb{R}$, then

$$S = \{(-\infty, x) : x \in D\}$$

is an open cover of $\mathbb{R}$ with no countable subcover.

(1.7.7) Every family S of open intervals covering a subset M of $\mathbb{R}$ has a countable subfamily which covers M.[9]

In its uncritical use of successive arbitrary choices, Young's proof closely resembled the one that Borel gave in 1895 for the Heine–Borel Theorem. Beginning with a fixed interval d in S, Young selected an interval d_1 which overlapped d to the left, then an interval d_2 overlapping d_1 to the left, and so on. This process could end, he argued, only by exhausting all the points of M to the left of d, or by the left endpoints of $d_1, d_2, \ldots$ approaching a sequential limit point p. Some interval e contained p, and so $d_{i+1}, d_{i+2}, \ldots$ could be discarded (where d_i was the interval of lowest index to overlap e). Then this procedure was repeated beginning with e. Choosing intervals in this fashion to the right of d as well, Young concluded that one obtained countably many intervals covering M [1903, 384–386].

Thus, in both France and Germany, future critics of the Axiom were freely employing sequences of arbitrary choices in real analysis before Zermelo's proof appeared. In England, Russell and Whitehead (both of whom, like Young, had strong ties to Cambridge) were using the Assumption implicitly in their critical investigations of the foundations of mathematics. At the First International Congress of Philosophy, held at Paris in August 1900, Russell and Whitehead had been greatly impressed by Peano, and soon thereafter they adopted his system of symbolic logic. By October, Russell began to extend Peano's logic to the theory of relations, which Peirce and Schröder had developed earlier. Subsequently, Russell published two articles on relations [1901, 1902] in Peano's journal *Rivista di matematica*.

Both articles involved the subtle use of arbitrary choices. In the first of them, Russell asserted and proved the following proposition, by means of which he defined the sum of finitely or infinitely many cardinal numbers:

(1.7.8) If A and B are classes of disjoint non-empty classes and if $f: A \to B$ is a bijection such that $f(a) = b$ implies $\overline{\overline{a}} = \overline{\overline{b}}$ whenever $a \in A$ and $b \in B$, then $\overline{\overline{\bigcup A}} = \overline{\overline{\bigcup B}}$.

To show (1.7.8), in effect Cantor's proposition (1.4.12), Russell selected a bijection from a onto $f(a)$ for each $a \in A$. Yet the symbolic form of Russell's proof concealed these arbitrary choices [1901, 124]. His second article used (1.7.8), where A and B were sets of order-relations, to define the sum of infinitely many order-types [1902, 19]. Closely related was his theorem that every dense ordered class A has a subclass of type ω between any distinct

[9] Lindelöf's demonstration utilized the Countable Union Theorem [1903, 698]; see also his [1905, 188]. In modern terminology, a topological space X is said to be *Lindelöf* if every family of open sets which covers X has a countable subfamily which covers X. If every subspace of X is Lindelöf, then X is said to be *hereditarily Lindelöf*. Thus (1.7.7) implies that $\mathbb{R}$ is hereditarily Lindelöf.

elements a and b of A. Here too the arbitrary choices were not immediately apparent [1901, 143–144].

By contrast, in 1901 Russell already sensed that some new assumption was required to deduce that a class is Dedekind-finite if and only if it is finite. Thus Russell introduced a postulate, similar to Burali-Forti's theorem (1.3.7), stating that the Dedekind-finite classes satisfied a kind of mathematical induction.[10] It followed at once that every Dedekind-finite class was finite. Since, however, he introduced fifteen other postulates in that article, it would be an error to place much emphasis on this one. By 1902, as was discussed in 1.1, Russell came to believe that he had proved this postulate by showing Cantor's proposition (1.1.4) that every infinite set has a denumerable subset. Russell's proof, containing arbitrary choices, appeared in an article by Whitehead [1902], to which he contributed one section.

Stimulated by his proof, Russell tentatively proposed a further postulate, which he regarded as a generalization of (1.1.4):

(1.7.9) Every infinite class u is the union of some family v of disjoint denumerable classes.

It may have been the following heuristic argument, which Russell gave in favor of (1.7.9), that led Hardy [1903] to his related argument for the proposition that every infinite cardinal is an aleph or is greater than all alephs (see 1.6):

> It has been shown that every infinite class contains classes whose cardinal number is α_0 [aleph-zero]. Thus if a class v of α_0's has been found which contains all but a finite number of u's, these may be added on to the v's; but if an infinite number of u's remain, new α_0's may be found out of them, and to this process there can be no end as long as any u's remain. It appears not impossible that a theorem may be discovered proving that this process can always be completed.[11]

In any case, Russell insisted on the importance of determining the truth or falsehood of (1.7.9). To this end, he deduced from it various propositions which he and Whitehead had not been able to establish directly. Relying on (1.7.8), he derived from (1.7.9) that for every infinite $\mathfrak{m}$ there is some $\mathfrak{n}$ such that

$$\mathfrak{m} = \aleph_0 \mathfrak{n}.$$

He then concluded that:

(1.7.10) If $\mathfrak{m}$ is infinite, then $\mathfrak{m} = \mathfrak{m} + \mathfrak{m}$.

---

[10] Russell 1901, 135. Like Burali-Forti and other followers of Peano, Russell used the term "primitive proposition" to mean a postulate.

[11] Russell in Whitehead 1902, 381–382.

Likewise, he derived the equivalent proposition (1.6.7) that

$$\text{if } \mathfrak{m} \text{ is infinite and } \mathfrak{n} \leq \mathfrak{m}, \text{ then } \mathfrak{m} + \mathfrak{n} = \mathfrak{m}.$$

Since $\mathfrak{m} = \aleph_0 \mathfrak{n} = \aleph_0^2 \mathfrak{n}$, he also obtained an apparently stronger result:

(1.7.11) If $\mathfrak{m}$ is infinite, then $\mathfrak{m} = \aleph_0 \mathfrak{m}$.[12]

Whitehead, on the other hand, regarded Russell's postulate (1.7.9) and its various consequences as "unproven theorems." Finally, Whitehead presented as an open problem the following proposition (1.6.8), which he did not believe to be a consequence of (1.7.9) but which paralleled (1.6.7):

If $\mathfrak{m}$ is infinite and $\mathfrak{n} \leq \mathfrak{m}$, then $\mathfrak{m}\mathfrak{n} = \mathfrak{m}$.[13]

All of these propositions, it was later recognized, depend on the Axiom.

The main thrust of Whitehead's article, which provided a general treatment of cardinal arithmetic, was to extend Cantor's cardinal multiplication to infinitely many factors. With this goal in mind, Whitehead introduced the multiplicative class $A^\times$ for every family A of disjoint non-empty classes. He defined $A^\times$ to be the class of all subclasses M of $\bigcup A$ such that for every S in A, $M \cap S$ has exactly one member [1902, 383]. Here he came close to stating the Axiom of Choice, in the form which Russell later named the Multiplicative Axiom. Yet at the time neither Whitehead nor anyone else recognized that such an axiom was needed (see 2.7 and 3.2).

Consequently, Whitehead's implicit uses of the Assumption did not appear as arbitrary choices but as instances of the unsuspected Multiplicative Axiom:

$A^\times$ is non-empty for every family A of non-empty classes.

Perhaps this proposition seemed so obvious to him that he felt no need to mention it, just as Cantor had not stated certain obvious theorems relying on the Assumption. In any case, Whitehead demonstrated several theorems which can easily be shown equivalent to the Multiplicative Axiom, such as the following:

(1.7.12) For every disjoint family A of non-empty classes, $\bigcup (A^\times) = \bigcup A$.

---

[12] Proposition (1.7.10) cannot be proved in **ZF**, since it implies that every Dedekind-finite set is finite (see 4.2). On the other hand, (1.7.10) is weaker than the Axiom, as Gershon Sageev later showed [1975], and in particular does not imply the Denumerable Axiom restricted to countable subsets of $\mathbb{R}$. Finally, Halpern and Howard [1970] established that (1.7.11) is equivalent to its consequence (1.7.10). It follows, since (1.7.11) clearly implies (1.7.9), that (1.7.9) is weaker than the Axiom.

[13] Whitehead 1902, 368, 392, 394.

(1.7.13) For all disjoint families A and B, if $A^x = B^x$, then $A = B$.

Another theorem had a family resemblance to the maximal principles, such as Zorn's Lemma and Hausdorff's Maximal Principle, which gained prominence decades later in algebra and analysis (see 3.4 and 4.4):

(1.7.14) Let A be a family of disjoint non-empty classes. If $S \subseteq \bigcup A$ and if no member of A contains more than one element of S, then $S \subseteq M$ for some M in A^x.

In other words, under the given hypotheses a class S could be extended to a member of A^x. To deduce these three theorems, Whitehead implicitly presupposed that certain multiplicative classes were non-empty [1902, 383–384].

After these preliminaries, Whitehead used his new notion of multiplicative class to define cardinal multiplication, for an arbitrarily large infinity of factors, and to relate multiplication to both addition and exponentiation. For a disjoint family A of classes, he defined $\sum_{a \in A} \bar{\bar{a}}$ as $\overline{\overline{\bigcup A}}$, and $\prod_{a \in A} \bar{\bar{a}}$ as $\overline{\overline{A^x}}$. The first of these definitions was similar to that given by Russell [1901]. In order to establish the uniqueness of cardinal addition and multiplication, Whitehead demonstrated that: If f is a bijection from A onto B, where A and B are families of disjoint non-empty classes, and if for every a in A and b in B, $f(a) = b$ implies that $\bar{\bar{a}} = \bar{\bar{b}}$, then $\bigcup A$ is equipollent to $\bigcup B$, and A^x to B^x. Then he connected cardinal addition, multiplication, and exponentiation by showing that

(1.7.15) $$\sum_{a \in A} \bar{\bar{B}} = \bar{\bar{A}} \cdot \bar{\bar{B}}$$

and that

(1.7.16) $$\prod_{a \in A} \bar{\bar{B}} = \bar{\bar{B}}^{\bar{\bar{A}}}.$$

As in the case of finite powers, cardinal exponentiation could be viewed as a special case of multiplication, which in turn became a special case of addition [1902, 383–388]. In all of these results Whitehead implicitly assumed multiplicative classes to be non-empty. Thus, as in Cantor's more limited attempt to treat sums of infinitely many cardinals as finite products (see 1.4), the Assumption was used implicitly in an unavoidable way.

Neither Russell nor Whitehead understood how much deductive strength lay hidden in Whitehead's theorems (1.7.12), (1.7.13), and (1.7.14). In fact, each of these three theorems implied both Russell's postulate (1.7.9) and Whitehead's unproved (1.6.8). Each also implied two propositions that Whitehead regarded as still undemonstrated in general: $\mathfrak{m}^2 = \mathfrak{m}$ and

$\mathfrak{m}^{\mathfrak{m}} = 2^{\mathfrak{m}}$ for every infinite $\mathfrak{m}$. Each of the three would turn out to be equivalent to the Axiom of Choice, as would the Multiplicative Axiom implicit in Whitehead's work, his (1.6.8), and the proposition that $\mathfrak{m}^2 = \mathfrak{m}$ for every infinite $\mathfrak{m}$. When Whitehead returned to such cardinality questions, in a paper written during 1903 and printed the following year, he took the Trichotomy of Cardinals as a hypothesis in order to deduce the proposition (4.3.4):

$$\text{If } \mathfrak{m}_1 < \mathfrak{n}_1 \text{ and } \mathfrak{m}_2 < \mathfrak{n}_2, \text{ then } \mathfrak{m}_1 + \mathfrak{m}_2 < \mathfrak{n}_1 + \mathfrak{n}_2. \text{ [1904, 31–32]}$$

Subsequently (4.3.4) too was shown to be equivalent to the Axiom (see 4.3). Yet on the eve of Zermelo's proof it remained quite unclear what role arbitrary choices and related assumptions played in cardinal arithmetic.

As the twentieth century opened, the Assumption was being used implicitly—in the guise of arbitrary choices as well as indirectly through such results as the Countable Union Theorem—even by mathematicians who were fundamentally unsympathetic to non-constructive methods. In analysis this occurred especially within the theory of real functions. The countable additivity of Lebesgue measure, the behavior of sets of first category, the structure of Baire's classification of real functions, and even that of Borel sets, all depended on the Axiom. Whereas the Denumerable Axiom or the Principle of Dependent Choices sufficed for such applications in real analysis, cardinal arithmetic was rife with propositions which turned out to be equivalent to the Axiom itself. Thus Russell and Whitehead, even more than the analysts, found themselves implicitly using the full strength of the Assumption.

When Zermelo's proof appeared, every one of the mathematicians discussed in this section rejected the Axiom of Choice on which the proof was based. The position of these critics was a delicate one, since the Axiom remained thoroughly entangled with their own researches. Subsequently, their reactions differed from individual to individual. In their magistral *Principia Mathematica*, Russell and Whitehead carefully made the Multiplicative Axiom an explicit hypothesis for those theorems depending on it, but they remained dubious of its truth (see 2.7 and 3.6). Young discarded the Axiom of Choice in favor of a nebulous procedure, due to Bernstein, which labeled the uncertainty of a proof by the cardinality of arbitrary choices used in it (see 2.5). At times Borel dismissed the Axiom as absurd and at other times he was inclined to accept the Denumerable Axiom, whereas Baire repudiated the Axiom unequivocally. Lebesgue, on the other hand, repudiated the Axiom equivocally (see 2.3, 3.6, and 4.11).

Why these future critics did not detect the Assumption hidden in their researches is a question to which there can be no final answer. During 1904, moreover, Russell did uncover Whitehead's implicit presupposition that the multiplicative class of a family of non-empty classes must be non-empty, and gradually realized its potency. As for analysts such as Borel, they had often extended to the denumerably infinite those processes already employed in the finite. In the wake of Zermelo's proof, all of these mathematicians faced a painful reappraisal of their researches, one which some of them carried out more fully, openly, and consistently than others.

1.8 Italian Objections to Arbitrary Choices

Despite the fact that by 1900 arbitrary choices were widely used in analysis, they attracted little explicit attention. Prior to Zermelo's proof, the only objections to making infinitely many arbitrary choices came from three Italian mathematicians: Peano [1890, 1902], Bettazzi [1892, 1896a], and Beppo Levi [1902]. All three were located at Turin, and there are strong reasons for believing that Peano's viewpoint spurred the others to reject arbitrary choices. Nevertheless, some historians have claimed that either Peano or Levi formulated the Axiom of Choice before Zermelo. These claims are best deferred until the articles in question have been examined.

In 1886 Peano published a new demonstration of the theorem, due to Cauchy, that the differential equation $y' = f(x, y)$ has a unique solution. Here Peano weakened Cauchy's hypotheses to require only that $f(x, y)$ be continuous. Four years later Peano returned to this theorem and generalized his proof to finite systems of first-order equations. When he arrived at a step that required a single element to be chosen from each set in a certain sequence $A_1, A_2, \ldots$ of subsets of $\mathbb{R}$, he remarked carefully:

> But since one cannot apply infinitely many times an *arbitrary* rule by which one assigns to a class A an individual of this class, a *determinate* rule is stated here, by which, under suitable hypotheses, one assigns to each class A a member of this class. [1890, 210]

To obtain his rule, he employed least upper bounds. Thereby he became the first mathematician who, while accepting infinite classes, categorically rejected the use of infinitely many arbitrary choices. Even though he was familiar with Cantor's researches, apparently it did not occur to him, or to anyone else at the time, that Cantor had often used such arbitrary choices. After Zermelo's proof appeared, Peano's suspicions were aroused, and he vigorously criticized the Axiom of Choice as well as earlier results depending on it implicitly (see 2.8).

In 1892 Bettazzi, who had just become Peano's colleague at the Military Academy in Turin, published an article on discontinuous real functions. In this context Bettazzi was led to distinguish between a limit point and a sequential limit point. While he did not cite Peano [1890] in arguing here against the use of infinitely many arbitrary choices, his terminology closely resembled Peano's. On the other hand, Bettazzi discussed the question of such choices in much more detail than Peano had done:

> A point can be taken arbitrarily from a given set of points [in ℝ], or from one of its subsets, or from a finite number of its subsets. But when one has to consider infinitely many of its subsets and to construct a subset formed by choosing in each of these subsets any point whatever (as will be the case in what follows), it does not suffice to say that one forms this set by taking a point *arbitrarily* in each of these subsets. For one cannot regard as determinate an *infinite* number of objects all chosen arbitrarily in given classes. This follows clearly when one notes that giving them arbitrarily is equivalent to defining them separately one at a time . . . . [1892, 176]

Bettazzi added that one could arbitrarily select a point from finitely many subsets of an infinite set of real numbers, so long as one stated "a rule... sufficient to define a *determinate* point in each subset remaining." Thus he agreed with Peano that the use of infinitely many arbitrary choices must be categorically rejected.

Since such choices could not be used, Bettazzi wished to investigate the conditions under which a rule of choice existed:

> In order that, given an arbitrary set G [in ℝ], one can form a set consisting of any point whatever from each of infinitely many determinate subsets of G, the following problem must be resolved: Given a set of points [in ℝ], indicate a rule by means of which a set determines a point in it, *whatever the given set may be.* In the present state of set theory, no solution is known for this problem. [1892, 176]

Despite the fact that no rule of choice was known for ℝ, he noted that such a rule was available for any countable set and, so far as he knew, only for countable sets. On the other hand, there were certain subsets of ℝ for which a rule of choice was known in special cases. Thus, for a closed set S, one could determine a point in any closed subset T of S lying in a given closed and bounded interval, namely, the least upper bound of T.

Bettazzi studied the existence of rules of choice because he wished to know under what conditions a limit point was also a sequential limit point. This question arose in the course of considering the limit points from the right of an arbitrary real function $y = f(x)$ along the line $x = a$. His first theorem was that, given a function $y = f(x)$ with domain G, if p is a limit point of f from the right along $x = a$, then p is also a sequential limit point of f from the right—*provided* that there is an interval $(a, a + \varepsilon)$ and a rule of choice for the domain of f restricted to this interval [1892, 181]. Then he generalized this result to obtain a second theorem: Under the same hypotheses, there is a countable subset C of the domain of f such that the set of all limit points of C from the right along the line $x = a$ is equal to a given closed subset A of the set B of all limit points of f along $x = a$ from the right. Since B was closed, the theorem applied in particular if A and B were equal [1892, 184–186].

Ironically, Bettazzi's second theorem used the Axiom of Choice indirectly, in the guise of the Countable Union Theorem, when he relied on one of his earlier results:

(1.8.1) Every closed subset A of $\mathbb{R}$ can be partitioned into a countable set C and another set E consisting of all those limit points of C not in C itself.

Likewise, he applied the Countable Union Theorem when he attempted to frame a restricted version of his second theorem, one that did not require the existence of a law of choice [1892, 187].

Four years later, reading a paper [1896a] before the Turin Academy of Sciences, Bettazzi again discussed arbitrary choices. There he questioned Dedekind's proof of 1888, that every infinite set S is Dedekind-infinite, precisely because it involved the arbitrary choice of a one–one function $f_n: \{1, 2, \ldots, n\} \to S$ for every n. Although Bettazzi hypothesized that one could postulate such choices, he immediately rejected the idea [1896a, 512]. Shortly thereafter, Burali-Forti persuaded him that Dedekind's theorem was correct. As a result, Bettazzi's acute observations concerning Dedekind's use of arbitrary choices stimulated no debate (see 1.3).

The third mathematician to oppose the use of infinitely many arbitrary choices was Beppo Levi. In 1900 Levi, who had received his doctorate at Turin four years earlier, published an article largely inspired by Baire's thesis of 1899. Levi sought to obtain properties satisfied by every real function (or by every subset of $\mathbb{R}$), since he believed that no non-trivial such properties were yet known. Without proof he proposed one such property as a theorem: Every subset A of $\mathbb{R}$ is equal to $(B \cup C)—D$, where B is some closed set and both C and D are nowhere dense sets.[1] In effect, he asserted that every subset of $\mathbb{R}$ has the Baire property.[2] Moreover, Levi insisted that his theorem implied one form of Cantor's Continuum Hypothesis: Every uncountable subset of $\mathbb{R}$ has the power of the continuum [1900, 75].

The relationship between the Axiom of Choice and Levi's results is intriguing. In particular, his proposition that every subset of $\mathbb{R}$ has the Baire property contradicts the Axiom of Choice, since the non-measurable set, later given by Vitali [1905] and shown to exist by means of the Axiom, lacks the Baire property (see 2.3). On the other hand, it is not clear why Levi's proposition should imply the above form of the Continuum Hypothesis. Although he stated [1900, 73n] that an article entitled "Researches on the Continuum and Its Power" would soon appear with the proof of his proposition and its corollaries, for unknown reasons the article was not published. Nevertheless, he carefully noted that $\mathbb{R}$ and the second number-

[1] Levi also noted that one could require C to be closed.

[2] See 1.7. The term *Baire property* was not introduced until some years later.

class might fail to be comparable.[3] Thus Levi distanced himself from Cantor's repeated uses of the Trichotomy of Cardinals.

In his thesis of 1901, Felix Bernstein established that the set of all closed subsets of $\mathbb{R}$ has the power of the continuum. A few pages later he remarked that the result from which Levi [1900] had deduced that every subset of $\mathbb{R}$ has the Baire property was mistaken.[4] By way of response, Levi published an analysis [1902] of Bernstein's thesis.

Levi began with a broad critique of set theory. First of all, he insisted, Cantor and his followers had frequently made the dubious assumption that certain properties known for finite sets hold for infinite ones as well. In this vein Cantor had asserted that every set can be well-ordered. Yet, Levi continued, several authors had come to doubt that well-ordering was possible in general. Even Cantor's student Bernstein had evidently abandoned the Well-Ordering Principle, since his thesis investigated the conditions under which two sets are comparable—a property already satisfied by any two sets which can be well-ordered [Levi 1902, 863].

Nevertheless, Levi added, Bernstein had made an assumption which "appeared, although stated rather obscurely, to be derived in essence from the same postulate of [well-]orderability in such a way that one might hesitate for a moment when trying to deny that this new assumption is self-evident" [1902, 863]. This assumption, which Levi regarded as a consequence of the Well-Ordering Principle, was the Partition Principle: If a set A is partitioned into a family S of disjoint non-empty sets s, then $\overline{\overline{S}} \leq \overline{\overline{A}}$.[5] After showing that the Partition Principle held whenever A was finite, Levi added that "this demonstration is applicable without change to any case where all the s are well-ordered or else, more generally, where we can distinguish a unique element in each s" [1902, 864]. However, he did *not* regard his comment as grounds for using the Partition Principle in general.

Levi devoted the remainder of his article to a new proof of Bernstein's theorem that the family of all closed subsets of $\mathbb{R}$ has the power of the continuum. Levi did so because he believed that Bernstein's proof was vitiated by using the Partition Principle to establish the following lemma:

(1.8.2) The set of all denumerable subsets of $\mathbb{R}$ has a power less than or equal to that of all infinite sequences of real numbers.

---

[3] In fact, although Levi did not know this, $\mathbb{R}$ and the second number-class must be incomparable if every subset of $\mathbb{R}$ has the Baire property and if every uncountable subset of $\mathbb{R}$ has the power of the continuum. For Vitali's set [1905] which lacks the Baire property was shown to exist by using the Axiom merely to well-order $\mathbb{R}$.

[4] Bernstein 1901, 48n. However, he had misread Levi's result and so his argument was not valid.

[5] Apparently Levi did not know that Burali-Forti [1896], employing unions rather than partitions, had postulated an incomplete version of the Partition Principle.

In his new proof Levi carefully avoided arbitrary choices and proceeded according to definite rules. Thus he was far from proposing the Axiom of Choice in his article. What he did was simply to recognize that the Partition Principle can be proved in those cases where a rule permits a unique element to be distinguished from each s in S.[6]

Consequently, we must reject the claims of those historians who, relying on the articles of Peano and Levi discussed above, have asserted that one or both of them formulated the Axiom of Choice explicitly before Zermelo.[7] What we can say is that Peano was the first to *reject* the use of infinitely many arbitrary choices. Levi, probably influenced by Peano, also endeavored to avoid proofs that employed such arbitrary choices.

One claim for Levi's priority is more difficult to evaluate. In 1958 Abraham Fraenkel wrote:

> According to a communication by letter from F. Bernstein, about 1901 G. Cantor and F. Bernstein tried to construct a one-to-one correspondence between the continuum [$\mathbb{R}$] and the set of all denumerable order-types . . . . When they met with an insurmountable difficulty, B. Levi proposed to solve the difficulty by introducing the principle of choice which he formulated in a general form. [Fraenkel and Bar-Hillel 1958, 48]

It is uncertain what Fraenkel meant by an "insurmountable difficulty," since by 1901 Cantor and Bernstein had succeeded in establishing the existence of such a correspondence (see 1.6). Yet they never referred to any hypothesis justifying arbitrary choices. It may be that inaccurate recollection by Bernstein or Fraenkel, decades after the event, credited Levi with something not truly his own. Whether or not he proposed some version of the Axiom of Choice to Cantor and Bernstein in 1901, Levi distinctly avoided anything of the sort in his article dated October 1902. Only in 1918 did he return to such questions by proposing an extremely restricted form of the Denumerable Axiom as a substitute for Zermelo's Axiom of Choice, which Levi still could not accept (see 4.7).

It should be emphasized that the mistrust of arbitrary choices, disseminated by Peano and his associates, was not typical of Italian mathematicians. Indeed, a particularly interesting use of arbitrary choices occurred in the generalized analysis that was developing in Italy late in the nineteenth century and that would eventually lead to functional analysis. One source for this generalized analysis was the work of Weierstrass. The particular result of interest to us here, which became known as Ascoli's Theorem, extended the Bolzano–Weierstrass Theorem to families of real functions.

---

[6] A similar view of Levi's position was reached independently by Barbara Moss [1979].

[7] Kennedy [1975, 209; 1980, 33] has pressed a claim for Peano. Beth has argued that the Axiom was stated by Peano but was first "seen, by Beppo Levi, to constitute an independent principle of proof" [1959, 376]. Van Heijenoort insisted that Levi's article [1902] recognized the Axiom "as a new mathematical principle" [1967, 139].

In 1884, inspired partly by the calculus of variations, Giulio Ascoli sought conditions under which a limit-curve would exist for a given family of curves. To this end he introduced the concept of equicontinuity. A family F of real functions, defined on a closed interval $[a, b]$, was said to be equicontinuous if, given any $\varepsilon > 0$, there was some $\delta > 0$ such that $|x - y| < \delta$ implied $|f(x) - f(y)| < \varepsilon$ for every f in F and for every x and y in $[a, b]$. Then Ascoli's Theorem stated that a uniformly bounded, equicontinuous sequence $\{f_n\}$ of functions, all defined on $[a, b]$, had a convergent subsequence [1884, 549].

To demonstrate his theorem, Ascoli noted that $\{f_n\}$ had a subsequence $\{g_n^{(1)}\}$ such that $g_1^{(1)}(a), g_2^{(1)}(a), \ldots$ converged to some point on $x = a$. Then $\{g_n^{(1)}\}$ had a subsequence $\{g_n^{(2)}\}$ such that $g_1^{(2)}(b), g_2^{(2)}(b), \ldots$ converged to some point on $x = b$. Next he chose a subsequence $\{g_n^{(3)}\}$ of $\{g_n^{(2)}\}$ such that $g_1^{(3)}((a + b)/2), g_2^{(3)}((a + b)/2), \ldots$ converged to some point on $x = (a + b)/2$. Proceeding in this manner, he chose sequences $\{g_n^{(4)}\}, \{g_n^{(5)}\}, \ldots$, each of which was a subsequence of all previous sequences and each of which converged pointwise to the previous points on $x = a$, $x = b$, and $x = (a + b)/2$, as well as to additional points on $x = (3a + b)/4$, $x = (a + 3b)/4$, and so forth, respectively. Since the abscissas of this set of points were dense in $[a, b]$, then the diagonal sequence $g_1^{(1)}, g_2^{(2)}, \ldots, g_n^{(n)}, \ldots$ was a convergent subsequence of $\{f_n\}$, as desired [1884, 545–549].

Ascoli used infinitely many arbitrary choices when he selected a subsequence $\{g_n^{(m)}\}$ at each stage m. Such a use was avoidable since for any sequence $g_1^{(m)}(c), g_2^{(m)}(c), \ldots$, the Bolzano–Weierstrass Theorem gave a uniquely specifiable limit point, which could then be used to define a subsequence. However, Ascoli made no attempt to state such a rule.

As later mathematicians generalized Ascoli's Theorem, it became more difficult to avoid the Axiom even if one wished to do so. In 1889 Cesare Arzelà, who taught at the University of Bologna, made the first such generalization: If F is an infinite set of real functions defined on $[a, b]$, uniformly bounded and equicontinuous, then F has a limit-function. Here f was a limit-function of F if for any $\varepsilon > 0$ and for all x in $[a, b]$ there were infinitely many functions g in F such that

$$f(x) - \varepsilon < g(x) < f(x) + \varepsilon.$$

Arzelà began his proof by selecting a sequence $\{f_n\}$ in F, and hence relied on Cantor's theorem (1.1.4) that every infinite set has a denumerable subset.[8] Whether or not one could state an alternative proof of Arzelà's Theorem

[8] Arzelà 1889, 344. Later he gave a demonstration of Ascoli's Theorem using a rule rather than arbitrary choices [1895, 230]. In a related article of the same year [1895a, 258] he cited Peano [1890], but not in regard to a rule of choice. On the other hand, Peano [1902, 8–9] criticized Arzelà for using infinitely many arbitrary choices in a textbook [1901] on the calculus.

avoiding the Axiom, for its later generalizations—such as that given by Maurice Fréchet in 1906 and discussed in 2.4—it would not be possible at all.

1.9 Retrospect and Prospect

That mathematics consists of constructions is a viewpoint whose ancestry can be traced at least as far as Euclid. During the nineteenth century, mathematicians both expanded and restricted what was permissible in such constructions. Which of the procedures valid for any finite number of steps, they inquired, could be extended to an infinite number? Among the procedures considered was the choice of an element from a set.

While even Euclid had, in effect, selected an element from each of finitely many sets, it appears that the second and third stages in the emergence of arbitrary choices did not take place until the nineteenth century: infinitely many choices made by a stated rule, or by an unstated rule, respectively. Certainly the third stage was already visible in Cauchy's proof of 1821 that a real function f, continuous on a closed interval, has a root there whenever the value of f has opposite signs at the endpoints. For Cauchy arbitrarily selected the terms of two convergent sequences in such a way that each term depended on those chosen previously. All the same, his arbitrary choices served only as a convenient shorthand for a rule that he could have supplied if he had wished to do so.

The watershed was reached in 1871 when Cantor made infinitely many arbitrary choices for which no rule was possible, and thereby initiated the fourth stage. This occurred in the proof of his theorem that a real function f is continuous at a point if and only if f is sequentially continuous there. Thus both the third and fourth stages originated in analysis. In 1877 Dedekind extended arbitrary choices to algebraic number theory when he made uncountably many of them in order to obtain representatives from congruence classes. Soon afterward both Cantor and Dedekind used the Assumption implicitly to characterize the boundary between finite and infinite sets. Yet they were no more aware than Cauchy had been, a half century earlier, that a new and important principle was involved.

In this regard Cantor's researches were especially significant since they served as the principal conduit for direct and indirect uses of the Assumption. Whereas Cantor introduced his topological notions—such as limit point, derived set, and perfect set—in terms of neighborhoods, a decade later Jordan defined many of these notions in terms of sequences. Treating these parallel notions as equivalent brought the Denumerable Assumption into play, as had already happened when Cantor equated continuity and sequential continuity. Moreover, his Countable Union Theorem and his tacitly used $\aleph_1$-Partition Principle occupied a central position within the theory of derived sets. In real analysis Baire, Borel, and Lebesgue used the Denumerable Assumption implicitly—both directly as arbitrary choices and indirectly as the Countable Union Theorem—although they would later criticize the Axiom when it appeared explicitly. Within set theory proper, Borel and Russell employed arbitrary choices in their respective proofs of Cantor's theorem that every infinite set has a denumerable subset. Finally, Whitehead deduced a number of theorems, later recognized as equivalent to the Axiom, by implicitly assuming that a certain multiplicative class was non-empty—the origins of Russell's later Multiplicative Axiom.

Independently of his use of arbitrary choices, Cantor proposed the Well-Ordering Principle in 1883 as a fundamental and fruitful law of thought. At the same time he published the form of the Continuum Hypothesis which implied that the real numbers can be well-ordered. Indeed, Cantor's correspondence suggests that it was this form of the Continuum Hypothesis which led him to the Well-Ordering Principle, as well as to a special case of the Equivalence Theorem. By 1895, however, he had become convinced that both the Well-Ordering Principle and its consequence, the Trichotomy of Cardinals, required a demonstration. Within two years he obtained such a proof (based on successive arbitrary choices), which was rediscovered by Jourdain in 1903. Although Cantor remained ill at ease with his proof, he encouraged Jourdain to publish his independently discovered version. Unknown to Cantor and Jourdain, in 1896 Burali-Forti had taken a postulational approach to the problem of deducing the Trichotomy of Cardinals. One of Burali-Forti's postulates was the Partition Principle, and the other turned out to be equivalent to the Axiom of Choice.

As the twentieth century opened, mathematicians had not arrived at a concensus regarding the Well-Ordering Problem. Borel and Russell remained sceptical of both the Well-Ordering Principle and the Trichotomy of Cardinals, whereas Schoenflies and Schröder favored the latter while doubting the former. Hilbert inclined to the belief that at least $\mathbb{R}$ could be well-ordered. As a step toward establishing or refuting the Well-Ordering Principle, Hardy used successive arbitrary choices to deduce both that $\aleph_1 \leq 2^{\aleph_0}$ and that every infinite cardinal either is an aleph or is greater than all alephs. From the latter theorem emerged Jourdain's attempted proof of the Well-Ordering Principle.

Why did future critics of the Axiom, such as Borel and Russell, fail to notice the use of arbitrary choices in their own research? In part, the answer is that they remained unaware (as did everyone else at the time) of the deductive strength of such arbitrary choices. Furthermore, the boundary between constructive and non-constructive methods continued to be vague. When the notion of a construction was gradually broadened to include denumerably infinite processes, the uniqueness of the object constructed did not immediately emerge as an important consideration. Nevertheless, three Italian mathematicians—Peano, Bettazzi, and Levi—asserted that infinitely many arbitrary choices were not permissible in mathematics but that, on the contrary, a rule must be given specifying the choices. Later both Peano and Levi expressed their opposition to the Axiom.

Thus, on the eve of Zermelo's proof that every set can be well-ordered, there was widespread use of arbitrary choices but equally widespread ignorance of their true deductive strength. No one had suggested the propriety of justifying such choices by an axiom. Like a stroke of lightning, Zermelo's proof suddenly illuminated the landscape and caused mathematicians to scrutinize the assumptions upon which they had been resting unawares.

Chapter 2

Zermelo and His Critics (1904–1908)

> Any argument where one supposes an *arbitrary choice* to be made a non-denumerably infinite number of times . . . [is] outside the domain of mathematics.
>
> Emile Borel [1905, 195]

When in 1904 Zermelo published his proof that every set can be well-ordered, many questions lay unresolved. In the wake of Russell's paradox, published in 1903, it was even uncertain what constituted a set. Moreover, Zermelo's proof itself raised a number of methodological questions: Was it legitimate to define a set A in terms of a totality of which A was a member, as Zermelo had done? Did the class W of all ordinals invalidate Zermelo's proof and entangle it in Burali-Forti's paradox? Most important of all, was his Axiom of Choice true? Was it a law of logic? Should one postulate simultaneous, independent arbitrary choices in preference to successive, dependent ones? Did the cardinality of the set of choices affect the validity of the Axiom, so that the Denumerable Axiom was true but not the Axiom of Choice in general?

All of these questions echoed a broader problem which had rarely been enunciated explicitly: What methods were permissible in mathematics? Must such methods be constructive? If so, what constituted a construction? What did it mean to say that a mathematical object existed? Normally mathematicians avoided such quasi-philosophical questions, and addressed them only when they felt their discipline to face a crisis. That Zermelo's proof precipitated such a crisis was shown by the extent of the resulting controversy. From 1905 to 1908 eminent mathematicians in England, France, Germany, Holland, Hungary, Italy, and the United States debated the validity of his demonstration. Never in modern times have mathematicians argued so publicly and so vehemently over a proof. This chapter is an exploration of their debate.

2.1 König's "Refutation" of the Continuum Hypothesis

During 1904 the Well-Ordering Problem underwent a series of reversals, and at first it seemed to be resolved negatively. In August a professor from Budapest, Julius König, delivered a lecture entitled "On the Continuum Problem" to the Third International Congress of Mathematicians at Heidelberg. König claimed to establish that the Continuum Hypothesis was false, since the power of the continuum was not an aleph, and so the real numbers could not be well-ordered. Listening to that lecture, Cantor became dismayed. For he continued to believe fervently that every set can be well-ordered and that $2^{\aleph_0} = \aleph_1$.[1]

Ironically, König's argument made sophisticated use of Cantor's cardinal arithmetic. As an initial step, König established that for each sequence $M_0, M_1, M_2, \ldots$ of infinite disjoint sets the following inequalities held:

$$\sum_{i=0}^{\infty} \overline{\overline{M}}_i \leq \prod_{i=0}^{\infty} \overline{\overline{M}}_i \leq \left(\sum_{i=0}^{\infty} \overline{\overline{M}}_i \right)^{\aleph_0}. \tag{2.1.1}$$

To derive the first inequality, König obtained a subset of $\prod_{i=\infty}^{\infty} \overline{\overline{M}}_i$ equipollent to $\sum_{i=0}^{\infty} \overline{\overline{M}}_i$ by making infinitely many arbitrary choices: "For this purpose one chooses from each M_i a definite element $\beta_i \ldots$" [1904, 145]. Next he established that for an increasing sequence of infinite cardinals the first inequality in (2.1.1) would be strict:

(2.1.2) If $\aleph_0 \leq \overline{\overline{M}}_i < \overline{\overline{M}}_{i+1}$ for all i in $\mathbb{N}$, then $\sum_{i=0}^{\infty} \overline{\overline{M}}_i < \prod_{i=0}^{\infty} \overline{\overline{M}}_i$.

From these two results, which used arbitrary choices in an essential way, he immediately deduced a third:

(2.1.3) If $\aleph_0 \leq \overline{\overline{M}}_i < \overline{\overline{M}}_{i+1}$ for all i in $\mathbb{N}$, then

$$\sum_{i=0}^{\infty} \overline{\overline{M}}_i < \left(\sum_{i=0}^{\infty} \overline{\overline{M}}_i \right)^{\aleph_0}.$$

To complete his argument, König relied on a proposition taken from Bernstein's recent dissertation:

(2.1.4) $\aleph_\alpha^{\aleph_0} = \aleph_\alpha \cdot 2^{\aleph_0}$ for every ordinal α.

---

[1] Kowalewski 1950, 202; see also Schoenflies 1922, 100.

If $\mathbb{R}$ could be well-ordered, König continued, then its power was $\aleph_\beta$ for some ordinal β. Letting $\overline{\overline{M}}_i$ be $\aleph_{\beta+i}$ in (2.1.3), he obtained

$$\text{(2.1.5)} \qquad \aleph_{\beta+\omega} < \aleph_{\beta+\omega}^{\aleph_0},$$

since $\sum_{i=0}^{\infty} \aleph_{\beta+i} = \aleph_{\beta+\omega}$. But then (2.1.4) implied that

$$\aleph_{\beta+\omega}^{\aleph_0} = \aleph_{\beta+\omega} \cdot \aleph_\beta = \aleph_{\beta+\omega},$$

contradicting (2.1.5). Therefore $\mathbb{R}$ could not be well-ordered.

König's argument stimulated some of those in his audience to explore the matter further. By the next day Zermelo had ferreted out the error in this argument.[2] As it happened, Bernstein's proof for (2.1.4) was inadequate when α was a limit ordinal—the crucial case.

After the Congress ended, this episode continued to intrigue mathematicians. A few of them—including Cantor, Hausdorff, Hilbert, and Schoenflies—soon met in Wengen, where König's argument and Bernstein's dubious result were the focus of discussion.[3] Later the same year, Hausdorff published an article in which he established a formula closely related to Bernstein's:

$$\text{(2.1.6)} \quad \aleph_{\beta+1}^{\aleph_\alpha} = \aleph_{\beta+1} \cdot \aleph_\beta^{\aleph_\alpha} \text{ for any ordinals } \alpha \text{ and } \beta.$$

On the other hand, Hausdorff questioned whether Bernstein's proposition was true in general: "Its correctness appears all the more problematical since from it would follow, as Herr J. König has shown, the paradoxical result *that the power of the continuum is not an aleph and that there are cardinal numbers which are greater than every aleph*."[4] Thus in 1904 Hausdorff thought that $\mathbb{R}$ could be well-ordered and that every infinite cardinal was an aleph—propositions of which he became dubious two years later, and then reaffirmed (see 2.5). Cantor, who felt relieved when Zermelo discovered the mistake in König's argument, continued to affirm the Well-Ordering Principle. Writing to Jourdain on 5 May 1905, Cantor claimed without proof that Bernstein's proposition was false in general, and added that, despite König's ill-considered attempt to refute the Well-Ordering Principle, his paper had its positive aspects.[5] In order to redeem his thesis, Bernstein insisted that he had only applied (2.1.4) when α was finite, a case for which his proof remained valid [1905a, 463–464].

---

[2] Kowalewski 1950, 202.

[3] Schoenflies 1922, 100–101.

[4] Hausdorff 1904, 571. Bernstein's proposition (2.1.4) is true for all $\aleph_n$ as well as for all $\aleph_\alpha$ less than or equal to $2^{\aleph_0}$; it is false for all $\aleph_\alpha$ of cofinality ω such that $2^{\aleph_0} < \aleph_\alpha$, by (2.1.3) and and (2.1.6). Thus (2.1.4) cannot be decided in **ZFC** since Cohen [1964, 108–109] has shown that $2^{\aleph_0}$ may take on arbitrarily large or small values among uncountable alephs.

[5] Cantor in Grattan-Guinness 1971a, 125–126.

When König's paper appeared in 1905 as part of the *Proceedings* of the Heidelberg Congress, he retracted his earlier claim that $\mathbb{R}$ cannot be well-ordered. Instead, he asserted that if Bernstein's result (2.1.4) is true in general, then $\mathbb{R}$ cannot be well-ordered. König also argued that the converse was true. If $\mathbb{R}$ cannot be well-ordered, then $2^{\aleph_0} > \aleph_\alpha$ for every ordinal α, from which Bernstein's proposition followed [1904, 147]. Hidden in König's argument for this converse was a case of the Trichotomy of Cardinals, in particular the assumption that the power of the continuum is comparable with every aleph. Yet a decade later Hartogs established that such a converse is impossible, since $2^{\aleph_0} > \aleph_\alpha$ cannot hold for every α [1915, 440–442].

This error in König's argument for the converse, like his use of arbitrary choices in his original proposition, reflects a fundamental misunderstanding of the relationship between Trichotomy, arbitrary choices, and the existence of a well-ordering for $\mathbb{R}$. Ironically, König's proof of his converse utilized a form of Trichotomy (for every α, $\aleph_\alpha \leq 2^{\aleph_0}$ or $2^{\aleph_0} \leq \aleph_\alpha$) sufficiently strong to guarantee that $\mathbb{R}$ can be well-ordered. Although his theorems (2.1.2) and (2.1.3) did not necessarily yield a well-ordering for $\mathbb{R}$, the arbitrary choices on which those theorems were based would provide such a well-ordering if permitted in general. Thus König's arguments were inherently contradictory, a fact which only emerged in the wake of Zermelo's proof. That a researcher as astute as König could so confuse these problems was a measure of Zermelo's insight in clarifying them.

Later mathematicians would accept König's theorems (2.1.2) and (2.1.3) as consequences of the Axiom of Choice. Instead of viewing his Heidelberg argument as refuting the Continuum Hypothesis, they regarded it as establishing that $2^{\aleph_0} \neq \aleph_{\alpha+\omega}$ for any ordinal α.[6] After Hausdorff's researches of 1906–1908, mathematicians eventually came to understand how König's argument also yields the stronger result that $2^{\aleph_0} \neq \aleph_\beta$ for any limit ordinal β cofinal with ω.[7] Nevertheless, König continued his attempts, ultimately unsuccessful, to refute the Continuum Hypothesis.

2.2 Zermelo's Proof of the Well-Ordering Theorem

Zermelo began his mathematical career in fields far removed from set theory. His doctoral dissertation of 1894 dealt with the calculus of variations, and soon thereafter he turned to mathematical physics, especially statistical

[6] Schoenflies claimed too much when he asserted [1908, 25] that König's proof showed that the power of the continuum cannot be $\aleph_\alpha$ for *any* limit ordinal α. Later Luzin and Sierpiński [1917] modified König's proof to deduce $2^{\aleph_0} \neq \aleph_\omega$ without using the Axiom.

[7] This was first pointed out by Lindenbaum and Tarski [1926, 310]. The *cofinality* of β is the least ordinal δ, order-isomorphic to a subset A of β, such that the least upper bound of A is β. Two ordinals are *cofinal* with each other if they have the same cofinality.

mechanics. After serving from 1894 to 1897 as Max Planck's assistant at the Institute for Theoretical Physics in Berlin, Zermelo moved to Göttingen. There he submitted his *Habilitationsschrift* on hydrodynamics in 1899, and began lecturing as a *Privatdozent*.

Reminiscing later about this period, Zermelo described how Hilbert had affected his career:

> Thirty years ago, when I was a *Privatdozent* at Göttingen, I came under the influence of D. Hilbert, to whom I am surely the most indebted for my mathematical development. As a result I began to do research on the foundations of mathematics, especially on the fundamental problems of Cantorian set theory, whose true significance I learned to appreciate through the fruitful collaboration of the mathematicians at Göttingen.[1]

It was in 1899 that Hilbert published the seminal book on the foundations of geometry in which he offered a purely formal axiomatization for Euclidean space. Next Hilbert turned to investigating the consistency of the system of real numbers. About this time Zermelo discovered in Schröder's algebraic logic what later became known as Russell's paradox, two years before Russell himself, and communicated it to Hilbert.[2] Three years later Zermelo discussed it with the philosopher Edmund Husserl as well, but did not publish it. Indeed, Zermelo viewed his argument as showing merely that any set which contains all its subsets as elements (such as the set of all sets) is self-contradictory.[3] This reticence to publish the paradox, or even to regard set theory as threatened by it, marks an important difference between Zermelo's foundational attitudes and those of Russell.

During the winter semester of 1900–1901 Zermelo delivered his first course of lectures on set theory. In general he relied heavily on Cantor's *Beiträge* of 1895 and 1897. However, section seven of Zermelo's lecture notes (on the comparability of sets) indicated that he did not regard the Trichotomy of Cardinals as proven.[4] That same section contained results soon published in his first set-theoretic paper, which Hilbert presented to the Göttingen Academy of Sciences on 9 March 1901, concerning the addition of infinite cardinals. Zermelo's paper used arbitrary choices both directly

---

[1] "Bericht an die Notgemeinschaft der Deutschen Wissenschaft über meine Forschungen betreffend die Grundlagen der Mathematik," p. 1. This unpublished report from Zermelo's *Nachlass*, kept at the Albert-Ludwigs-Universität (Freiburg im Breisgau), appears as an appendix in Moore 1980, 130–134.

[2] Zermelo 1908, 118–119. On 7 November 1903 Hilbert wrote to Frege that Zermelo had discovered this paradox three or four years earlier. Hilbert's letter is printed in English translation in Frege [1980, 51–52]. When in May 1903 Zermelo spoke on Frege's work [1884, 1903] to the Mathematical Society of Göttingen, he seems to have been primarily concerned with the concept of number in Cantor, Dedekind, and Frege (see **DMV 12** (1903), 345–346).

[3] Husserl made notes of his conversation with Zermelo, which were recently rediscovered. On this subject see Rang and Thomas [1981].

[4] Zermelo's notes for this course can be found in his *Nachlass*, but most of them are in his idiosyncratic version of Gabelsberger shorthand.

and indirectly. Their direct use occurred when he selected denumerably many bijections in order to prove that if

$$\mathfrak{m} = \mathfrak{m} + \mathfrak{p}_n$$

for each n, then

$$\mathfrak{m} = \mathfrak{m} + \sum_{n=0}^{\infty} \mathfrak{p}_n.$$

On the other hand, their indirect use ensured that the sum of denumerably many cardinals was well-defined. In neither case did he remark the importance of such arbitrary choices [1901, 35–37].

Then Zermelo's perspective shifted. In August 1904, after discovering the flaw in König's attempt to refute the Continuum Hypothesis, he turned his attention to the Well-Ordering Problem. During conversations with Erhard Schmidt, his thoughts crystallized, and on 24 September his proof was complete. Sent to Hilbert, it quickly appeared in *Mathematische Annalen* as a three-page article entitled "Proof that Every Set can be Well-Ordered" [1904].

Zermelo's demonstration of the Well-Ordering Theorem began with an arbitrary non-empty set M for which a well-ordering was desired. Letting S be the set of all non-empty subsets M' of M, he made the first explicit statement of the principle later called the Axiom of Choice: "*To each subset M' one associates any element m'_1 which occurs in M' itself and which may be named the 'distinguished' element of M'.*"[5] In other words, Zermelo postulated the existence of a function $\gamma : S \to M$ such that $\gamma(M') \in M'$ for every M' in S. Cantor had termed any function $f : A \to B$ as a covering of A by elements of B,[6] and Zermelo adopted this terminology when he described a function $\gamma : S \to M$ given by the Axiom as a covering of a special kind:

> The number of these coverings γ is equal to the product $\Pi\mathfrak{m}'$ extended over all [non-empty] subsets M' [of power $\mathfrak{m}'$ respectively] and hence is different from 0 in any case. In what follows, any covering is considered and from it a definite well-ordering of the elements of M is derived. [1904, 514]

(How such a covering—or, as we now say, a choice function—could actually be obtained was the fundamental objection of many mathematicians, including Baire, Borel, and Lebesgue as well as Peano and Russell; see 2.3, 2.7, and 2.8.)

At this point some definitions are in order. Let $s(b, x)$ be the segment of x determined by b, that is, the set of all elements preceding b in the well-

[5] Zermelo 1904, 514; his emphasis.

[6] Cantor 1895, 486.

ordered set x. Zermelo defined a γ-set to be any well-ordered subset M_γ of M such that b was the distinguished element of $M-s(b, M_\gamma)$ for every b in M_γ. He described any member of a γ-set as a γ-element.

Zermelo began his proof by establishing certain properties of γ-sets. First of all he noted that γ-sets existed, for instance $\{m_1\}$, where m_1 was the distinguished element of M. Next he demonstrated that for any two distinct γ-sets, one was a segment of the other. Hence if two γ-sets had a common element, they agreed on the ordering of all preceding elements.

In fact, the set L_γ of all γ-elements was a γ-set, as Zermelo proceeded to show. First, he defined the order $<$ of any two elements a and b in L_γ to be their order relative to any γ-set A of which they were both members, and this order was seen to be independent of the choice of A. Thus for any distinct a and b in L_γ, $a < b$ or $b < a$ but not both. A similar argument yielded that for any elements a, b, c in L_γ, if $a < b$ and $b < c$, then $a < c$. Hence he concluded that L_γ was an ordered set. Second, to show L_γ well-ordered, Zermelo considered any non-empty subset B of L_γ and any member b of B. Since b belonged to some γ-set M_γ and since $B \cap M_\gamma$ was a non-empty subset of the well-ordered set M_γ, then $B \cap M_\gamma$ had a first element c; in fact, c was also a first element of B. Finally, that L_γ was a γ-set followed from the last sentence of the previous paragraph. (To Henri Poincaré, the very definition of L_γ was objectionable; see 2.4.)

Zermelo completed his proof of the Well-Ordering Theorem by establishing that L_γ equaled M. Since L_γ was clearly a subset of M, it remained to show that L_γ contained every member of M. Otherwise $M-L_\gamma$ would contain a distinguished element m', with the consequence that $L_\gamma \cup \{m'\}$ would be a γ-set and m' would be a γ-element not in L_γ—clearly absurd. (It was this final step which German Cantorians such as Bernstein and Schoenflies found objectionable, fearing that Burali-Forti's paradox had slipped into the proof; see 2.5.)

Here Zermelo remarked three consequences of his proof—the cardinal of every infinite set is an aleph, the Trichotomy of Cardinals is true, and the following equalities hold:

$$\mathfrak{m} = 2\mathfrak{m} = \aleph_0 \mathfrak{m} = \mathfrak{m}^2.$$

In this manner the fundamental propositions true for alephs were extended to all infinite cardinals.

By way of conclusion, Zermelo discussed his Axiom at greater length:

> The preceding proof is based on the assumption that coverings γ exist in general, therefore on the principle that even for an infinite totality of [non-empty] sets there always exist mappings by which each set corresponds to one of its elements, or, formally expressed, that the product of an infinite totality of sets, each of which contains at least one element, is different from the empty set. Indeed, this logical principle cannot be reduced to a still simpler one, but is used everywhere in mathematical deductions without hesitation. So for example the general validity

> of the theorem [Partition Principle], that the number of subsets into which a set is partitioned is less than or equal to the number of its elements, cannot be demonstrated otherwise than by assigning to each subset one of its elements.[7]

Thus he stated the Axiom of Choice in full generality and gave evidence in its favor—its simplicity, the previous widespread use of arbitrary choices, and an elementary result (the Partition Principle) whose proof required the Axiom.

Since the quoted form of the Partition Principle had first been mentioned by Levi, who rejected it in general, Zermelo may already have suspected that his Axiom would encounter opposition. Yet the extent of this opposition, which arose at once and quickly spread throughout Europe, surpassed all expectations. Soon the critics of Zermelo's proof far outnumbered its supporters.

The controversy took place mainly within communities of mathematicians which were local or national rather than international. In France (see 2.3 and 2.4), Germany (2.5), and England (2.7), his proof gave rise to extensive debates. In Hungary (2.6), Italy (2.8), Holland (2.9), and the United States (2.10), public reaction was vehement but sporadic. Nevertheless, in each country mathematicians searched for the distinctions—partly mathematical and partly philosophical—needed to understand why the Axiom created such deep divisions in their ranks.

2.3 *French Constructivist Reaction*

Whether one describes Baire, Borel, and Lebesgue as French semi-intuitionists[1] or as French Empiricists,[2] it is clear that they shared strong constructivistic sympathies and profoundly influenced each other's research. Yet it is hard to characterize their common beliefs. In their work one finds not a detailed philosophy of mathematics but rather a piecemeal response, differing from individual to individual, to Cantorian set theory and to the questions that it raised about existence proofs in mathematics.

Their responses were distinctly ambivalent. On the one hand, all three employed Cantorian concepts and theorems in their analytic investigations. In fact, their researches would have been severely limited without the Can-

[7] Zermelo 1904, 516. By the product of sets, Zermelo probably conceived of something analogous to their Cartesian product, a notion which had not yet been clearly formulated. However, he may have intended the product of sets to be interpreted as Whitehead's multiplicative class, and then Zermelo's principle would constitute the Multiplicative Axiom.

[1] Heyting 1955, 6.

[2] Murata 1958, 93. Concerning these three mathematicians, see also Monna 1972 and Medvedev 1976.

torian notions of denumerability, cardinality, derived set, and countable ordinal. On the other hand, Baire, Borel, and Lebesgue had misgivings about set theory and even about the Cantorian concepts that they used repeatedly. Their misgivings grew much more pronounced after Zermelo's proof appeared.

Of the three, Borel aired his ambivalence the earliest, and doggedly returned to these questions again and again over the next fifty years. His *Leçons sur la théorie des fonctions* of 1898, devoted half to set theory and half to its applications in the theory of complex functions, distinguished sharply between the merely "philosophical interest" of abstract set theory and the "practical utility" of set theory applied to other branches of mathematics [1898, ix]. For the pragmatic Borel, this meant confining attention to those sets which were denumerable or had the power of the continuum or, at least, could be defined uniquely by a countable number of conditions. Thus he mistrusted any set of functions with power greater than that of the continuum. Moreover, he was beginning to consider even the notion of uncountable set as a fundamentally vague and negative one—a tendency which time would strengthen.[3]

On 1 December 1904 Borel finished a brief article, requested by David Hilbert as an editor of *Mathematische Annalen*, on the question of Zermelo's proof. What Zermelo had done, Borel wrote, was to show the equivalence of the problems of:

(A) well-ordering an arbitrary set M, and
(B) choosing a distinguished element from each non-empty subset of M.

What Zermelo had *not* demonstrated, Borel insisted, was that the equivalence of A and B provides

> a general solution to problem A. In fact, to regard problem B as resolved for a given set M, one needs a means, at least a theoretical one, for determining a distinguished element m' from an arbitrary subset M' [of M]; and this problem appears to be one of the most difficult, if one supposes, for the sake of definiteness, that M coincides with the continuum [$\mathbb{R}$].[4]

He did not consider Zermelo's argument to be any more valid than well-ordering a set M by selecting a first element, then a second, and so on transfinitely until all the elements of M had been exhausted. No mathematician, he felt certain, could accept such reasoning. Particularly objectionable was Zermelo's use of uncountably many arbitrary choices, a use which Borel exiled "outside mathematics." After dismissing Zermelo's proof, he left open the possibility that denumerably many arbitrary choices (and hence the Denumerable Axiom) might be permissible.

---

[3] Borel 1898, 6–20, 109–110, 122.

[4] Borel 1905, 194.

In a footnote, Borel cited a letter which Baire had written him on this subject:

> Personally, I doubt that a common measure can ever be found between the continuum (or what comes to the same thing here, the set of all sequences of positive integers) and the well-ordered sets; for me these two things are defined only in potentiality, and it may well be that these two potentialities cannot be reduced to each other.[5]

Although Borel stated his agreement with this letter, Baire appears to have gone appreciably beyond Borel by doubting that $\mathbb{R}$ could ever be well-ordered. Soon afterward, Baire restricted even further what he allowed into mathematics.

The next mathematician to add to the correspondence discussing Zermelo's proof was the analyst Jacques Hadamard. After he read Borel's article in *Mathematische Annalen*, Hadamard wrote Borel a dissenting letter. First of all, Hadamard distinguished Zermelo's proof sharply from reasoning which required an infinite number of successive choices, each of which depended on those made previously. Zermelo's proof was acceptable, he emphasized, precisely because the choices were *independent* of each other. Second, Hadamard saw no essential difference between denumerably many and uncountably many choices; such a difference would exist for dependent choices but not for independent ones.[6]

Was it possible, Hadamard inquired, to make such independent choices effectively, that is, in a way that someone could actually perform? Certainly Zermelo had given no method for doing so, and it seemed unlikely that anyone could provide one. What Zermelo had done was to state an existence proof. The essential distinction, which Hadamard credited to Jules Tannery,[7] was between establishing:

(i) that a function exists and
(ii) that it can be specified uniquely.

Hadamard added that many mathematical questions would have a completely different meaning, and different solutions, if (i) were replaced by (ii). Furthermore, he continued, even Borel had used functions which he proved to exist but which could not be defined uniquely, especially in certain theorems on the convergence of complex series. As for the notion of unique definability,

[5] Baire in Borel 1905, 195.

[6] Hadamard in Baire *et alii* 1905, 261–262. This article, consisting of five letters exchanged between Baire, Borel, Hadamard, and Lebesgue, is translated in Appendix 1. It may be that the four of them decided to publish these letters at the meeting of the *Société Mathématique de France* held on 4 May 1905. On that occasion, mentioned in the fifth letter, Borel gave a paper "On the Principles of Set Theory" and Lebesgue acted as commentator; see **SMF 33** (1905), 157.

[7] Tannery 1897, 132–139. In case (i) Tannery said that the function can be defined, and in case (ii) that it can be described. Thus he used the term "describe" for what was later designated by the term "define."

it was, to borrow Borel's phrase, "outside mathematics" since it belonged to the psychology of the human mind. Finally, Hadamard argued that the existence of the function γ used by Zermelo was equivalent to the possibility of choosing an element from any given set.[8] In this regard, Hadamard was mistaken. His conflation of the Axiom with the choice of a single element appears to have led Arnaud Denjoy [1946, 111] and Paul Lévy [1950, 23] to repeat the same error.

Borel forwarded this letter to Baire, who quickly replied to Hadamard. On the whole, Baire agreed with Borel but went even further. Baire regarded any infinity as a potential infinity, merely a matter of convention. Even in his thesis he had treated Cantor's infinite ordinals as only a *façon de parler* [1899, 36]. He now asserted as well that if one is given an infinite set, "*I consider it false to regard the subsets of this set as given.*"[9] *A fortiori*, it made no sense to assume, as Zermelo had done, that from each subset an element was chosen—though it was not contradictory to do so. What Zermelo had proved, Baire insisted, was merely that

> we do not perceive a contradiction in supposing that, in each set which is defined for us, the elements are positionally related to each other in exactly the same way as the elements of a well-ordered set. In order to say, then, that one has established that every set can be put in the form of a well-ordered set, the meaning of these words must be extended in an extraordinary way and, I would add, a fallacious one.[10]

Zermelo's result was consistent but meaningless. In the last analysis, Baire concluded, everything in mathematics must be reduced to the finite.

Borel next requested Lebesgue's opinion. Responding with a lengthy letter, Lebesgue stood at first on middle ground between Borel and Hadamard; but from Lebesgue's cautious reasoning there gradually emerged a distinct position. He considered the central question to be this: "*Can one prove the existence of a mathematical object without defining it?*"[11] In particular, Lebesgue inquired whether an existence proof is legitimate if it does not specify uniquely an object of the type purported to exist. While he recognized that it was a matter of convention whether one restricted existence proofs in this way, and admitted that he himself had deviated at times from such usage, he remained convinced that one could prove the existence of a mathematical object only by defining it uniquely. What Lebesgue rejected, in other words, were proofs that show the existence of a non-empty class of objects of a certain kind rather than a specific object of that kind.

This constructivist attitude, which Lebesgue described as close to Kronecker's, permeated the rest of his letter. Zermelo's proof had not given a way of determining the covering γ uniquely, and yet one needed to be

---

[8] Hadamard in Baire *et alii* 1905, 261–262; *cf.* footnote 36 of 4.8 re the ε-axiom.

[9] Baire *et alii* 1905, 264; Baire's emphasis.

[10] *Ibid.*

[11] Lebesgue in Baire *et alii* 1905, 265; Lebesgue's emphasis.

sure that γ remained the same object throughout the proof. How, Lebesgue wondered, could one be certain? Even for a given non-empty subset M' of M, one might not be able to choose an element in the sense of determining this element uniquely.[12]

After due consideration Lebesgue concluded that, Borel notwithstanding, a denumerable infinity of arbitrary choices was just as objectionable as an uncountable one. Consequently, Lebesgue regarded as unproved not only the proposition (1.6.1) that every uncountable set has a subset of power aleph-one, but also—in contrast to Borel—the proposition (1.1.4) that every infinite set has a denumerable subset.[13] Lebesgue added: "Although I seriously doubt that a set will ever be named which is neither finite nor infinite, it has not been proved to my satisfaction that such a set is impossible."[14] In sum, he found Zermelo's proof, like many general arguments in set theory, "too little Kroneckerian to have meaning" as a theorem of existence.[15]

Borel forwarded Baire's and Lebesgue's letters to Hadamard. Surprised by their adoption of Kronecker's views on existence proofs, Hadamard affirmed that a mathematical object can exist even if we cannot define it uniquely. The way in which existence was established did not matter, for, once proved, the object's existence was a fact like any other. Otherwise the object did not exist. He recognized that his three opponents interpreted the Well-Ordering Problem differently than he did: "I would say rather—Is a well-ordering possible?—and not even—Can *one* well-order a set?—for fear of having to think who this *one* might be. Baire would say: Can *we* well-order it?"[16] Consequently, Hadamard found their interpretation completely subjective, psychological, and hence contrary to the nature of mathematics.

As Hadamard emphasized, two different conceptions of mathematics were at stake. While granting his opponents' right to retain theirs, he viewed the debate as similar to the one which had divided Riemann from his critics as to what functions should be allowed into mathematics. Indeed, the rule that Lebesgue demanded to determine each value of Zermelo's covering γ resembled the analytic expression demanded by Riemann's adversaries. Hadamard understood the central issue in the debate over Zermelo's proof to be this:

> From the invention of the infinitesimal calculus to the present, it seems to me, the essential progress in mathematics has resulted from successively annexing notions

---

[12] *Ibid.*, 266–269.

[13] Thus at times Lebesgue restricted non-constructive proofs more severely than Borel did, despite the contrary claim by Monna [1972, 73].

[14] Lebesgue in Baire *et alii* 1905, 269. It is unclear what Lebesgue meant by this. Perhaps he intended to express that it was unproved whether every set was either finite or Dedekind-infinite.

[15] *Ibid.*, 267.

[16] Hadamard in Baire *et alii* 1905, 270.

> which, for the Greeks or the Renaissance geometers or the predecessors of Riemann, were "outside mathematics" because it was impossible to define them.[17]

Thus he firmly rejected the constriction of mathematics on philosophical or psychological grounds.

While concluding his letter, he returned to the details of Zermelo's proof. It did not perturb him that a covering γ was not uniquely determined by some characteristic property, for the fact that there existed many such coverings certainly justified the choice of one. Nor was he concerned by Bernstein's objection based on Burali-Forti's paradox and on the collection W of all ordinals (see 2.5): "We have the right to form a set only from previously existing elements, and it is easily seen that the definition of W supposes the contrary."[18] To solve the paradox, Hadamard insisted, one must recognize that the collection W is self-contradictory. As a parting shot, he took aim at his opponents' belief that one should investigate only those functions uniquely determined by a finite number of words. Hence the set of all real functions, which for Hadamard had the uncountable cardinality $\aleph^{\aleph}$ (where $\aleph$ was the power of the continuum), was only denumerable for his opponents —as, indeed, were all infinite sets!

The fifth and final letter in this exchange passed from Borel to Hadamard. Borel echoed Lebesgue's concern with Zermelo's references to the same distinguished elements throughout his proof, since Zermelo could not characterize such elements even for himself. As for Hadamard's sally on cardinality, Borel replied: "I prefer not to write alephs. Nevertheless, I willingly state arguments equivalent to those which you mention, without many illusions about their intrinsic value, *but intending them to suggest other, more serious, arguments*."[19] Borel considered Cantor's reasoning analogous to certain theories in mathematical physics which did not express reality but which led to new physical phenomena. In effect, he limited Cantor's arguments to their heuristic value in finding new theorems in analysis, which must then be proved constructively.

As an example Borel cited an argument, given in his book *Leçons sur les fonctions de variables réelles* of 1905, which had been generated by Cantor's ideas. There, Borel had claimed to define effectively a real function not in Baire class 0, 1, ..., n for any previously given finite n, and had carried out the construction for n equal to two [1905a, 156]. Yet this same example illustrates how deeply the Axiom was entangled with Borel's researches in analysis. As Sierpiński pointed out in 1918, the Axiom entered Borel's proof when he relied on the proposition (1.7.4): Any function $f(x)$ of Baire class two is representable as $\sum_{i=1}^{\infty} \sum_{j=1}^{\infty} P_{i,j}(x)$ where for every i and j, $P_{i,j}(x)$ is

---

[17] *Ibid.*

[18] *Ibid.*, 271.

[19] Borel in Baire *et alii* 1905, 272; Borel's emphasis.

some real polynomial. Moreover, Sierpiński established that if one could define effectively such a representation for each f in Baire class two, then one could determine effectively an example of a non-measurable set—a conclusion which was very implausible.[20] Despite Borel's desire to avoid the Axiom or at least to limit it to denumerable families of sets, here he had unwittingly used a stronger form.

This sequence of letters between Baire, Borel, Hadamard, and Lebesgue—published in the *Bulletin de la Société Mathématique de France* during 1905—remains a classic statement of the grounds for accepting or rejecting the Axiom. Afterward, the Axiom's three critics grew still more intent on allowing into mathematics only those objects which could be uniquely defined by a finite number of words. Their position remained vague, however, since they lacked exact criteria for definability in natural languages and since they did not consider symbolic languages such as those of Frege and Peano. In fact, the Axiom still found its way into their researches. As was discussed above, this happened to Borel almost simultaneously with the correspondence over the Axiom. For Lebesgue the situation was more complicated. While in his book on integration he appeared ambivalent about infinities of arbitrary choices, avoiding them explicitly at one point but using them at another [1904, 63, 131], by the time that he wrote his letter to Borel he had clearly retreated from such choices.

Nevertheless, in Lebesgue's article of 1905 which connected his researches to those of Baire and Borel, the Axiom appeared indirectly several times. There Lebesgue termed a function $f: \mathbb{R}^n \rightarrow \mathbb{R}$ analytically representable if one can give a definite rule for constructing f by means of countably many additions, multiplications, variables, constants, and limits of variables. In particular, any analytic expression could be brought into the form of an analytically representable function [1905, 145, 197]. By means of the latter notion, Lebesgue linked his ideas on real functions to Baire's:

(2.3.1) A function is analytically representable if and only if it is in Baire class α for some countable α.

While demonstrating this result, Lebesgue relied implicitly on the Denumerable Axiom via Baire's theorem that the limit of a sequence $f_1, f_2, \ldots$ of functions in Baire's classification was also in Baire class α for some countable α [1905, 152]. Without the Axiom, it is unclear whether some analytically representable function might fail to belong to Baire's classification. Likewise Lebesgue used Baire's proposition (1.7.5), that the family of sets of first category in $\mathbb{R}$ is closed under countable unions, to deduce that every Borel set in $\mathbb{R}$ has the Baire property [1905, 187]—a result requiring the Axiom.

---

[20] Sierpiński 1918, 135. Previously Hardy [1906, 15] had pointed out that Borel's proof required the Axiom, but was mistaken about the step in the proof which required it.

Lebesgue's paper culminated in three theorems which showed that his measurable sets and measurable functions yielded a richer classification than those of Baire and Borel:

(2.3.2) For every countable α, there exists a function in Baire class α.

(2.3.3) There exists a measurable function that is not in Baire's classification and hence is not analytically representable.

(2.3.4) There exists a measurable set that is not a Borel set.

In each of these theorems Lebesgue actually claimed a stronger result, namely, that the relevant function or set not only exists but is definable. The proofs that he offered all used the Axiom implicitly. To establish (2.3.2), he chose arbitrary functions $f_{i,n}$ such that for each n, $f_{1,n}, f_{2,n}$, ... converged to a given function f_n [1905, 211]. For (2.3.3) he required the Axiom to ensure that there existed just $2^{\aleph_0}$ functions in Baire's classification—a cardinality result which in 1917 Sierpiński proved sufficient, without the Axiom, to yield a non-measurable set.[21] As (2.3.4) depended on (2.3.3), the Axiom appeared implicitly there as well.[22] Of course in 1905 Lebesgue remained unaware of the Axiom's role in these proofs, although it is surprising that he did not notice his arbitrary selections in (2.3.2). However, even the Axiom could guarantee only that the functions and sets mentioned in these three theorems existed, not that they were definable. Eventually it became clear that the notion of definability required one to specify the formal system rather than remaining within a natural language.

During 1905 Lebesgue also wrote a monograph, not published until 1971, which illuminates his philosophical views in the aftermath of Zermelo's proof. Lebesgue believed the proper battleground to be the theory of functions, and considered the fundamental question to be what functions one could legitimately introduce. His response was subtle. He found Dirichlet's general definition of a function vague since it did not indicate by what means one could name or specify a function. In a critique similar to Brouwer's later intuitionism, he remarked that even for the analytically representable real function whose value is zero at all rational numbers and one at all irrationals, we do not know how to calculate the function's value for, say, Euler's constant. Faced with these difficulties, he continued, one might be tempted to avoid irrational numbers, in as much as no calculation can be completed with them. Yet he did not wish to eliminate irrationals, since they arose naturally in many questions of analysis [1971, 37–39]. One might be tempted to ask Lebesgue if the Axiom of Choice did not also arise naturally in many questions of analysis.

[21] Lebesgue 1905, 212–215; Sierpiński 1917a, 883–884.

[22] Lebesgue 1905, 215–216.

At this juncture Lebesgue divided mathematicians into two camps, the Idealists and the Empiricists—names previously introduced by the German analyst Paul du Bois-Reymond.[23] The Empiricists only admitted those functions which could be uniquely defined, while the Idealists accepted others as well. Their attitudes toward mathematical reasoning also diverged. The proof that every continuous function is integrable was, for the Empiricist, only "a form devoid of meaning" but which acquired meaning when one restated the argument for a specific function. On the other hand, such a general proof was

> the definitive and complete line of reasoning for the Idealist, since for him a function is [uniquely] determined when he affirms it is . . . . When an Idealist wants to determine a function, he does not seek a characteristic property which would permit him, as well as others, to be sure of always thinking of the same function; [rather] he contents himself with saying that he chooses this function . . . ; he affirms . . . that he is always thinking of the same function. This affirmation, which the Idealist recognizes and declares unverifiable, appears meaningless to the Empiricist, who places these [functions] peculiar to the Idealist outside mathematics. [1971, 39]

Thus Lebesgue made definability the touchstone for his Empiricist philosophy of mathematics.

Here we have a variation in a philosophical mode of the correspondence which Lebesgue and his colleagues exchanged over Zermelo's proof. Clearly Lebesgue regarded Hadamard as an Idealist, while considering Baire, Borel, and himself to be Empiricists. What Zermelo's proof had done, Lebesgue asserted, was to revive the dispute between Idealists and Empiricists.

Nevertheless, as in his letter, Lebesgue remained more ambivalent than his fellow Empiricists. He recognized that some eminent mathematicians were Idealists and even that the Idealist position might have practical consequences some day. Furthermore, in the past those who wished to extend the concept of function had always been in the right, and perhaps the same would hold true for the extension proposed by the Idealists. In any case, the matter would probably be resolved not by theoretical arguments but by the degree to which the Idealists' functions proved useful [1971, 41–44].

Soon such Idealist functions proved useful in a way that was distasteful to Lebesgue. For in 1905 Giuseppe Vitali utilized a choice function to demonstrate the existence of a non-measurable subset of $\mathbb{R}$. However, Vitali had a broader goal: to establish that Lebesgue's Measure Problem, discussed in 1.7, had no solution. In other words, there existed no translation-invariant, countably additive, positive real measure defined on all bounded subsets of $\mathbb{R}$ such that the unit interval had measure one.

---

[23] Lebesgue 1971, 38; du Bois-Reymond 1887, 64–143. In modern philosophical parlance, however, Lebesgue's Idealists would be termed realists, while his Empiricists would be called idealists.

Vitali began by supposing that there was such a measure. For any real x, he let

$$A_x = \{x + b: \quad b \text{ is rational}\}$$

and then defined H as the set of all such A_x. After choosing an element p in $A_x \cap (0, \frac{1}{2})$ for each distinct A_x in H, he termed G_0 the set of all such p. (Here the Axiom served to justify the choices.) Then he let

$$G_q = \{p + q: \quad p \in G_0\}$$

for each rational number q. Thus all G_q were disjoint congruent sets, and so all had the same measure. In particular $G_0, G_{1/2}, G_{1/3}, \ldots$ were subsets of the interval [0,1] and hence the sum of their measures was at most one. Since there were denumerably many of them, the measure of each had to be zero by the countable additivity of the measure. Consequently, the measure of G_q was zero for each rational q, and the measure of $\mathbb{R}$, which was the union of all G_q, was likewise zero—a contradiction. Therefore Lebesgue's Measure Problem had no solution. Lebesgue's particular measure did not apply to all bounded subsets of $\mathbb{R}$ and so did not fulfill all the conditions which he had demanded of it. Those who did not grant that $\mathbb{R}$ can be well-ordered, Vitali added, could conclude instead that there cannot exist both such a measure and a well-ordering of the real numbers [1905, 3–5].

Lebesgue's philosophical presuppositions affected his response to this non-measurable set, as became evident in 1907 when he published two articles on related questions. The first contained a solution to a problem posed in 1891 by Carrado Segre: Do there exist complex functions f such that for all complex z_1 and z_2,

(i) $f(z_1 + z_2) = f(z_1) + f(z_2)$ and
(ii) $f(z_1 z_2) = f(z_1) f(z_2)$,

other than the three functions $f(z) = 0$, $f(z) = z$, and $f(z) = \bar{z}$? Lebesgue established that no other solutions f are measurable functions. Yet if one accepted Zermelo's proof that $\mathbb{R}$ can be well-ordered, Lebesgue argued, then there were infinitely many other functions f satisfying the conditions (i) and (ii). Hence, for an Idealist accepting that proof, there existed non-measurable sets, as Vitali had already demonstrated. But for an Empiricist, Lebesgue insisted, non-measurable sets had not been shown to exist and there remained only three solutions to Segre's problem [1907a, 32–39].

In a second article of 1907, Lebesgue at last abandoned his moderate stance and attacked Zermelo's proof mathematically. After reiterating his doubts about the proof's adequacy, Lebesgue focused on the difficulties inherent in a specific case of the Axiom of Choice: choice functions on the family of all denumerable subsets of $\mathbb{R}$. He deduced that any such function must be non-measurable and hence not analytically representable. For

Lebesgue, this indicated how recondite and essentially undefinable Zermelo's choice functions must be—if one could say that they existed at all. Lebesgue continued to doubt that a non-measurable set would ever be defined uniquely [1907, 202–203].

By way of conclusion, Lebesgue again noted the gulf separating Empiricists from Idealists. Although he willingly used definitions framed in an Idealist manner, he restricted his proofs to Empiricist reasoning. For him, Burali-Forti's paradox concerning the class of all ordinals, as well as Richard's paradox of definable real numbers (see 2.4), illustrated the pitfalls inherent in Idealist reasoning, and hence Empiricists had to be vigilant in scrutinizing such Idealist "proofs." Nevertheless, Lebesgue concluded, to the Idealist he had shown via the Axiom of Choice that

(i) there exists a non-measurable set, and
(ii) there exists a set which, like its complement, is of second category on every interval and hence lacks the Baire property.[24]

Borel, like Lebesgue, continued to grow more constructivistic as a result of Zermelo's proof. In 1908, when Borel delivered a lecture at the International Congress of Mathematicians held in Rome, he extended his attack on Cantorian set theory. There he treated the notion of uncountable set as a "purely negative" one. The continuum, he claimed, was given by geometric intuition. From an arithmetic point of view the continuum was never given in its entirety, since no real number was known to be undefinable and yet only denumerably many real numbers would ever be defined. As for arbitrary choices, Borel added:

> The complete arithmetic notion of the continuum requires that one admit as legitimate a denumerable infinity of successive arbitrary choices. This legitimacy appears very debatable, but one must distinguish it in essence from the legitimacy of a non-denumerable infinity of choices (successive or simultaneous). The latter notion appears to me . . . entirely meaningless. As for a denumerable infinity of choices, clearly one cannot effect them all, but . . . one is assured that any given choice will be effected at the end of a finite time. [1908, 16]

Thus Borel's position had changed, since 1905, only by becoming still more restrictive. Contrary to his claim, an arithmetic definition of the continuum $\mathbb{R}$ (such as that given by Cantor or Dedekind) did not require the

[24] Lebesgue 1907, 211–212. Later Solovay [1965, 1970] established that if there is a model of **ZF** and the proposition that there exists a strongly inaccessible cardinal, then there is a model of **ZF** in which the Principle of Dependent Choices is true but every subset of $\mathbb{R}$ is measurable and has the Baire property. Recently, Shelah [1980] proved that the inaccessible cardinal is essential to Solovay's result, in the sense that if there is a model of **ZF** and the Principle of Dependent Choices in which every subset of $\mathbb{R}$ is measurable, then there is a model of **ZF** and the proposition that there exists a strongly inaccessible cardinal. See also footnote 6 of 5.2.

Axiom. As for the Denumerable Axiom, he wrote in another article that whether or not one accepts a denumerable infinity of successive arbitrary choices "is a *metaphysical* question, in the sense that a positive or negative answer will never have any influence on the development of mathematics" [1908, 447]. For better or worse, he could hardly have been more mistaken.

In one respect Borel refined his earlier position. No longer did he regard the relevant distinction as that between denumerable and non-denumerable sets. From a practical point of view, the principal distinction was now between those sets effectively enumerable and those not. According to Borel, a set was effectively enumerable if one could state, "by means of a finite number of words, a definite process for attributing unambiguously a [unique natural number as] rank to each of its elements" [1908a, 446–447]. Some denumerable sets were not effectively enumerable, since to attribute a rank to each element might require an infinity of arbitrary choices. Moreover, an infinite subset of an effectively enumerable set need not be effectively enumerable. At last Borel had produced a positive concept, effective enumerability, in response to his qualms about set theory and Zermelo's Axiom. Related notions of effectivity for subsets of $\mathbb{N}$, such as recursive enumerability, were investigated by recursion theorists beginning in the 1930s.

Despite this positive contribution, Borel's work continued to involve tacit or indirect uses of the Axiom, as did the work of Baire and Lebesgue. Indeed, Lebesgue later acknowledged that he had used the Axiom in his two papers [1907, 1907a] concerning non-measurable sets.[25] Nevertheless, he did not understand how profoundly his researches, as well as those of Baire and Borel, depended on the Axiom.

In the most extreme case, if the Denumerable Axiom were false, then $\mathbb{R}$ could be a countable union of countable sets.[26] If so, then the hierarchy of Borel sets was trivialized, as was Baire's classification of functions. Every set of real numbers belonged to the first four levels of Borel's hierarchy, and was $F_{\sigma\delta\sigma} \cap G_{\delta\sigma\delta}$ in Hausdorff's notation or $\mathbf{\Delta}_4^0$ in modern notation. Moreover, Lebesgue measure failed to be countably additive, and every measurable set was a Borel set. On the other hand, there could be a Borel set that was non-measurable, lacked the Baire property, and, though it was uncountable, had no perfect subset.[27] Thus, in the absence of the Axiom, the theory of real functions could diverge radically from the theory that Baire, Borel, and Lebesgue had developed. In the final analysis, their mathematics and their philosophy of mathematics proved to be incompatible. Little by little it became evident how fundamental this incompatibility really was.

[25] Lebesgue 1922, 61.

[26] Feferman and Levy 1963; Cohen 1966, 143–146.

[27] Juris Steprans has recently pointed out to the author that a Borel set with these three properties occurs in the model of Feferman and Levy [1963], where the Countable Union Theorem fails.

2.4 *A Matter of Definitions: Richard, Poincaré, and Fréchet*

The first French journal to mention Zermelo's proof was the *Revue générale des sciences pures et appliquées.* There on 15 November 1904 an anonymous author reported that, at the Heidelberg Congress, König had shown the impossibility of well-ordering $\mathbb{R}$, whereas Zermelo had soon afterward proved the opposite![1] The anonymous author promised to discuss this paradoxical situation in a later issue of the *Revue.* On 30 March 1905 he did so. Many mathematicians, he began, objected to Zermelo's use of arbitrary choices, but he did not share their objections. The crucial distinction, due to Jules Tannery [1897, 132], was between proving that some function with a given property exists and defining such a function uniquely. Because the alphabet is finite, one could uniquely define only denumerably many mathematical objects. In particular, the choice functions postulated by Zermelo could not be defined uniquely. However, as one could choose an element from a given set, so one could choose an element from each set in a finite or infinite family of sets.[2]

On the other hand, the anonymous author continued, the paradox obtained by juxtaposing König's and Zermelo's results remained unresolved, as did the Burali-Forti paradox. Indeed, the introduction of the transfinite seemed to produce various contradictory consequences, which reminded the anonymous author of the paradoxes that had arisen when mathematicians first introduced irrational, negative, and imaginary numbers—the sign of a fruitful theory which should be pursued vigorously. Here the views of the anonymous author, actually Hadamard, resembled those expressed in his first letter to Borel.[3]

Less than two months later Jules Richard, an instructor at a *lycée* in Dijon, responded to this anonymous account by sending a letter to the *Revue*, and the anonymous author printed it in his column. It contained Richard's paradox, now well-known, which relied on the diagonalization procedure that Cantor [1891, 76] had used to show $\mathbb{R}$ uncountable: Arrange all two letter combinations in alphabetic order, then all three letter combinations, and so on. Delete all combinations which do not define a real number. Then all the real numbers definable in a finite number of words form a denumerable well-ordered set $E = \{p_1, p_2, \ldots\}$. Let the real numbers $s = .a_1a_2\ldots$ between 0 and 1 be such that a_n is one more than the nth decimal of p_n if this decimal is not 8 or 9, while a_n is 1 otherwise. Although s is defined by a finite number of words, s is not in E, a contradiction. How,

---

[1] "Le troisième Congrès international des Mathématiciens," **RG 15** (1904), 961.

[2] "La théorie des ensembles," **RG 16** (1905), 241–242.

[3] See 2.3 and Appendix 1. Hadamard acknowledged authorship in Baire *et alii* 1905, 270. Van Heijenoort [1967, 270] mistakenly attributed authorship to Louis Olivier.

Richard asked, was this paradox to be resolved? The definition of s, he noted, actually required a definition of E, which would consist of infinitely many words.[4]

In the same column the anonymous author (again Hadamard) mentioned the long-awaited publication of König's proof that $\mathbb{R}$ cannot be well-ordered. Since König's argument depended on Bernstein's dubious proposition (2.1.4), the author concluded that Zermelo appeared to be vindicated.[5]

Richard's paradox also influenced Henri Poincaré, at that time the patriarch of French mathematics and an astute critic of research in the foundations of mathematics. Poincaré's articles of 1905–1906 attacking such research, particularly Russell's and Hilbert's, included an analysis of Zermelo's proof. In contrast to Russell, Poincaré believed that any attempt to prove or disprove Zermelo's Axiom from other postulates was illusory:

> The axioms in question [the Axiom of Choice and Russell's Multiplicative Axiom] will always be propositions which some will admit as "self-evident" and which others will doubt. Each person will believe only his intuition. Yet there is one point on which everyone will agree: The Axiom is "self-evident" for finite classes. But if it is unprovable for infinite classes, it is doubtless unprovable also for finite classes, which are not yet distinguished from the former at this stage of the theory. Thus it [the Axiom of Choice] is a synthetic *a priori* judgment without which the "theory of cardinals" would be impossible, for finite as well as infinite numbers. [1906a, 313]

Two aspects of Poincaré's comment deserve emphasis. The first is that he believed intuition to play a central role in mathematics. The series of articles from which this quotation was taken sought to legitimize the place of intuition in mathematics and to refute the logicist thesis that all mathematics is reducible to logic. The centrality of intuition was an integral part of Poincaré's Kantian perspective, equally visible in the second aspect: synthetic *a priori* judgments. For Poincaré, the principle of complete induction on $\mathbb{N}$ was such a judgment and could not be reduced to logic.[6] Certain other such judgments were likewise essential to mathematics. Consequently, it is quite significant that Poincaré regarded the Axiom of Choice as such a synthetic *a priori* judgment. That he believed the Axiom necessary for the theory of finite cardinals may have induced him to accept it, since he was very explicit in rejecting the actual infinite [1906a, 316]. Although cognizant of Borel's and Russell's criticisms, Poincaré became the second French mathematician to approve of the Axiom publicly.

On the other hand, Poincaré added: "Thus, although I am favorably disposed to accept Zermelo's Axiom, I reject his proof, which for an instant had made me believe that aleph-one could actually exist" [1906a, 315].

---

[4] Richard in Hadamard 1905a, 541. The letter was reprinted in Richard 1905.

[5] Hadamard 1905a, 543.

[6] Poincaré 1905, 318; 1906a, 302. For a detailed discussion of Poincaré's philosophy of mathematics, see Mooij 1966.

Poincaré considered Zermelo's proof to be flawed—like Bernstein's, Burali-Forti's, and Whitehead's demonstrations of other set-theoretic propositions—by an "impredicative definition." The definition of a mathematical object A was impredicative if A was defined in terms of a class B of which A was a member. In particular, Zermelo had defined L_γ by using the class of all γ-sets, although L_γ was itself a γ-set. Such impredicative definitions, Poincaré insisted, were born from the actual infinite and in turn sired the paradoxes of set theory. Through Richard's resolution of his own paradox, Poincaré came to realize that impredicative definitions contained a vicious circle and that by banning them one could circumvent other paradoxes such as Burali-Forti's [1906a, 315, 307]. As a result of reading Poincaré's article, Zermelo began a correspondence with him (see 3.2).

When Richard returned to his paradox in 1907, he had been influenced by Poincaré's articles of 1905–1906 and especially by their negative view of the actual infinite. At first this was not apparent, however, for Richard intended to prove the following proposition, essentially (1.1.4):

(i) Every uncountable set A has a denumerable subset.

One could select $a_1, a_2, \ldots$ arbitrarily from A, but there was an "*embarras du choix*." In order to avoid an infinity of arbitrary choices, one needed a rule. Similar difficulties arose, he remarked, if one attempted to prove another proposition:

(ii) For any sequence $A_1, A_2, \ldots$ of disjoint non-empty sets there is a sequence $a_1, a_2, \ldots$ such that a_n is in A_n for every positive integer n.

Zermelo's Axiom entered in an unorthodox fashion when Richard wrote:

> For us these propositions [(i) and (ii)] constitute *Zermelo's Axiom*. They do not appear to be provable in the most general case. They are provable in a very extensive case, the only one which arises in practice. Let us examine the first proposition. Let us suppose here that "In the set A there are elements definable in a finite number of words." By making this supplementary hypothesis we do not restrict the problem very much. One cannot really conceive of a set which does not possess this property. With our restriction *Zermelo's Axiom is immediately demonstrated*. [1907, 97]

In order to prove (i), Richard gave a lexicographic well-ordering, similar to the one found in his paradox, for the definitions of elements in A. Then he did the same for elements in the union of all A_n so as to prove (ii). Actually, for (i) and (ii) he needed to assume that every non-empty set contains a definable element. Such an assumption bore a distant resemblance to the constructible sets that Gödel later used to establish the relative consistency of the Axiom (see 4.10).

Thus in 1907, Richard succeeded in deducing the Denumerable Axiom from a hypothesis about definable elements. Why he restricted the Axiom to the denumerable case is clear from his hostility towards uncountable sets and even toward the Axiom itself. His article concluded: "I regard these questions as interesting but as absolutely useless in mathematics. The true mathematics, which helps us to understand the external world, has nothing to do with non-denumerable sets or with objects not definable in a finite number of words" [1907, 98]. By making mathematics the handmaiden of physics, Richard believed that he revealed how inapplicable set theory was to the real world.

Ironically, by the time that Richard cast these aspersions on set theory, a number of treatises had been published showing how useful set theory could be to the theory of functions.[7] In the same vein, one of Hadamard's doctoral students, Maurice Fréchet, had relied heavily on set theory in his thesis of 1906. There he generalized analysis and developed what he called the functional calculus. Influenced heavily by Arzelà (see 1.8), Fréchet laid the foundations for an abstract treatment of functional analysis.[8]

Fréchet used limits of sequences to approach problems which we now consider topological, and so he required the Denumerable Axiom (*cf.* 1.2). Like Borel when introducing Borel sets and like Lebesgue when proposing his Measure Problem, Fréchet used axioms to characterize the objects in which he was interested. In particular, Fréchet broadened the analytic concept of limit by axiomatizing it via his notion of L-class: A given set B was an L-class if there was a function F on some sequences of members of B such that

(i) $F(b, b, \ldots) = b$ for every b in B, and
(ii) if $F(b_1, b_2, \ldots) = b$, then any subsequence of $b_1, b_2, \ldots$ had the same F-value b.

The F-value was called the "limit" of the sequence. Likewise, an element c of B was termed a "limit-element" of B if there existed some sequence of distinct elements $c_1, c_2, \ldots$ in B such that c was the limit of $c_1, c_2, \ldots$. If C was a subset of B such that for every infinite subset of C there was a limit-element of B, then C was called compact. (We would now say, sequentially conditionally compact.) He next established the following generalization of a topological theorem of Cantor's:

(2.4.1) If $C_1, C_2, \ldots$ are all sequentially closed, non-empty subsets of a compact set C and if $C_{i+1} \subseteq C_i$ for $i = 1, 2, \ldots$, then the intersection of all C_i is non-empty.

The proof, which Fréchet described as typical of those in his thesis, relied on denumerably many arbitrary choices to pick an element from each C_i [1906, 1–7].

---

[7] These included Borel 1898 in France, Schoenflies 1900 in Germany, and Young 1906 in England.

[8] On the early development of functional analysis, see Siegmund-Schultze 1978.

Indeed, the Denumerable Axiom aided Fréchet in many of his results. One important example concerned his concept of extremal set. He defined a subset S of B to be extremal (we would now say, sequentially compact) if S was compact and sequentially closed; a function $f: B \rightarrow \mathbb{R}$ was said to be sequentially continuous on S if, for any s in S and for any sequence $s_1, s_2, \ldots$ in S whose limit was s, $\lim_{n \rightarrow \infty} f(s_n) = f(s)$. Of course, this generalized the sequential continuity of a real function, discussed in 1.2. At this point Fréchet established the following extension of a classical theorem of real analysis:

(2.4.2) If $f: S \rightarrow \mathbb{R}$ is a sequentially continuous function on an extremal set S, then f is bounded on S and attains both its maximum and minimum values there.

The Denumerable Axiom was likewise essential when Fréchet extended Arzelà's Theorem to extremal sets [1906, 30].

Despite his extensive use of arbitrary choices and despite the fact that his thesis was completed in 1906, Fréchet nowhere referred to Zermelo's Axiom or to his proof. This seems surprising in view of the fact that Hadamard, who had vigorously defended the Axiom in 1905, supervised the thesis.[9] Later Fréchet tended to avoid discussing the Axiom, even when he used it implicitly.[10] Despite this attitude his investigations, one of the principal pathways leading to general topology, corroborate how the Axiom increased in importance as analysis grew more abstract and topological.

2.5 *The German Cantorians*

Zermelo's proof received mixed reviews in Germany. While its French critics had tended to reject the Axiom of Choice, its German critics hardly noticed the Axiom, and argued instead that Burali-Forti's paradox had crept into the proof. At issue was the question of how high a cardinality a well-ordered set could attain without generating a contradiction. In Germany, unlike France, there were mathematicians who both accepted the Axiom and openly explored its consequences. Thus in 1905 Georg Hamel used a well-ordering of $\mathbb{R}$ to obtain discontinuous real functions f such that

---

[9] Fréchet 1906, 7–8, 71. On the other hand, he presented to the Paris Academy of Sciences the first results from his thesis, namely his axiomatization of limits, on 21 November 1904 (see Fréchet 1904). This was presumably before he was aware of Zermelo's proof.

[10] See Fréchet 1934, 15. He used (1.1.4) and the Countable Union Theorem among other results [1934, 19–20].

$f(x + y) = f(x) + f(y)$ for all real x and y—a problem similar to that of Segre discussed in 2.3. From 1906 to 1908 Felix Hausdorff, who at first was sceptical of the Axiom, consciously applied it to obtain many theorems about ordered sets. All in all, Germany received the Axiom more favorably than any other country, until the Warsaw school of mathematics emerged in 1918 (see 4.1).

On 29 October 1904, after he had received Zermelo's proof, Hilbert presented to the Göttingen Academy of Sciences a paper by Bernstein. This was a response to Levi's criticism of 1902 that Bernstein's thesis inadvertently assumed the Partition Principle, which Levi stated as follows: If a set A is partitioned into a family S of disjoint sets, then $\overline{\overline{S}} \leq \overline{\overline{A}}$. After remarking the need to assume as well that every member of S is non-empty, Bernstein added: "I regard this principle as one of the most important in set theory, and I see no objection to using it" [1904, 558].

In order to clarify the logical significance of the Partition Principle, Bernstein introduced his concept of many-valued equivalence. He termed two sets A and B "many-valued equivalent" if there exists a non-empty family C of bijections from A onto B such that no member of C is "distinguished." Apparently he meant that no member of C could be uniquely defined, and hence that C was infinite. If a theorem asserted that A and B were equipollent, then the theorem was said to have "multiplicity" $\overline{\overline{C}}$. The logical significance of such a theorem was less, he granted, than that of a theorem where a member of C could be uniquely specified. In particular, his theorem that the set of all closed subsets of $\mathbb{R}$ has the power of the continuum had multiplicity $2^{\aleph_0}$. Recognizing that a theorem's multiplicity depended on present knowledge, he concluded with the hope that future research would reduce it as far as possible [1904, 558]. Except for W. H. and G. C. Young, then at Göttingen, Bernstein's notion of many-valued equivalence found no adherents (see 2.7).

Within a few months Hilbert, who had encouraged the controversy over Zermelo's proof by presenting Bernstein's paper to the Göttingen Academy, wrote his colleague Adolf Hurwitz about that controversy:

> Schoenflies and I have discussed Zermelo's proof, recently published in the *Annalen*, that it is possible to well-order every set. Against Zermelo, Schoenflies puts forward objections which, however, I do not find any more valid than those which Bernstein (Halle) makes. The whole polemic will appear in one of the next issues of the *Annalen*.[1]

Indeed it did. Printed in 1905, volume 60 of the *Annalen* contained articles on Zermelo's proof by Borel (see 2.3), J. König (2.6), and Jourdain (2.7) as well as Bernstein, Schoenflies, and Hamel. Except for Hamel, none of these authors accepted the validity of Zermelo's proof.

---

[1] Hilbert in Dugac 1976, 271.

Using Burali-Forti's paradox, Bernstein attacked the arguments of Cantor, Jourdain, and Zermelo that every set can be well-ordered. Bernstein interpreted this paradox in the following fashion: Since the set W of all ordinals is well-ordered, it has an order-type, say β, which must be the largest ordinal; but then the set $W \cup \{\beta\}$ is well-ordered and has as its ordinal $\beta + 1$, giving $\beta + 1 > \beta$, a contradiction [1905, 187–188]. Bernstein knew that both Cantor and Jourdain had described W as an inconsistent set and had argued that no well-defined set could have W as a subset; they had concluded that every consistent set can be well-ordered.

However, Bernstein used similar means to arrive at opposite ends. The first step was to modify Burali-Forti's paradox in order to show that there exists no well-ordered set S of which W is a segment, since otherwise S would be order-isomorphic to a segment of itself, an absurdity. Consequently, Bernstein rejected Cantor's principle that for any ordinal α there is a next ordinal $\alpha + 1$. Henceforth, that principle would apply only to segments of W, not to the ordinal β of W itself. In this way Bernstein claimed to circumvent Burali-Forti's paradox. Yet W could still be a proper subset of another set V, Bernstein insisted, as long as the well-ordering on W was not extended by any element of $V - W$. With this limitation, he argued that the set W was both consistent and legitimate.[2]

At this point Bernstein proposed a set which, he claimed, could not be well-ordered. For every ordinal α, $\aleph_{\alpha+1}$ was the power of some segment of W, and so W could not have the power $\aleph_\alpha$. Hence there existed an infinite set, namely W, whose power was not an aleph, and therefore Cantor's Aleph Theorem was false. Furthermore, he continued, the set Z of all subsets of W could not be well-ordered. For Cantor had established that every set has a smaller cardinality than its power set, and so $\overline{\overline{W}} < \overline{\overline{Z}}$; if Z could be well-ordered, then Z would be order-isomorphic to W or a segment of W, yielding the absurd consequence that $\overline{\overline{Z}} \leq \overline{\overline{W}}$. Indeed, Bernstein considered Z to be the simplest example of a set that cannot be well-ordered [1905, 192]. The sets Z and $\mathbb{R}$ might even be equipollent, he noted, in which case there would exist no well-ordering of the real numbers.

It is significant that Bernstein deviated from his teacher Cantor on such a critical point. Except briefly in 1884 and at König's lecture of 1904, Cantor had great faith in the validity of the Continuum Hypothesis and the Well-Ordering Principle. Bernstein's article, on the other hand, stressed how little was really known about the Continuum Problem, since no one had been able to refute the proposition that

$$2^{\aleph_\alpha} = 2^{\aleph_0}$$

[2] Bernstein 1905, 188–190. Jourdain objected to this limitation in a letter of 25 April 1905 to Russell, who agreed that it was nonsense in his reply three days later (see Grattan-Guinness 48). Both Hadamard and Zermelo [1908a] were highly critical as well; see 2.3 and 3.1.

for every ordinal α [1905, 192]. In particular, he added, it was not even known if

$$2^{\aleph_0} < 2^{\aleph_1}.$$ [3]

Evidently Bernstein abandoned the Well-Ordering Principle rather than renounce two others. Like Jourdain, he might have admitted that some well-ordered sets, such as W, do not possess ordinal or cardinal numbers. This Bernstein explicitly refused to do. More radically, he might have dropped the Principle of Comprehension, which stated that for every well-formed proposition P with one free variable there exists a set whose members are all those objects satisfying P. Although Bernstein does not appear to have entertained this possibility, Zermelo gave up the Principle of Comprehension as a step toward his axiomatization of set theory (see 3.2).

On what grounds did Bernstein reject Zermelo's proof of the Well-Ordering Theorem? The answer lies in how Bernstein interpreted the set W of all ordinals. Zermelo had not eliminated the possibility, he claimed, that for some set M the set L_γ of all its γ-elements equaled W. Although Zermelo's proof depended on extending L_γ by an element of $M - L_\gamma$ if $M \neq L_\gamma$, this extension was not possible if L_γ should equal W. On the other hand, Bernstein thought the Axiom of Choice superfluous if one used his notion of many-valued equivalence. While no single choice function could be specified, he insisted that no axiom was required to ensure that the set of such choice functions was non-empty [1905, 193]. Posterity was to disagree.

Schoenflies' critique, published next to Bernstein's in the *Annalen*, resembled it in many respects. As early as 1900 Schoenflies had doubts about Cantor's claim that every set can be well-ordered (see 1.6). In 1905 Schoenflies remained neutral toward this claim, but his view of W paralleled Bernstein's. In particular, Schoenflies treated W as a legitimate set and avoided the Burali-Forti paradox by disallowing any extension of W to a more inclusive well-ordered set [1905, 181].

Nevertheless, Schoenflies stressed the distinction, which Bernstein had not, between what he called the general and special theories of well-ordered sets. The general theory rested on the definition of well-ordered set and on its immediate consequences. To prove the Well-Ordering Theorem on this basis, Schoenflies asserted, it was necessary and sufficient to show the following proposition:

(2.5.1) There is no infinite sequence $\mathfrak{m}_1, \mathfrak{m}_2, \ldots$ of cardinals such that $\mathfrak{m}_{i+1} < \mathfrak{m}_i$ for $i = 1, 2, \ldots$.

Zermelo had not followed this path but had relied instead upon the special theory, namely, Cantor's principles for generating ordinals. However, it

[3] In 1963 Feferman pointed out that, in Cohen's model of **ZFC** and $2^{\aleph_0} = \aleph_2$, we also have $2^{\aleph_0} = 2^{\aleph_1}$; see Cohen 1964, 110.

remained uncertain how far those principles could be used to generate Cantor's number-classes without also generating a contradiction. For the moment Schoenflies regarded only the first and second number-classes as secure. If one introduced any others, he insisted on the need to postulate them individually and then to legitimize their existence [1905, 182].

Like Bernstein, Schoenflies found Zermelo's proof faulty since L_γ might be order-isomorphic to W, in which case the well-ordering on L_γ could not be extended by a new element. If there existed well-ordered sets of every power, then one could always extend a well-ordered set by another element. But then, Schoenflies insisted, one was really postulating that every set can be well-ordered, and consequently a proof was redundant. Here he failed to recognize the role that the Axiom of Choice played in the proof, since he neither objected to the Axiom nor found it novel. For him the problem of well-ordering every set remained open.[4] His misunderstanding, both in regard to the Axiom and to (2.5.1), was partly caused by his belief that the Trichotomy of Cardinals was true, though as yet unproved, while the Well-Ordering Theorem was probably false. In fact, Trichotomy and the Well-Ordering Theorem turned out to be equivalent.

In contrast to Bernstein and Schoenflies, Georg Hamel's article of 1905 in the *Annalen* not only approved of Zermelo's demonstration but contained significant new theorems deduced from it. Hamel, a *Privatdozent* at Karlsruhe, was primarily concerned with vector analysis and the foundations of mechanics. By utilizing Zermelo's result that $\mathbb{R}$ can be well-ordered, he obtained a basis B for the real numbers as a vector space over the field of rational numbers.[5] Later termed a Hamel basis, B was defined via transfinite recursion from the given well-ordering of $\mathbb{R}$ by the following criterion: If the real number x was not a linear combination of basis elements which occurred earlier in the well-ordering, then x belonged to B; otherwise, x was not in B. Zermelo soon conjectured that the Axiom was indispensable to obtain such a Hamel basis [1908a, 114]. In the absence of the Axiom, it was later shown, a vector space could fail to have a basis or could have two bases of different cardinalities.[6]

As Hamel was aware, Cauchy had proved that every continuous real function f, such that $f(x + y) = f(x) + f(y)$ for all x and y, was of the form $f(x) = kx$ for some constant k. On the other hand, Hamel's basis B provided him with discontinuous solutions f of a different form. First, he let $f(b)$ be any non-zero real number for each b in B; next, for each real x (where $x = q_1 b_{i_1} + \cdots + q_n b_{i_n}$ for some n, some rationals $q_1, \ldots, q_n$, and some basis elements $b_{i_1}, \ldots, b_{i_n}$), he defined $f(x)$ to be $q_1 f(b_{i_1}) + \cdots + q_n f(b_{i_n})$. Moreover, Hamel was able to show that every such discontinuous solution f

[4] Schoenflies 1905, 182–186.

[5] Hamel 1905, 459–460. A *basis* for a vector space V over a field F is a maximal linearly independent subset of V.

[6] Jech and Sochor 1966, 354; see also Jech 1973, 145.

could be obtained from some Hamel basis in this way [1905, 462]. Fréchet, using the countable additivity of Lebesgue measure and hence indirectly the Axiom, later extended Cauchy's result from any continuous function to any measurable function [1913]. A few years afterward, Sierpiński established without the Axiom that any of Hamel's solutions f must be non-measurable and, with the Axiom, that some Hamel bases are measurable.[7]

While Hamel had no philosophical scruples over Zermelo's proof, Gerhard Hessenberg proved more cautious. In a lengthy article, published in 1906 in the philosophical journal *Abhandlungen der Friesschen Schule* which Hessenberg had helped to found, he analyzed the fundamental concepts of set theory. There he examined the use of arbitrary choices and the problem of well-ordering any given set.

Since Zermelo read the proof sheets of that article when it was reprinted the same year as a book,[8] one might expect that his proof would receive favorable treatment. Yet Hessenberg did not criticize the proof's opponents, as Hadamard had done, nor did he commit himself to the Axiom of Choice. Rather, he scrutinized a number of possible principles of choice related to the Axiom. The most limited was the postulate (i): For any consistently defined non-empty set A, an element of A can be given. He considered this postulate to be non-trivial because it was not satisfied by the set consisting of those real numbers that were not finitely definable. Indeed, influenced by Richard's paradox, Hessenberg left open the question of how far (i) remained valid [1906, 151].

Another such principle was (ii): From any infinite set one can choose a denumerable subset. In contrast to Richard, Hessenberg found (ii) unfruitful as a postulate, since it eliminated an important problem in set theory—to establish that every infinite set is Dedekind-infinite. At first Hessenberg had even believed that (ii) could be deduced from (i), but Zermelo had shown him how a stronger form of the Axiom of Choice was required for (ii).[9]

One might also assume, Hessenberg continued, the stronger principle (iii): If a choice is possible at all, it is possible aleph-zero times. This principle, he insisted, led to Burali-Forti's paradox. For reasons that remain unclear, Hessenberg tried to preserve both the assertion that the collection W of all ordinals is a set and the principle that every set can be well-ordered, by the following ad hoc procedure: If, while attempting to well-order a set M, we were to exhaust W but not M, we should place the remaining members of M at the beginning of the well-ordering. How far such a process might be continued, he preferred not to discuss! This confusion illustrates how murky the principles underlying set theory were at the time.

The most far-reaching principle of choice examined by Hessenberg was the assumption (iv): One can make arbitrarily many successive, dependent

---

[7] Sierpiński 1920a and 1920d; see also Blumberg 1918.

[8] Hessenberg 1906, v.

[9] *Ibid.*, 155–156, v. Note that here Hessenberg temporarily made the same error as Hadamard.

choices. To accept this principle was in effect, he acknowledged, to assume that every set can be well-ordered. The final alternative (v) was that of simultaneous, independent choices, as postulated by Zermelo's Axiom, which Hessenberg stated in a form provided by Zermelo and later known as the Multiplicative Axiom: If S is a family of disjoint non-empty sets, then there is a set M such that for every A in S, $A \cap M$ contains exactly one element.[10]

After reporting Zermelo's proof in detail, Hessenberg cautiously stated what Zermelo had demonstrated: If it is possible to distinguish an element in each non-empty subset of a set M, then M can be well-ordered. Indeed, Hessenberg was well aware of the controversy surrounding the Axiom. No one had effected such choices for all subsets of $\mathbb{R}$, he remarked, and so Zermelo's theorem brought one no closer to resolving Cantor's Continuum Problem. In a comment apparently directed at König, Hessenberg noted that Zermelo's proof removed the grounds for believing that $\mathbb{R}$ cannot be well-ordered. For one could never specify a non-empty set in which it was impossible to choose an element.[11] Here Hessenberg tacitly presupposed his principle of choice (i).

As he concluded his discussion, Hessenberg remained neutral toward Zermelo's Axiom and the Well-Ordering Theorem. Both involved open questions as to which sets were non-contradictory and what it meant for a mathematical object to exist [1906, 218]. However, in 1908 Zermelo's axiomatization of set theory answered these questions to Hessenberg's satisfaction, and he became a staunch supporter of Zermelo's system (see 3.2 and 3.3).

Like Hessenberg, Felix Hausdorff fluctuated in his attitude toward the Well-Ordering Theorem and the Axiom. As was seen in 2.1, prior to Zermelo's proof Hausdorff thought that it would be paradoxical if some cardinal were greater than every aleph. When that proof appeared, he was sceptical at first, then lauded it. On the other hand, having come to set theory from astronomy, Hausdorff had little enthusiasm for foundational questions. Indeed, he regretted the intelligence, wasted on the paradoxes and similar matters, that could have produced genuine mathematical results.

The extremely original and fruitful papers on ordered sets, which Hausdorff published between 1906 and 1908, chronicle his changing attitude toward Zermelo's Axiom and proof. In 1906 Hausdorff explicitly set aside the question whether every infinite cardinal is an aleph, since mathematicians had not reached a consensus over this Aleph Theorem. Consequently, he refused to employ the corollary of the Aleph Theorem that

$$\mathfrak{m} = 2\mathfrak{m} = \mathfrak{m}^2$$

for every infinite $\mathfrak{m}$. Yet, like so many other mathematicians, Hausdorff inadvertently used the Axiom. Here this occurred when he defined a set B

[10] Hessenberg 1906, 158, 154. Regarding Russell and the Multiplicative Axiom, see 2.7.

[11] *Ibid.*, 212, 216–217.

of order-types and a subset A of B such that $\bar{\bar{B}} = 2^{\aleph_0}$ and $\bar{\bar{A}} = \aleph_1$, giving $\aleph_1 \leq 2^{\aleph_0}$.[12]

By the following year Hausdorff's attitude had shifted, and he came to accept Zermelo's proof. In particular, Hausdorff used a well-ordering of $\mathbb{R}$ to show that there exists a pantachie, a notion adopted from Paul du Bois-Reymond's study of monotonically increasing real functions. In place of such functions, Hausdorff investigated sequences of real numbers with the following partial order: For any sequences f and g, $f < g$ if g is eventually greater than f, *i.e.*, there exists some m such that for all $n \geq m$, $f_n < g_n$. Hausdorff defined a pantachie as a maximal set P of such sequences ordered by $<$; that is, P cannot be extended by any other sequence and still remain ordered by $<$ [1907, 118]. This concept of maximality, not clearly stated in 1907, turned out to be extremely significant. In 1909 it emerged as a form of Hausdorff's Maximal Principle, later shown equivalent to the Axiom of Choice (see 3.4).

Hausdorff's investigations of pantachies stimulated him to define certain sophisticated order-types called H-types, which generalized Cantor's order-type η of the rational numbers and its universal property: Every ordered set of power aleph-zero is order-isomorphic to a subset of η. Similarly, for an H-type τ, every ordered set of power aleph-one is order-isomorphic to a subset of τ. To demonstrate this fundamental property of H-types, Hausdorff relied on the Axiom [1907, 127]. The following year, he generalized the H-types as well.

In a second paper published in 1907, Hausdorff explicitly used the full strength of Zermelo's Well-Ordering Theorem to obtain two further results on ordered sets. That the rationals and the irrationals are both dense in $\mathbb{R}$ may have suggested to Hausdorff the first of these:

(2.5.2) Every dense set M can be partitioned into two sets, each of which is dense in M.[13]

The second result applied his new notion of cofinality to obtain a classification of infinite cardinals. A cardinal was said to be singular if it was cofinal with some smaller cardinal; otherwise it was regular. This definition made sense precisely because Hausdorff assumed the Well-Ordering Theorem, from which it followed that every infinite cardinal was an aleph. Though stated in 1907, his second result was not proved until the following year:

(2.5.3) $\aleph_{\alpha+1}$ is regular for every α.[14]

---

[12] Hausdorff 1906, 113, 117.

[13] An ordered set M is *dense* if between any two distinct elements of M there is a third element of M. A subset A of an ordered set M is *dense in M* if between any two distinct elements of M there is an element of A.

[14] Hausdorff 1907a, 542. Feferman and Levy [1963] gave a model of **ZF** in which $\aleph_1$ was singular and had cofinality ω. Gitik [1980], using a large cardinal hypothesis, obtained a model of **ZF** in which every infinite cardinal other than $\aleph_0$ was singular with cofinality ω.

In a lengthy article of 1908, codifying his extensive researches on ordered sets, Hausdorff protested against foundational discussions—especially constructivistic and finitistic objections to set theory. From his perspective such objections fell into three categories:

(i) the need to sharpen the concept of set, perhaps axiomatically;
(ii) the criticisms, raised against set theory, which applied to mathematics in general;
(iii) the absurdities of "Scholastics" who were clinging to literalisms.

With the first type of critic, Hausdorff hoped, it would eventually be possible to reach an understanding; those of the second type could be ignored, while the third type deserved the most unequivocal rejection. Nevertheless, Hausdorff consciously avoided any set which contained every ordinal or every cardinal [1908, 437]. Although he named no critics individually, the previous year he had defended Hardy's proof of 1903 that $\aleph_1 \leq 2^{\aleph_0}$ against E. W. Hobson's objections.[15]

After he accepted the Well-Ordering Theorem, Hausdorff refused to note when he used it or to consider as significant that some of his theorems could dispense with it. In fact, the Axiom was thoroughly entangled with the results of his paper of 1908 and especially with the e_α-types (the order-types of what are now called the η_α-sets). A generalization of the H-types to any power, the e_α-types were characterized by two theorems which relied on the Axiom:

(2.5.4) If E is an e_α-type and B is an ordered set of power $\aleph_\alpha$, then B is order-isomorphic to a subset of E.

(2.5.5) There is at most one e_α-type of power $\aleph_\alpha$.

The existence of an e_α-type with power $\aleph_\alpha$ was implied by the Generalized Continuum Hypothesis, which Hausdorff first formulated here: The sum of all $\aleph_\alpha^{\mathfrak{m}}$, where $\mathfrak{m}$ ranges over all cardinals less than $\aleph_\alpha$, equals $\aleph_\alpha$. As a corollary he derived the form of the Generalized Continuum Hypothesis more familiar to later mathematicians:

$$2^{\aleph_\alpha} = \aleph_{\alpha+1}$$

for all ordinals α [1908, 487, 494].

In the second installment of his report on set theory, where Hausdorff's work in particular was analyzed, Schoenflies expanded his earlier discussion of Zermelo's proof. The proof remained inadequate because Zermelo had failed to show that L_γ is never equal to W, a set which Schoenflies now feared to be self-contradictory. Influenced by Borel's article [1905], Schoenflies believed that Zermelo had shown only the equivalence of the Axiom

[15] Hausdorff 1907, 155–156. See 1.6 and 2.7.

of Choice and the proposition that every set can be well-ordered. In fact, Schoenflies insisted, Zermelo's Axiom merely shifted the Well-Ordering Problem to a different perspective. At best Zermelo had provided a pure existence proof rather than a constructive one. Yet for particular sets such as the continuum $\mathbb{R}$, the desideratum was always a constructive proof.[16]

Nevertheless, Schoenflies acknowledged that the Well-Ordering Problem remained one of the most fundamental questions in set theory. Equally fundamental from his standpoint was the Trichotomy of Cardinals, which he believed many authors to have used implicitly. In place of Zermelo's Axiom, he argued that one should adopt Trichotomy as a postulate. Schoenflies felt considerably more comfortable with Trichotomy, believing it to be essentially weaker than the Axiom, but his argument for this claim was necessarily vague: Trichotomy is weaker because it concerns only the whole set while the Axiom involves all the subsets. From the "weaker" postulate of Trichotomy, he argued, one could still deduce, for every infinite cardinal $\mathfrak{m}$, that $\mathfrak{m}$ is either an aleph or greater than all alephs and that $\mathfrak{m} = \mathfrak{m}^2$. However, he offered no proof.[17]

On the other hand, Schoenflies recognized the need not just to assume Trichotomy as a postulate but to axiomatize set theory as a whole. Such an axiomatization appeared more difficult to him than in the case of geometry since set theory lacked any comparable visual intuition. Moreover, the foundations of arithmetic, logic, and even cognition were involved with set theory. In the final analysis, he felt certain, the real foundations of set theory must be psychological. Though no progress had yet been made towards axiomatizing set theory, he believed that the Principle of Comprehension must constitute one such axiom [1908, 37]. When Zermelo's axiomatization appeared later that year, Schoenflies was dissatisfied with it and soon proposed a system of his own (see 3.3).

At the same time that Schoenflies' report appeared, Bernstein was using the Axiom in a way that requires us to consider some earlier developments. In 1884 Ludwig Scheeffer, then at Munich, wished to determine those sets A such that $f(x) = g(x) + C$, for some constant C and for all x in A, whenever the real functions f and g had equal Dini derivatives D^+ on A. After establishing that any countable A was such a set, which we shall term a fundamental set, Scheeffer established that an uncountable A failed to be fundamental whenever A had a perfect subset. Moreover, he doubted that there exists an uncountable set of real numbers without a perfect subset [1884, 287].

In the context of trigonometric series and their sets of uniqueness, the subject that had led Cantor to set theory, Bernstein extended Scheeffer's results by showing that a subset A of $\mathbb{R}$ is fundamental if and only if A has no

---

[16] Schoenflies 1908, 27, 31, 34–36.

[17] *Ibid.*, 31–32, 34, 36. That Trichotomy is weaker than the Axiom remained a common view until Hartogs [1915] proved them equivalent.

perfect subset. To establish that this was actually an extension, Bernstein proved the following result:

(2.5.6) There is an uncountable set of real numbers without a perfect subset.

Bernstein remained unaware that this result, which used only the proposition that $\aleph_1 \leq 2^{\aleph_0}$, required the Axiom. On the other hand, he claimed that (2.5.6) implies the existence of a non-measurable set, but supplied no proof.[18]

In sum, five German mathematicians published their views on Zermelo's proof between 1904 and 1908. Two opposed the proof, one remained neutral, and two were in favor. Moreover, Hilbert supported the proof in private but ventured no public opinion. This was the most positive initial reaction that the proof received in any country.

The opponents, Bernstein and Schoenflies, objected to the role that they thought the "set" W played in the proof. Because of Burali-Forti's paradox, they were convinced that W could not be extended to a more inclusive well-ordered set, which they believed Zermelo to require. Bernstein rejected the proposition that every ordinal has a successor, whereas Schoenflies stressed the difference between the general theory of well-ordered sets and the special theory. In particular, Schoenflies asked whether one could define Cantor's third or higher number-classes without producing a contradiction. Surprisingly, Bernstein claimed to refute the Well-Ordering Theorem by giving a "set" that could not be well-ordered, the power set of W.

About the Axiom of Choice, Bernstein and Schoenflies were confused. At first, Schoenflies did not recognize the Axiom's significance at all, and Bernstein thought his own notion of many-valued equivalence made the Axiom superfluous. Later Schoenflies proposed the Trichotomy of Cardinals as an alternative postulate, which he regarded as essentially weaker than the Axiom. On the other hand, Hessenberg, who interpreted Zermelo's proof as showing that the Axiom of Choice implies the Well-Ordering Theorem, voiced no opinion about the Axiom's validity.

The two mathematicians who approved of Zermelo's Axiom and proof put both into service. Hamel showed that there exists a basis for $\mathbb{R}$, considered as a vector space over the field of rational numbers, and hence that there

[18] Bernstein 1908, 329–330. It should be noted that the negation of (2.5.6) yields the Continuum Hypothesis (1.5.1): Every uncountable subset of $\mathbb{R}$ has the power of the continuum. Solovay [1970] showed that if there is a model of **ZF** with an inaccessible cardinal, then there is a model of **ZF** in which the Principle of Dependent Choices is true but (2.5.6) is false. It is a recent but difficult result, due to Shelah, that in **ZF** one can prove the existence of a non-measurable set from $\aleph_1 \leq 2^{\aleph_0}$; thus it seems unlikely that Bernstein possessed such a proof. In all probability a modification of the methods in [Solovay 1970] will suffice to show that, contrary to Bernstein's claim, the existence of a non-measurable set does not follow from (2.5.6) (personal communication from Solovay, 14 April 1982). On the other hand, Lyubekii [1970] established that if there is a non-measurable $\mathbf{\Sigma}_2^1$ set of real numbers, then there is a $\mathbf{\Pi}_1^1$ set satisfying (2.5.6).

are discontinuous real functions f such that $f(x + y) = f(x) + f(y)$. At first avoiding the Axiom but then accepting it, Hausdorff used the Well-Ordering Theorem extensively in his innovative papers on ordered sets. He consciously refused to state when he applied the Axiom and when he could dispense with it, an act which reflected his disdain for constructivistic criticisms of set theory. Nevertheless, the Axiom clearly played an essential role in his theorems on pantachies, on the partition of dense sets, and on e_α-types. The Axiom was equally visible when he showed $\aleph_{\alpha+1}$ to be regular for all ordinals α and when he formulated the Generalized Continuum Hypothesis. Without any doubt, Hausdorff utilized Zermelo's Axiom and Well-Ordering Theorem more fully than anyone else before the rise of the Warsaw school in 1918.

Yet Hausdorff's ad hoc proposal, to avoid any collection containing all the ordinals or all the alephs, revealed the uncertain state of set theory. Although he objected to fruitless philosophical discussions on foundational questions, he recognized, as did Schoenflies, that an axiomatization of set theory might well be needed. Hausdorff refused to speculate on what axioms it ought to include, but Schoenflies proposed the Trichotomy of Cardinals and the Principle of Comprehension. Neither author was aware of the two axioms, discussed in 1.5, which Burali-Forti had suggested in 1896.

Certainly the Axiom of Choice had to be considered vis-à-vis any such axiomatization, since after 1904 Zermelo's proof was on the minds of set-theorists quite as much as the paradoxes. In fact, when Hausdorff and Schoenflies wrote in 1908, Zermelo's axiomatization was about to be published, together with a new proof of the Well-Ordering Theorem (see 3.1 and 3.2). Ironically, this axiomatization would provoke a new controversy rather than quiet the old one.

2.6 *Father and Son: Julius and Dénes König*

These two Hungarian mathematicians, who were strongly influenced by Cantorian ideas, diverged in their approaches to mathematics. The father Julius moved increasingly into philosophical waters while his son Dénes remained on firmer mathematical ground. At first this difference was not apparent, for J. König's Heidelberg lecture of 1904 (reprinted in volume 60 of *Mathematische Annalen* near the critiques of Zermelo's proof) was concerned with mathematics alone [1905a]. Yet after his lecture J. König began to redirect his attack on the Continuum Problem. Whereas he had previously approached it through the Cantorian apparatus of alephs, in another paper of 1905 he introduced other methods. There he insisted that, as an exact science, set theory could not dispense with axioms, which were necessarily somewhat arbitrary. However, the axioms that he proposed were quasi-psychological, such as the existence of mental processes satisfying the formal laws of logic.[1] In this way his philosophical presuppositions came to affect his view of mathematics and especially of the axiomatic method.

At this point in his paper [1905], J. König returned to his earlier preoccupation: to prove that $\mathbb{R}$ cannot be well-ordered. After introducing a notion of finite definability, he argued somewhat along the lines of Richard's paradox, though independently (*cf.* 2.4). Although $\mathbb{R}$ was uncountable, there were just countably many definable real numbers. Consequently, he noted, some real numbers cannot be finitely defined. If $\mathbb{R}$ could be well-ordered, then there would be a least real number r such that r was not finitely definable—a contradiction since r was finitely defined in this way. Therefore the set of all real numbers cannot be well-ordered. Furthermore, he added, the same argument would apply to any uncountable well-ordered set, and so every well-ordered set must be countable. On the other hand, he knew that Cantor's second number-class was such an uncountable well-ordered set, consistently defined by the phrase "the totality of all denumerable ordinals" [1905, 159].

How was one to escape from this dilemma? J. König attempted to do so by distinguishing sharply between two meanings of the word "set": completed sets, and sets in the act of becoming. The word "set" (*Menge*) was reserved for the former, and the term "class" (*Klasse*) for the latter. In particular, he considered the continuum $\mathbb{R}$ to be a "set" whereas the second number-class remained only a "class." Thus the second number-class was well-ordered, but $\mathbb{R}$ could not be. Finally he claimed that his distinction sufficed to resolve such set-theoretic paradoxes as Burali-Forti's. But he did not elaborate [1905, 160].

When J. König published the second part of this paper [1906], he moved even further into philosophy. He stressed the connection of set theory with logic and epistemology, as well as the need to develop a theory of logical evidence. As revised, his paradox was now essentially Richard's: The set of all finitely definable real numbers is denumerable, and yet by Cantor's diagonal process one can define another real number. To avoid this contradiction, J. König adopted an ad hoc device which he called "pseudofinite definitions." From these considerations he then deduced that the second number-class was denumerable, a surprising conclusion since, in agreement with Cantor, he had asserted the opposite a year earlier.[2]

Soon afterward, his son Dénes König investigated related questions in a more traditionally mathematical manner. In an article of 1908, D. König proposed to establish the propositions that

$$\mathfrak{m} = 2\mathfrak{m} \quad \text{and } \mathfrak{m} = \aleph_0\mathfrak{m}$$

[1] J. König 1905, 156–157.

[2] König's papers [1905, 1906] were severely criticized by Vivanti [1908].

for every infinite cardinal $\mathfrak{m}$. If one did not accept Zermelo's proof, and he added that many set-theorists did not, then these propositions remained undemonstrated. To remedy this situation, he wished to derive both propositions from the following hypothesis, which he considered very simple:

(2.6.1) If a set M has more than one element, then there is a bijection $f: M \to M$ such that $f(s) \neq s$ for all s in M. [1908]

Contrary to his claim, he did not deduce $\mathfrak{m} = \aleph_0 \mathfrak{m}$ from (2.6.1) without the Axiom. Rather, his proof tacitly used the Principle of Dependent Choices when he obtained a certain sequence of partitions of an infinite set M into countable sets. This sequence yielded proposition (1.7.9): Every infinite set M is the union of some family A of denumerable sets.[3] Then he relied implicitly on the Denumerable Axiom, via (1.4.1), to conclude that $\overline{\overline{M}} = \aleph_0 \overline{\overline{A}}$. In effect he had shown that, without the Axiom, (2.6.1) and the Principle of Dependent Choices imply $\mathfrak{m} = \aleph_0 \mathfrak{m}$ for all infinite $\mathfrak{m}$.

To sum up, both Julius and Dénes König viewed Zermelo's proof sceptically. Arguing that $\mathbb{R}$ cannot be well-ordered, the father shifted from the arithmetic of alephs to finite definability, epistemology, and a generally constructivistic approach. Moreover, Julius König increasingly came under the sway of Poincaré's philosophy of mathematics.[4] On the other hand, his son doubted Zermelo's proof because of its mathematical uncertainties. Yet like many other mathematicians, Dénes König implicitly used the Axiom. Ironically, by 1914 both father and son came to accept Zermelo's Well-Ordering Theorem (see 3.6).

2.7 *An English Debate*

The English mathematicians who debated Zermelo's proof had connections with Cambridge, and all but one of them had previously investigated Cantorian notions. In particular, Hardy had published a demonstration that $\aleph_1 \leq 2^{\aleph_0}$ and that every infinite cardinal either is an aleph or is greater than all alephs. Influenced by Hardy's article, Jourdain had proposed his own argument that every set can be well-ordered. Russell and Whitehead had investigated cardinal arithmetic in the context of the foundations of mathematics, while W. H. Young had been occupied with point-set topology. The only non-Cantorian was the analyst Ernest Hobson, who initiated the attack on Zermelo's proof.

[3] Apparently he did not know of Russell's work on (1.7.9) and related propositions; see 1.7.

[4] See, for example, J. König 1906a.

These English critics of the proof focused primarily on the Axiom of Choice, despite the fact that all but Hobson had used the Assumption implicitly (see 1.7). Russell was dubious of the Axiom, which he had discovered independently in a different form. Presumably Whitehead's views resembled Russell's, since their *Principia Mathematica*, written over the next few years, echoed Russell's scepticism. Hobson dismissed the Axiom on grounds similar to Borel's, and Young preferred Bernstein's notion of many-valued equivalence. At first Jourdain thought the Axiom unnecessary, though he later came to defend it. Only Hardy accepted the Axiom at once, but he then rejected Zermelo's proof. Thus, as a whole, English reaction to the proof was quite negative.

Before turning to the English debate over Zermelo's proof, we must understand how Russell's views evolved. During the summer of 1904, he was sporadically concerned with the Well-Ordering Problem. On 5 July, Jourdain wrote Cantor that Russell believed he could resolve his own paradox if he could be certain that the Aleph Theorem (every infinite cardinal is an aleph) was true; yet Russell remained dubious of this theorem.[1] During the same period, Jourdain and Russell were corresponding about Jourdain's argument for the Aleph Theorem. That argument had failed to convince Hardy, Whitehead, and Russell.[2] It was from a different direction that Russell then discovered what he later called the Multiplicative Axiom. To satisfy Jourdain's curiosity, in 1906 Russell gave an account of this discovery:

> As for the multiplicative axiom, I came on it so to speak by chance. Whitehead and I make alternate recensions of the various parts of our book [*Principia Mathematica*], each correcting the last recension made by the other. In going over one of his recensions, which contained a proof of the multiplicative axiom, I found that the previous proposition used in the proof had surreptitiously assumed the axiom. This happened in the summer of 1904. At first I thought probably a proof could easily be found; but gradually I saw that, if there is a proof, it must be very recondite.[3]

---

[1] Jourdain in Grattan-Guinness 1971a, 118. It is not known how Russell intended to circumvent his paradox by using the Aleph Theorem.

[2] Grattan-Guinness 1977, 26, 29.

[3] Letter of 15 March 1906 in Grattan-Guinness 1977, 80. Some of Russell's work in 1904 or early 1905 on this question can be found in his "Plan for a General Introduction to Vol. II." These are his worksheets, kept in the Russell Archives, for the second volume of the *Principles*, which eventually turned into *Principia Mathematica*. On page 633 of these worksheets he wrote: "It appears that, both for $2^\alpha > \alpha$ and for $\exists^{|} x^{|} K$ [the Multiplicative Axiom], it is important to study the conditions under which

$$(\exists F) : v \in K \,.\, \supset_v .\, F^{|} v \in v$$

..., *i.e.*, for any class v of the set considered, there must be some propositional function $\phi^{|}(v)^{|}z$ which is satisfied only by one member of v ...." Here he wanted to know, in effect, when a choice function F existed; see also footnote 7 below.

Thus both Whitehead and Russell became aware of the Multiplicative Axiom shortly before Zermelo formulated the Axiom of Choice. As the quotation makes clear, neither Whitehead nor Russell originally regarded the Multiplicative Axiom as an *axiom*, but rather as a *theorem* to be demonstrated—little by little realizing the difficulty involved in such a demonstration. This attitude contrasts sharply with that of Zermelo, who recognized in September 1904 that the Axiom of Choice was a simple, fundamental principle and must be postulated, not proved [1904, 516].

Russell reached the Multiplicative Axiom by a route quite different from Zermelo's path to the Axiom of Choice. Whereas Zermelo had been preoccupied with the Well-Ordering Problem, Russell's Multiplicative Axiom arose from considering the infinite product of disjoint sets, *i.e.*, their multiplicative class, in order to define the product of infinitely many cardinals. During 1902 Whitehead, who was already collaborating with Russell, had defined the multiplicative class K^x of a disjoint family K of non-empty classes: K^x is the class of all those subclasses M of the union of K such that for every S in K, $M \cap S$ has exactly one member. When finally published, Russell's Multiplicative Axiom took the following form [1906, 49]:

(2.7.1) If K is a disjoint family of non-empty classes, then K^x is non-empty.

In effect, Whitehead had implicitly used (2.7.1) in his article of 1902 (see 1.7).

At what point did Russell come to regard (2.7.1) as an axiom? One must understand that when (2.7.1) resisted his attempts to prove it, he became increasingly sceptical of its validity. For Russell, (2.7.1) became an axiom in the sense of a fundamental unproved assumption but *not* in the sense of a self-evident truth. Certainly he treated (2.7.1) as an undemonstrated assumption by early 1905 and may have done so by late 1904.[4]

Thanks to the recently rediscovered correspondence between Russell and the French philosopher Louis Couturat, we can see how Russell's thought developed vis-à-vis the Multiplicative Axiom and Zermelo's proof.[5] On 5 September 1904, Couturat informed Russell that König had refuted the Continuum Hypothesis. According to H. Fehr, Couturat's informant, no one at the Heidelberg Congress had discovered an error in König's proof, and even Cantor had asked to think about it further.[6] Couturat inquired if Russell had deduced in his logical calculus that the continuum can be well-ordered, thereby rebutting König. On 22 September, Russell replied that he

[4] Russell in Grattan-Guinness 1977, 48.

[5] Anne-Françoise Schmid is preparing a critical edition of this correspondence, which is kept at the Bibliothèque de la Ville, La Chaux-de-Fonds, Switzerland.

[6] Fehr's account differs from that of Kowalewski [1950, 202], who claimed that Zermelo found the error; see 2.1.

had not. The subject was closed until 18 December when Couturat conveyed "some important news (which I owe to my friend Borel)":

> König's refutation of G. Cantor's theorem was mistaken, and he recognized it himself. Moreover, Zermelo has published a proof of this same theorem. . . . But since this proof is based on the calculus of "alephs," some mathematicians do not find it any more valid than König's refutation. When you have the time to study the proof, I would be curious to learn . . . whether you can resolve the question, one way or the other, in your logical calculus.

Gradually, Zermelo's proof assumed fundamental importance for Russell, who wrote on New Year's Day: "I believe I can refute the theorem in question." Relying on Whitehead's version of the proof, which he still had not seen, Russell insisted on 5 February 1905 that Zermelo made an assumption analogous to (2.7.1): "But this [(2.7.1)] is questionable. I have been trying to prove it for a long time, without the least success." As a result, Russell confessed himself unable to show the proposition (1.7.8), which justified the addition of infinitely many cardinals.

On 26 April, Russell informed Couturat that he had finally read Zermelo's proof and had translated it into his symbolic logic. He noted that Zermelo's Axiom—stating that for every class M there is a function f such that $f(A) \in A$ for each non-empty subclass A of M—implied (2.7.1), which he described as

> a theorem without which much of cardinal arithmetic would be impossible . . . . I do not know if it is true or false, and I find it too complicated for a Pp [postulate] . . . . One could deduce everything necessary from the axiom
>
> $$(\exists f): \exists{}^{|}u\,.\supset_u.\, f{}^{|}u \in u \qquad \text{Pp?}$$
>
> But I do not know if this axiom is true.[7] I have not found any fallacy in Zermelo's argument, and I consider the result that he proves to be very interesting; but it is not $C^{||}\Omega = Cls$ [every class can be well-ordered] since one should hardly admit an axiom that is so complicated and so dubious.

Russell expressed similar sentiments in a letter to Jourdain two days later: "Zermelo's proof is a failure. . . ." But when Russell rephrased the proof as stating that the Axiom of Choice implies that every set can be well-ordered, he described it as "an important and interesting theorem."[8] Similarly, he insisted that Hardy's proof (every infinite cardinal is either equal to some aleph or greater than all of them) was fallacious, since it relied implicitly on (2.7.1).

---

[7] This proposition, now known as the *Axiom of Global Choice*, asserts that there exists a choice function f such that $f(u) \in u$ for *every* non-empty set u. Easton [1964, 1964a] showed that in **ZF** the Axiom of Global Choice is stronger than the Axiom of Choice; see also 4.8 and A. Levy 1961.

[8] Russell in Grattan-Guinness 1977, 46–48.

When he wrote to Couturat on 26 April, Russell also mentioned the following weaker version of (2.7.1):

(2.7.2) If K is a disjoint family of non-empty classes, all of which can be well-ordered, then K^x is non-empty.

Russell wondered if it were possible to demonstrate at least (2.7.2):

> At first I believed that I could show it, when each of the classes could be well-ordered, for then one could pick the first term in each of them. Yet to do so, we must choose, for each class, *one* of the [well-ordered] relations of which it is the field. Thus, for these relations, we are using the very principle that is to be proved.

Here and elsewhere, Russell grasped the subtlety of the Axiom more deeply than anyone else at the time, with the possible exception of Zermelo.[9]

Russell next turned the correspondence to the two definitions of finite set: finite and Dedekind-finite. On 12 May he informed Couturat that he knew of no satisfactory proof showing these two definitions to be equivalent, since the usual proof relied on Zermelo's Axiom. "I do not see any reason," he added, "to believe that this axiom is true." If the Axiom was false, he concluded, then one could not demonstrate that every infinite set has a denumerable subset.

Russell's letter shocked Couturat, who wrote on 28 June how distressed he was to learn that these two definitions could not be shown equivalent. On 10 July, Couturat returned to this matter:

> That you have not yet been able to prove the proposition [(2.7.1)] is possible. But what astonishes me is that you seem to doubt its truth. I find it self-evident that under the hypothesis [K is a family of disjoint non-empty classes] we can extract an individual from each class in K and hence form a new class. Thus, seeing that you cannot prove something which appears so simple and easy, I wonder if the fault lies with the definitions or the principles of your system. . . . It seems to me that a logical system should be able to provide all that is given by *common sense*.

Russell granted, in his reply of 23 July, that common sense affirmed (2.7.1), but added that it affirmed many fallacious propositions and hence was an unreliable guide. On the other hand, he described as "rather simple," and hence perhaps acceptable as a postulate, the assumption that there exists a function f such that $f(u) \in u$ for every non-empty class u: "But for the moment I do not know if we can admit such an axiom without generating contradictions." As he saw it, the problem was that this assumption yielded the Axiom of Choice, from which Zermelo had deduced that every class can be well-ordered. Yet Russell continued to believe he could show that some classes cannot be well-ordered.

---

[9] It should be noted that (2.7.2) does not imply the Axiom or even the Ordering Principle (every set can be ordered). On the other hand, the Ordering Principle does not yield (2.7.2); see Pincus 1968, 234, and 1972, 736, as well as Jech 1973, 105, 115.

Russell gave a more polished form to his thoughts on Zermelo's Axiom, expressed in correspondence with Couturat and Jourdain, when he began to draft his reply to an article [1905] by Hobson. Delivered to the London Mathematical Society in February, this article had vehemently attacked Cantorian set theory. Hobson used Burali-Forti's paradox to inquire how far the theory of transfinite ordinals and cardinals was sound. He answered: not very far. While accepting ω and $\aleph_0$, he questioned the existence of the first uncountable ordinal ω_1 and the first uncountable cardinal $\aleph_1$. In this vein he rejected Hardy's argument of 1903, which exhibited a set A of real numbers with power $\aleph_1$, on the grounds that Hardy used arbitrary choices instead of stating a "norm" for A. Indeed, Hobson doubted that there exists *any* uncountable well-ordered set.[10]

By a norm for a set A, Hobson meant a finite set of conditions which make it "logically determinate" whether any particular object belongs to A. It would be unduly restrictive, he granted, to require an algorithm deciding whether any given object belonged to A. On the other hand, he considered an infinity of arbitrary choices as too vague to be legitimate. For they provided no norm and so could not define a specific set [1905, 173, 182].

At this point, Hobson analyzed various set-theoretic arguments in terms of norms. He discarded Cantor's proof of 1895 that every infinite set has a denumerable subset. Nor was he satisfied by Hardy's extension of Cantor's proof to show that every infinite cardinal either is an aleph or is greater than all alephs. Similarly, he contended, one must reject Jourdain's and Zermelo's arguments that every set can be well-ordered. For all of these proofs relied on arbitrary choices and failed to state a norm. Viewing Jourdain's and Zermelo's arguments as fundamentally similar, he also faulted Zermelo for failing to deal with the possibility of inconsistent sets. Finally, he added, one could not assume *a priori* that the Trichotomy of Cardinals is true or that the continuum can be well-ordered [1905, 182–185].

During June 1905, Russell and Hardy began to correspond about Hobson's article, which they viewed quite differently.[11] On 22 June, Russell wrote to Hardy that "Hobson is not clear, but I think he is in the main right.... Much of what I have to say is only putting his views more clearly." While he was pleased that Hobson's particular criticism of Hardy's proof of $\aleph_1 \leq 2^{\aleph_0}$ was mistaken, Russell carefully pointed out how arbitrary choices appeared in that proof. What was involved, he insisted, was (2.7.1), which "may be true, but I know of no reason for believing it. The point is, that there is not necessarily any definition of such a class as we want. The results are very awkward." One such result, Russell observed, was that (2.7.1) may fail even for an infinite class of pairs.

[10] Hobson 1905, 170–171, 178, 182.

[11] These letters are kept in the Russell Archives.

Hardy's letter of 30 June largely concerned (2.7.1), his proof of $\aleph_1 \leq 2^{\aleph_0}$, and Hobson's notion of norm. In general, Hardy was sceptical of Hobson's criticisms:

> I have thought about what you said in your letter and discussed the points with Jourdain, whom I saw recently.... I see that the 'norm' which I suggested in my former letter does not meet the real difficulty....
>
> My present position is rather like this.
>
> I admit that it seems impossible to prove the existence [non-emptiness] of the multiplicative class in general, and don't see how to do it in the particular case I want. . . .
>
> On the other hand, to deny the [non-emptiness of the multiplicative] class, (1) seems paradoxical, as an idealist would say 'does not satisfy the intellect'; (2) does not seem logically necessary; (3) does not seem to me ... to solve in an obvious way difficulties otherwise insoluble; (4) seems to make hay of a lot of the most interesting mathematics.

Responding to Hardy on 2 July, Russell remained agnostic about the truth of (2.7.1). He remarked that only denumerably many objects can be defined with a finite number of symbols, then added: "This, so far as it goes, favors $\exists{}^{\mid}x^{\mid}K$ [(2.7.1)], for it shows that a class may have members without our being able to mention any...." After pointing out that (2.7.1) can be proved in many special cases, such as when K is a well-ordered family of well-ordered classes, he returned to Zermelo's axiom: "I agree that much interesting mathematics depends on this axiom; and if it really strikes anyone as obvious, I know no reason for refusing to admit it." On 5 July, in the last letter of the correspondence, Hardy returned to this point:

> I should not like to say that anything so abstract and general as Zermelo's axiom strikes me as obvious, exactly; but . . . the more one thinks of it the more paradoxical the contrary seems, so that unless it appeared to lead to contradictions I should. . . be disposed to assume it and hope for the best.

By early in 1906 the London Mathematical Society published articles from Hardy, Jourdain, and Russell which replied to Hobson's critique. Acknowledging Hobson's general attack on set theory, Hardy restricted himself to a few points of rebuttal. After he refuted Hobson's technical criticisms of his proof that $\aleph_1 \leq 2^{\aleph_0}$, he turned his attention to Zermelo's Axiom and the Multiplicative Axiom. The latter axiom, he admitted, was assumed implicitly in his proof that $\aleph_1 \leq 2^{\aleph_0}$ as well as in J. König's propositions of 1904 on cardinal exponentiation (see 2.1). On the other hand, he found it interesting that Borel objected to uncountable arbitrary choices but used them in an example, given in 1905, of a real function not in Baire class 0, 1, or 2 (see 2.3). Hardy, who regarded Borel's objection as similar to Hobson's, saw no essential difference between denumerably and uncountably many arbitrary choices [1906, 10–15].

For Hardy, these examples illustrated the importance of the Axiom of Choice and the Multiplicative Axiom, both of which remained unproved: "It is, of course, quite possible that the existence [non-emptiness] of the multiplicative class may not be universal, but may hold in extensive particular cases ..." [1906, 16]. While the Multiplicative Axiom (2.7.1) certainly held when each member of the family K was well-ordered, he thought that it might be shown to hold when K was well-ordered even though the members of K were not. In any case, he provisionally accepted the Multiplicative Axiom in full generality [1906, 17].

Nevertheless, Hardy rejected Zermelo's proof. Although Hardy did not state a reason in print, he seems to have shared Hobson's concern that Zermelo had not excluded inconsistent sets from the proof.[12] Writing to Russell on 5 July, Hardy gave his opinion on this matter: "I suppose that Jourdain's and Zermelo's arguments are not unassailable even if $\exists ! x ! K$ [(2.7.1) holds], depending as they do on Burali-Forti's contradiction."

Jourdain devoted his reply [1906] to Hobson's critique largely to defending the Multiplicative Axiom (2.7.1). This was a significant shift from Jourdain's article in *Mathematische Annalen* [1905], where he had failed to see any need for Zermelo's Axiom.[13] In his letters to Russell, such as that of 6 May 1905, Jourdain had argued similarly that one could dispense with the Axiom in his proof that every set can be well-ordered. Russell disagreed, insisting that even to show the proposition (1.1.4), that every infinite set has a denumerable subset, requires the Multiplicative Axiom. On 13 November Jourdain wrote Russell again with a proof of (1.1.4). In fact, Jourdain showed only that every Dedekind-infinite set has a denumerable subset, since he did not distinguish clearly between infinite and Dedekind-infinite sets. Four days later Russell accepted this argument but dismissed Jourdain's claim that, without the Multiplicative Axiom, one could show (1.6.1): Every uncountable set has a subset of power $\aleph_1$.[14] As a result of his correspondence with Russell, by early in 1906 Jourdain became resigned to the need for the Multiplicative Axiom.

In his reply to Hobson, Jourdain cited as evidence for the Axiom the fact that almost every writer on set theory had used it implicitly. Among other examples, he mentioned his own implicit use while demonstrating the proposition (1.6.6) that an ordered set with no subset of type $^*\omega$ is well-ordered. Even the fact that theorems formerly shown by the Axiom could now be proved without it (such as Bernstein's result that there are exactly $2^{\aleph_0}$ closed subsets of $\mathbb{R}$) constituted further evidence for the Axiom. Similarly, Jourdain rejected Hobson's claim that making infinitely many arbitrary choices was like determining an infinite decimal by a throw of the dice for each digit. That chance process, borrowed from Paul du Bois-Reymond, could not

[12] Hardy 1906, 17; Hobson 1905, 184.

[13] There Jourdain had also claimed that neither his proof nor Zermelo's ensured that the power of the continuum is an aleph, since neither had shown $\mathbb{R}$ to be a consistent set (*i.e.*, to have no subset order-isomorphic to the class W of all ordinals); see Jourdain 1905, 465, 468–469.

[14] Grattan-Guinness 1977, 52–53, 56–59.

define a decimal by means of logical constants. On the other hand, Jourdain insisted, infinitely many arbitrary choices were "justified by an axiom (whose truth can hardly be denied seriously) which is expressible in terms of the logical constants..." [1906, 276]. Thus Jourdain used logicist views to argue for the Axiom.

Influenced by Russell's belief [1906, 49] that the Multiplicative Axiom might be weaker than Zermelo's, Jourdain next attempted to strengthen the former axiom so that they would be equivalent. Jourdain gave his strengthened axiom the following form: If K is a family of non-empty classes, then there exists a subset S of the union of K such that for every A in K, $A \cap S$ has exactly one member [1906, 283]. In effect, he had merely dropped from the Multiplicative Axiom (2.7.1) the hypothesis that K be disjoint. On 14 March 1906 he sent an argument that his new axiom and the Axiom of Choice are equivalent to Russell, who soon accepted this claim.[15] Nevertheless, Jourdain's axiom is false. If $K = \{\{1\}, \{2\}, \{1, 2\}\}$, then the set S required by his axiom contains both 1 and 2, and so cannot contain a unique member of $\{1, 2\}$. It remains uncertain why Jourdain and Russell overlooked this error, unmentioned in the literature, which resembled Burali-Forti's omission of 1896 when stating the Partition Principle.

On the other hand, Russell's reply to Hobson illuminated the role of both Zermelo's Axiom and his own Multiplicative Axiom. While agreeing with Hobson on several points, Russell distinguished sharply between the Axiom of Choice and the problem of inconsistent sets such as W. Doubts about the Axiom arose, Russell insisted, from the impossibility of determining a norm to justify the selection of one element from each of infinitely many sets. Yet the inconsistent sets had a perfectly definite norm, and thus a norm (or a propositional function, as Russell preferred to call it) was a necessary but not a sufficient condition for determining a set.[16] By developing his "no-classes theory," which relied on substitutions and propositional functions in place of classes, he hoped to avoid the set-theoretic paradoxes.[17] In any case, he claimed, none of these paradoxes is "essentially arithmetical, but all belong to logic, and are to be solved, therefore, by some change in current logical assumptions" [1906, 37]. To determine what these changes should be was Russell's research program.

In his reply, as in his letters, Russell recognized clearly that much of cardinal arithmetic depended on his axiom and Zermelo's. One needed the Multiplicative Axiom, he remarked, to demonstrate that the union of a denumerable family of pairs is denumerable, as well as to define the addition and multiplication of an infinite number of cardinals. In the same vein the Multiplicative Axiom guaranteed that for all cardinals $\mathfrak{m}$ and $\mathfrak{n}$, the cardinal

[15] *Ibid.*, 81–82.

[16] Here, in effect, Russell abandoned the Principle of Comprehension and was the first to do so in print. Yet he soon reinstated this principle in his theory of types.

[17] Russell 1906, 29–30, 45–47, 53.

sum of $\mathfrak{m}$ disjoint sets, each having $\mathfrak{n}$ members, is $\mathfrak{mn}$ and that the product of $\mathfrak{m}$ factors each equal to $\mathfrak{n}$ is $\mathfrak{n}^{\mathfrak{m}}$, as Whitehead [1902] had supposed.[18]

Russell also pointed out in his reply that one required the Multiplicative Axiom to obtain the proposition (1.3.3) that every Dedekind-finite set is finite. In a letter of 10 July 1905, Couturat had reminded Russell that Burali-Forti [1896] deduced this proposition from a postulate which "has an analogy with that of Zermelo." Burali-Forti's postulate, an incomplete version of the Partition Principle, stated that if A is a family of non-empty classes, then A is equipollent to a subclass of the union of A. On 23 July, Russell described this postulate as interesting but remained uncertain of its truth. Couturat broached the subject again on 17 December, asking if Russell had found a satisfactory proof that finiteness and Dedekind-finiteness are equivalent.[19] Two days later Russell wrote that he had not. However, using the set $\{\{x\}, \{y\}, \{x, y\}\}$, where $x \neq y$, he showed that Burali-Forti's postulate was false unless the family A was assumed disjoint. There would be such a proof of equivalence, Russell added, if one could demonstrate the proposition (1.3.4) that the power set of every Dedekind-finite class is Dedekind-finite.[20]

Soon afterward, Couturat informed Mario Pieri, a member of Peano's school of logicians, that Burali-Forti's postulate was false. Consequently, when Pieri composed an article on the consistency of arithmetic, he suggested the following alternative postulate [1906, 205]:

(2.7.3) If A is a Dedekind-infinite family of non-empty classes, then the union of A is Dedekind-infinite.

Writing to Couturat on 5 March 1906, Russell granted that Pieri's postulate implies that a class is finite if and only if it is Dedekind-finite, and pointed out that this postulate is equivalent to (1.3.4).

During 1906 Russell finished a manuscript, entitled "Multiplicative Axiom," where he went beyond his article by studying different forms of the Axiom of Choice.[21] In this manuscript, he recognized that the Multiplicative Axiom holds if and only if the following proposition holds as well:

(2.7.4) Every function f includes a one-one function with the same range as f.

---

[18] *Ibid.*, 48–50.

[19] Couturat added that, following Russell, he had pointed out in his book [1905] how this equivalence depended on Zermelo's Axiom. There Couturat also rejected Zermelo's proof because it relied on the Axiom. See Couturat 1905, 65, 53.

[20] Russell acknowledged in a letter of 17 January 1906 that his earlier postulate (1.7.9)—every infinite class is the union of some family of disjoint denumerable classes—"probably depends on Zermelo's axiom," and hence he abandoned (1.7.9). In fact, it easily yields that every Dedekind-finite set is finite.

[21] This work, which remains unpublished, is kept in the Russell Archives; see pp. 4–7 of it.

Similarly, he claimed that Zermelo's Axiom is equivalent to a proposition that he regarded as more general than (2.7.4):

(2.7.5) Every relation S includes a function with the same domain as S.

On the other hand, he had not yet realized that the Multiplicative Axiom and Zermelo's Axiom were equivalent, as were (2.7.4) and (2.7.5). Offering no proof, he asserted the equivalence of (2.7.4) and the Partition Principle.[22] Finally, he remained ambivalent about assuming either (2.7.4) or (2.7.5) as a postulate:

> It is obvious *a priori* that there are existent [non-empty] classes of which no member can be specified: *e.g.*, the class of indefinable ordinals.
>
> If one could generalize the multiplicative axiom, one might find something worthy to be an axiom . . . . Thus we could do with a general axiom giving us members of classes in cases where we cannot specify any members. Such an axiom might well be legitimate.

Here Russell inclined towards accepting the Axiom more than he was to do in any of his public pronouncements.

Why did Russell remain sceptical of both the Axiom of Choice and his own axiom (2.7.1) when he understood so well their importance in the theory of cardinals? To exclude his own paradox and that of the largest cardinal, he knew that logic (which for him included set theory) would have to be restricted [1906, 33]. As a letter to Jourdain illustrates, he was prepared to abandon some of Cantor's set theory in order to be assured of the consistency of logic.[23] It seems that he was equally willing to do without those theorems based on the Axiom, which he characterized as follows: "The axiom asserts that we can find some rule by which to pick out one term from each existent [non-empty] class contained in [a given class] w."[24] Here Russell's word "rule" must be emphasized. Tannery in 1897 and Hadamard in 1905 had distinguished sharply between functions which exist abstractly and those which can be determined by a definite rule (see 2.3). It is this distinction which Russell ignored. In effect, he assumed that the abstract existence of a choice function, supplied by the Axiom, was legitimate only if a definite rule could be given as well. Thus his view of mathematical existence paralleled that of such French Empiricists as Borel and Lebesgue.

How, precisely, did Russell's philosophical presuppositions shape his view of the Axiom? Perhaps he would have argued that they had no influence, for he had written:

> All philosophical differences . . . ought not to affect the detail of mathematics, but only the interpretation. Mathematics would be in a bad way if it could not

[22] It is still not known whether this equivalence holds. On the other hand, both Russell [1908, 258–259] and Zermelo [1908a, 274] established that the Multiplicative Axiom is true if and only if the Axiom of Choice is true as well.

[23] Letter of 2 November 1905 in Grattan-Guinness 1977, 56.

[24] Russell 1906, 47; *cf.* 52.

> proceed until the dispute between idealism and realism has been settled. When a new entity is introduced, Dr. Hobson regards the entity as *created* by the activity of the mind, while I regard it as merely *discerned*; but this difference of interpretation can hardly affect the question whether the introduction of an entity is legitimate or not, which is the only question with which mathematics, as opposed to philosophy, is concerned. [1906, 41]

Nevertheless, Russell's realist interpretation, wedded to a constructivist insistence on rules, led him to believe that Zermelo's Axiom and his own were true only under restricted conditions: "To discover the conditions subject to which Zermelo's axiom and the multiplicative axiom hold would be a very important contribution to mathematics and logic, and ought not to be beyond the powers of mathematicians" [1906, 52]. Russell failed to grasp the vagueness of his proposal, probably because he regarded axioms (at least in logic) as self-evident truths. Indeed, he expected the truth or falsehood of Zermelo's Axiom to emerge from research in logic, rather than from technical results in mathematics, and not to be settled until it became clear which classes existed [1906, 53].

Russell's article [1906] did not influence Hobson's views any more than the replies of Hardy and Jourdain. In 1907 Hobson still objected to Hardy's claim that every infinite cardinal is either an aleph or greater than all alephs, as well as to Jourdain's and Zermelo's arguments that every set can be well-ordered. Since his previous article, Hobson had become sympathetic with Borel's critique [1905] but, unlike Borel, rejected even denumerably many arbitrary choices. On the other hand, Hobson implicitly used such choices when he argued that every infinite set has a denumerable subset. Once again, a critic of the Axiom had great difficulty in disentangling himself from it.[25]

W. H. Young, who had liberally used successive arbitrary choices, did not participate in the debate instigated by Hobson. Nevertheless, W. H. and G. C. Young felt compelled to discuss such choices in their textbook of 1906, *The Theory of Sets of Points*. Apparently the Youngs were the only mathematicians to support Bernstein's interpretation of arbitrary choices in terms of "many-valued equivalence" (*cf.* 2.5). For the Youngs, Cantor's theorem that an infinite set M has a denumerable subset was "weighted" with multiplicity $\overline{\overline{M}}^{\aleph_0}$; and the larger the multiplicity, so they claimed, the weaker the theorem. On the other hand, they accepted neither Jourdain's nor Zermelo's argument for the Well-Ordering Theorem, and they regarded the Trichotomy of Cardinals as an open problem.[26] While reviewing the Youngs' book, Jourdain took them to task for accepting Bernstein's many-valued equivalence, and insisted that the Axiom was required for an infinity of arbitrary choices to possess any meaning [1906c].

---

[25] Hobson 1907, 207–210, 155.

[26] Young and Young 1906, 290, 147, 288, 150.

Over the next few years Jourdain, whose belief in the Axiom remained strong, published several papers delimiting those theorems which needed it. He recognized that the product of an infinity of order-types, like that of an infinity of cardinals, required the Axiom, but that the Heine-Borel Theorem did not.[27] In 1907 he stated but did not prove that the Trichotomy of Cardinals implies the Axiom, and hence that both are equivalent—reversing his earlier assertion that Trichotomy is weaker. Moreover, he indicated how one could deduce the Equivalence Theorem without the Axiom.[28] Finally, he proved in 1908, with no use of the Axiom, a theorem that he had previously shown by arbitrary choices: $\aleph_\alpha \cdot \aleph_\alpha = \aleph_\alpha$ for every ordinal α.[29] On the other hand, he erroneously believed the Axiom necessary to prove that $\mathfrak{m}^{\mathfrak{m}} > \mathfrak{m}$ for every infinite cardinal $\mathfrak{m}$ [1908a, 376].

Thus in England, as in France and to a lesser extent in Germany, Zermelo's proof was rejected. Hobson dismissed the Axiom, as well as Cantor's uncountable ordinals and cardinals, for reasons similar to Borel's. Although Russell also objected to Zermelo's proof because it relied on the Axiom, he still explored the consequences of his own related assumption, the Multiplicative Axiom, in cardinal arithmetic. He remained suspicious of both axioms. Hardy, who accepted the Multiplicative Axiom for its usefulness, objected to Zermelo's proof because he thought it involved the inconsistent class W of all ordinals. During the next decade, these mathematicians did not substantially alter their initial reaction to Zermelo's Axiom and proof.

Jourdain, on the other hand, reversed his position. Originally he had published an argument for the Well-Ordering Theorem. When Zermelo's appeared, he thought it less informative than his own. Through correspondence with Russell, he gradually understood how deeply the Axiom of Choice was embedded in his own thinking, and so he accepted the Axiom as valid. On the other hand, he adopted Russell's view that a rule was needed to make the requisite choices. Over the next decade, there lingered in Jourdain a passion to prove the Axiom of Choice and thereby to render it unnecessary. In 1918 he believed that at last he had discovered a proof supplying the desired rule. However, his argument ran afoul of arbitrary choices as it had so often before (see 3.8).

The Axiom remained a very subtle assumption that could be disentangled from set theory and analysis only with the greatest care. Both Russell and Jourdain undertook to determine where the Axiom was necessary and where it could be avoided. In Russell's case, distrust of the Axiom, together with a desire (later embodied in *Principia Mathematica*) to ascertain precisely what assumptions were needed in mathematics, led him to explore the Axiom in depth. Yet until set theory had been axiomatized, it remained uncertain which theorems required the Axiom. Even after the emergence of

---

[27] Jourdain 1906a, 15; 1906b, 65.

[28] Jourdain 1907, 366–367, 356.

[29] Jourdain 1908, 506–511; 1904a, 298–301.

Zermelo's axiomatization in 1908, one could demonstrate only that a proposition was equivalent to the Axiom or could dispense with it. There existed no technique for establishing that a proposition required the Axiom but was weaker than it, and one could say only that all *known* proofs of the given proposition used the Axiom. Much of the later research on the Axiom would be devoted to developing such a technique.

2.8 Peano: Logic vs. Zermelo's Axiom

The only Italian mathematician who responded to Zermelo's proof before 1908 was Giuseppe Peano. In August 1906, while surveying the recent literature on the foundations of mathematics, Peano analyzed both Zermelo's proof and Richard's paradox. Although Peano found both quite paradoxical, he was not disturbed by them, nor by Burali-Forti's and Russell's paradoxes. In every age, he claimed, antinomies arose in mathematics and then were resolved by discovering a hidden error in the reasoning. As for Richard's paradox, Peano believed the difficulty to lie in natural language rather than in mathematics, for natural languages lacked the precision possessed by a formal system of symbolic logic [1906, 157].

Peano knew of the debate that had taken place over Zermelo's proof, especially the articles in *Mathematische Annalen* (by Bernstein, Borel, Jourdain, J. König, and Schoenflies), the published letters (between Baire, Borel, Hadamard, and Lebesgue), and Poincaré's article of 1906. Yet Peano's approach differed from theirs in that he criticized the Axiom as a principle of logic within a formal system [1906, 144–145]. More precisely, he treated the Axiom in terms of the principles of logic found in his lengthy treatise, *Formulaire de mathématiques.*[1] From these principles one could deduce theorems by arbitrarily choosing any finite number of elements from non-empty sets, a technique which he then described in detail. In the *Formulaire* the introduction of n successive arbitrary elements required $n + 2$ propositions, and thus infinitely many arbitrary choices would require infinitely many propositions. Since no proof could have an infinite number of steps, he insisted, then one could not use an infinite number of arbitrary choices. He concluded that Zermelo's Axiom must be rejected [1906, 145–147]. Two years later Zermelo was to criticize this argument severely (see 3.1).[2]

---

[1] Peano 1895–1908. Regarding the history of this work, see Kennedy 1980, 44–50, 64–65, 76–77.

[2] Despite Peano's prohibition, later mathematicians explored formal systems in which proofs could have infinite length; see in particular Zermelo 1932 and 1935, as well as Rosser 1937. Concerning the prehistory of infinitary logic, see Barwise 1980 and Moore 1980.

In his article Peano recalled his injunction of 1890 against infinitely many arbitrary selections (see 1.8), and stressed that in the *Formulaire* there were deduced, without the Axiom, propositions which others had shown through its implicit use. It was possible to avoid the Axiom, not only for any finite number of arbitrary choices but even in certain infinite cases, by using a rule. Indeed, he added, this was what he had done in his article of 1890.[3]

On the other hand, Peano emphasized that if it was not known how to eliminate the Axiom from the proof of a proposition, as occurred in a number of cases, then its proof was invalid. By way of example, he cited Cantor's proposition (1.1.4) that every infinite class has a denumerable subclass. Here he singled out for censure Borel's *Leçons sur la théorie des fonctions* [1898], where (1.1.4) was demonstrated by implicitly using the Axiom. Unlike Borel and Richard, Peano made no attempt to save (1.1.4). Towards its truth or falsehood, he was indifferent. For him, it remained merely another unsolved problem, like Goldbach's conjecture that every even integer greater than two is the sum of two primes [1906, 148].

2.9 *Brouwer: A Voice in the Wilderness*

In 1907 L. E. J. Brouwer completed his doctoral dissertation at the University of Amsterdam. In it he began to formulate intuitionism, the constructivist philosophy of mathematics that he would elaborate throughout his career. Except for a critical analysis of Kant's and Russell's [1897] views on the philosophical foundations of geometry, Brouwer's thesis consisted of a polemic against Cantorian set theory and the logical foundations of mathematics. Whereas Russell [1903], as a logicist, had insisted that mathematics is a part of logic, Brouwer claimed rather that logic depends upon mathematics.[1] Although he granted the existence of each denumerable ordinal, he denied the existence of the set of all such ordinals as well as of ω_1, the first uncountable ordinal. Furthermore, he argued that the Equivalence Theorem lacked real meaning and that no conceivable cardinal number was greater than the power of the continuum.[2]

In the same spirit Brouwer investigated the Well-Ordering Problem. Contrary to Cantor, he could find no grounds for regarding the principle that every set can be well-ordered as a law of thought. As for Zermelo's supposed

---

[3] Peano 1906, 145. He had already made many of these remarks in a letter of 24 July 1906 to Russell; see Peano in Kennedy 1975, 209, 216.

[1] Brouwer 1907, 127–131. Additions to his thesis can be found in Brouwer 1917.

[2] Brouwer 1907, 144, 151, 155.

proof, he interpreted Borel's criticism of it to mean that one might just as well accept the Well-Ordering Principle as accept Zermelo's Axiom. He agreed, and accepted neither. While he granted that every countable set can be well-ordered, he was equally certain that this was *not* true of the continuum $\mathbb{R}$, since most of its elements must remain unknown. Indeed, he added that the continuum could not be well-ordered because every well-ordered set was countable. The Well-Ordering Problem, he concluded, was illusory [1907, 152–153].

Brouwer saw the Continuum Problem from a perspective quite different from classical set theory. For he denied the existence, not only of the second number-class, but of the continuum as a system of individual, mathematically existing points. Consequently, he treated the stronger form of the Continuum Hypothesis (the continuum $\mathbb{R}$ and the second number-class are equipollent) as logically possible but devoid of mathematical content. If one introduced as logical entities the second number-class S and the set $\mathbb{R}$ of all real numbers, then he believed it consistent to assume that S and $\mathbb{R}$ were of the same power. In his eyes, this was legitimate since both S and $\mathbb{R}$ were "denumerably unfinished" sets (denumerable but such that new elements could always be adjoined by a previously defined mathematical process); and all denumerably unfinished sets had the same power.[3]

Brouwer regarded as mathematically more reputable Cantor's original form (1.5.1) of the Continuum Hypothesis: Every uncountable subset A of the continuum has the power of the continuum [1907, 151]. To establish this form, Brouwer offered a rather obscure argument. In the set A one selected a well-defined, and hence countable, subset T such that all other points of A were approximations to the everywhere dense set T, and then one chose a similar subset T' of $\mathbb{R}$. The undefined points of A and $\mathbb{R}$ were made to correspond with each other by means of the analogous parts in T and T'. From this, Brouwer concluded that the Continuum Hypothesis (1.5.1) followed [1907, 66–67].

For unknown reasons, perhaps because he wrote in Dutch, no one appears to have cited Brouwer's remarks on the Well-Ordering Problem and the Continuum Problem during this period. Even Borel and Lebesgue, who shared Brouwer's constructivistic tendencies, did not draw on his observations. Yet within a few years, he would create an intuitionistic set theory with its own problems, distinct from but often parallel to those of Cantorian set theory, and acquire a school of followers in Holland. As a result, intuitionism continues to flourish today as one of the major schools of constructivist thought in the philosophy of mathematics, and has given rise as well to its own body of technical mathematics.[4] It is now recognized

[3] *Ibid.*, 149–150. Brouwer considered all infinite sets to be either denumerable, denumerably unfinished, or of the power of the continuum.

[4] For a discussion of later developments in intuitionism, see Kleene and Vesley 1965, Heyting 1966, and Gauthier 1977.

that some forms of the Axiom, confined to number-theoretic functions, are true in intuitionistic logic because of how the quantifiers are interpreted there.[5]

2.10 *Enthusiasm and Mistrust in America*

Across the Atlantic, Zermelo's proof generated far less discussion than it had in Europe. Among the relatively small community of American mathematicians, only one accepted the proof in print. Two others knew of the controversy surrounding the Axiom, and a fourth investigated the related question of non-measurable sets.

Cassius Keyser, a professor at Columbia University, had a much broader perspective than the majority of American mathematicians. In 1905 he published some reflections on set theory in a philosophical journal. Enjoining cooperation between mathematics and philosophy, he challenged the practitioners of both disciplines to recognize their common interest in set-theoretic problems.

Keyser reacted more enthusiastically to Zermelo's proof than did any other mathematician at the time except Hadamard, and fully accepted Zermelo's solution to the Well-Ordering Problem: "The fundamental character of the problem, the simplicity of the means employed in its solution and especially the bearings of the solution on kindred dependent questions, justify a . . . report upon the matter in this place" [1905, 208]. Indeed, he effusively praised "the marvelous character, the awful comprehensiveness, of Zermelo's proposition" [1905, 210]. While he cited the Trichotomy of Cardinals as one of the proof's consequences, he found nothing unusual in Zermelo's Axiom and mentioned it only in passing.

By contrast, Keyser emphasized that Cantor had lacked an acceptable proof for the Well-Ordering Theorem, a lack which had caused Borel in 1898 to question whether every set can be well-ordered. "Concepts and proofs," Keyser insisted, "are essentially *social* affairs. They must be intelligible to at least two minds, or, what is tantamount, to one person at least twice" [1905, 211]. In fact, although Cantor had possessed an argument intelligible to at least three persons (Harward, Jourdain, and himself), it never gained wider currency. What is important for a proof, or a concept, as a social affair is that it become interesting to a mathematical community who then exploit its potential. This is what occurred to Zermelo's Axiom, though very slowly at first (see Chapter 4).

---

[5] Dummett 1977, 52–54, 315, and Troelstra 1977, 986, 1028, 1033–1035. There is some dispute as to which forms of the Axiom are intuitionistically valid; see Gielen *et alii* 1981, 123–124, 126.

Around the turn of the century, a small group of mathematicians, sometimes known as the American postulate theorists, formulated various axiomatic systems under the influence of Hilbert's *Grundlagen der Geometrie*. Two members of this group were Edward Huntington, who became a professor at Harvard in 1901, and Oswald Veblen, who left the University of Chicago in 1905 to join the faculty at Princeton. To the end of an article [1905] on the continuum as an order-type, Huntington appended a discussion of well-ordered sets. He was aware of the debate over Zermelo's proof, especially the article by Hobson [1905] and those in volume 60 of *Mathematische Annalen*. As a result, he maintained a studied neutrality on the question whether every set can be well-ordered, noting only that "the proof given by Zermelo . . . has not been generally accepted" [1905, 41n]. Likewise, he added that the Trichotomy of Cardinals may be false [1905, 39]. Veblen, on the other hand, knew of Zermelo's proof from reading an article [1907] by Johannes Mollerup of Copenhagen concerning $\aleph_1$-limit points.[1] Yet Veblen ventured no opinion on the proof, even though he implicitly used the Axiom to obtain $\aleph_1 \leq 2^{\aleph_0}$ [1908a, 235]. During the same period, recognizing that Hardy [1903] had not succeeded in defining a subset of $\mathbb{R}$ with power $\aleph_1$, he attempted to extend Hardy's methods constructively.[2]

A fourth American mathematician, Edward Van Vleck, was distinctly reserved toward the Axiom. Van Vleck, a professor at the University of Wisconsin, did not mention the Axiom directly in 1908 when he discussed sets that are not Lebesgue-measurable, but he did note how they depended on arbitrary choices. Moreover, he was acquainted with Lebesgue's article [1907] on Zermelo's choice functions and non-measurable sets. Ignoring Lebesgue's philosophical scruples, Van Vleck gave an example of a non-measurable set which he had discovered before learning of that article. To each irrational x in the open interval (0, 1), he assigned the pair of sets

$$P_x = \left\{\left\{\left(\frac{x}{2^{\pm p}}\right) \pm \left(\frac{m}{2^n}\right): m, n, p \in \mathbb{N}\right\}, \left\{1 - \left(\left(\frac{x}{2^{\pm p}}\right) \pm \left(\frac{m}{2^n}\right)\right): m, n, p \in \mathbb{N}\right\}\right\}.$$

Then the irrationals in (0, 1) were partitioned into two sets A and B. Choosing one set in each pair P_x, he placed all of its members in A, while he put all members of the other into B. Then he showed that the resulting sets A and B are non-measurable and lack the Baire property [1908, 239–241]. In effect, his argument established that the Axiom, restricted to families of power $2^{\aleph_0}$ whose members are pairs, yields a set that is neither measurable nor a Baire set.

Van Vleck hoped to eliminate the arbitrary choices from his example of a non-measurable set. His motives remain vague, since he did not refer

---

[1] A point p is an $\aleph_\alpha$-*limit point* of a subset A of $\mathbb{R}$ if there is some interval I such that, for every subinterval S of I containing p, $S \cap A$ has power $\aleph_\alpha$. Cantor [1885] referred to such a limit point as *of the* $(\alpha + 1)$*th order*; see 1.5.

[2] Veblen 1908; *cf.* 4.7.

to any criticism of the Axiom found in Lebesgue's article or elsewhere. Because of this article, he suspected that there exists no rule for picking an element from each set $\{(x/2^{\pm p}) \pm (m/2^n): m, n, p \in \mathbb{N}\}$. Nevertheless, he insisted, there might be another way to eliminate the choices if one could state a rule for selecting one member from each of an uncountable number of pairs:

> Thus it seems to me possible, and perhaps not difficult, to remove the arbitrary element of choice in my example by confining one's attention to a proper subset of the continuum, though as yet I have not succeeded in proving that this is possible.[3]

However, Van Vleck's hope has not been borne out, since it is now known that the existence of a non-measurable set cannot be deduced in **ZF**.[4]

2.11 *Retrospect and Prospect*

When Zermelo explicitly formulated the Axiom of Choice in 1904, mathematicians responded by posing an important question: Is the use of infinitely many arbitrary choices a legitimate procedure in mathematics? Prior to that time, many mathematicians had used such choices with a greater or lesser degree of awareness in set theory, analysis, and algebraic number theory, but no one had recognized that a significant new axiom was involved. On the other hand, from 1890 to 1902 three Italian mathematicians had found instances of such choices and had rejected them. Even these critics remained unaware how widely the use of arbitrary choices had diffused and how these choices had been absorbed indirectly through results such as the Countable Union Theorem. Their scruples went unnoticed until Zermelo demonstrated the Well-Ordering Theorem.

Coming in the wake of J. König's argument that the real numbers cannot be well-ordered, Zermelo's proof that every set can be well-ordered generated a major controversy, which soon involved mathematicians in many European countries as well as the United States. Some mathematicians, such as Hadamard, even spoke of König's and Zermelo's results as a paradox—a view which König himself propagated in a second attempt to refute the Continuum Hypothesis. Nevertheless, the controversy over Zermelo's proof arose in good part because a widely used procedure which had been regarded as innocuous—infinitely many arbitrary choices—turned out to be equivalent to Cantor's Well-Ordering Principle, of which many mathematicians remained sceptical. Thereby doubt was cast on arbitrary selections as embodied in the Axiom of Choice.

---

[3] Van Vleck 1908, 241. For the relevance of his work to probability theory, see Novikoff and Barone 1977.

[4] See footnote 24 of 2.3 and footnote 6 of 5.2.

In addition to the seeming paradox of König's and Zermelo's results, paradoxes affected the proof in other ways. German Cantorians such as Bernstein and Schoenflies believed that Zermelo's proof was invalid precisely because of Burali-Forti's paradox, a claim which illustrates the fluidity of set-theoretic assumptions during the period 1904–1908. Bernstein was prepared to abandon the principle that every ordinal has a successor, but not the principles that every set has a cardinal and that every ordered set has an order-type. In addition, he discarded Cantor's Well-Ordering Principle since he had found a set, so he believed, that could not be well-ordered. On the other hand, Schoenflies held that one must introduce a new postulate for each number-class, and felt certain only of the countable ordinals. He inclined towards accepting the Trichotomy of Cardinals, which he erroneously claimed to be weaker than both the Axiom and the Well-Ordering Theorem. Borel in France and Hobson in England were ready to confine set theory to the countable. More radically, Baire regarded all infinite sets as established by convention, and espoused the view, also adopted by Poincaré, that every infinite set was merely a potential infinity. These criticisms directed against Zermelo's proof formed part of a more fundamental attack on Cantorian set theory.

None of Zermelo's critics was prepared to give a precise formulation to the term "definable," and yet it occurred repeatedly in their writings. Analyzing Richard's paradox, Poincaré believed the error in Zermelo's proof to reside in the use of an impredicative definition. Indeed, Russell's Vicious Circle Principle, which provided the basis for his theory of types, originated from prohibiting such impredicative definitions.

Definability played a different role when Hadamard exchanged letters on Zermelo's proof with Baire, Borel, and Lebesgue. There the principal question concerned existence proofs: Can one establish the existence of a mathematical object with a certain property P by showing that the class of objects satisfying P is non-empty, or must one define a unique object in this class? The problem was at once a philosophical and a mathematical one, but it was made more difficult by the lack of precise criteria for definability relative to a given language. Moreover, Hadamard insisted that Borel's and Lebesgue's requirement of definability reduced the set of all real functions (and every other infinite set) to a denumerable set.

Thus the reactions to Zermelo's proof, though predominantly negative, were quite diverse. Hadamard, Hausdorff, and Keyser accepted the proof in full generality, while Hardy and Poincaré accepted the Axiom but disputed the proof. Borel, who remained sceptical of both the Axiom and the Well-Ordering Theorem, granted their equivalence; tentatively he accepted the Denumerable Axiom alone. On the other hand, Baire, Lebesgue, Hobson, and Peano rejected every form of the Axiom.

This very diversity illustrates how difficult it would be to argue convincingly for national characteristics as a decisive factor in the debate. Neither the French, the Germans, nor the English—the only nationalities with several mathematicians discussing the proof among themselves—were in agreement. In each of these countries mathematicians argued both for and against the proof, both for and against the Axiom. Nevertheless, the French criticisms of the proof flowed largely from constructivistic beliefs, whereas the German opponents concerned themselves primarily with Burali-Forti's paradox. The English shared both concerns.

In spite of the debate, the Axiom served those who accepted it as a powerful means of obtaining new mathematical results. Hausdorff, in his studies of ordered sets, used the Axiom repeatedly to investigate the order-types of η_α-sets and the Generalized Continuum Hypothesis, as well as to establish that $\aleph_{\alpha+1}$ is regular for every α. In real analysis, both Hamel and Vitali relied on Zermelo's Well-Ordering Theorem to obtain a well-ordering for the real numbers. From such a well-ordering Hamel defined a Hamel basis for $\mathbb{R}$, as well as discontinuous real functions f such that $f(x + y) = f(x) + f(y)$ for all x and y. Likewise, Vitali produced a non-measurable set and established that Lebesgue's Measure Problem had no solution. Eventually, such non-measurable sets would lead to Hausdorff's paradox and then to the question whether there existed a measurable cardinal (see 3.7 and 4.11).

On the other hand, even some of those mathematicians who opposed the Axiom wished to determine its consequences, either as a means of protecting themselves against dubious results or as a way of showing what improbable conclusions followed from the Axiom. Within set theory, Russell did not attempt so much to deduce new results from the Axiom as to ascertain those earlier results which had relied on it. Thus he recognized that the Axiom was required to define the addition and multiplication of infinitely many cardinals. He also uncovered the Axiom's role in establishing that every Dedekind-finite set is finite and that every infinite set has a denumerable subset. Independently of Russell, Zermelo also recognized that these three results depended on the Axiom. Finally, Russell noted that the Axiom is essential even to prove that the union of denumerably many sets, each containing two elements, is denumerable. Like everyone else at the time, he lacked a method for transforming his excellent intuition into a *proof* that the Axiom was needed for these results. Such a method had to await the work of Abraham Fraenkel and Andrzej Mostowski, as well as the more powerful method of Cohen (see 4.9 and the Epilogue).

Chapter 3

Zermelo's Axiom and Axiomatization in Transition (1908–1918)

> Many mathematicians still stand opposed to the Axiom of Choice. With the increasing recognition *that there are questions in mathematics which cannot be decided without this axiom*, the resistance to it must increasingly disappear.
>
> Ernst Steinitz [1910, 170–171]

As early as 1896, even before the discovery of set-theoretic paradoxes, a few mathematicians had suggested that set theory ought to be axiomatized. Yet interest in formulating such an axiomatization remained very faint even in 1903, when Russell restated Burali-Forti's result of 1897 as a paradox, and published his own paradox as well. Hilbert, for example, viewed Russell's paradox as revealing that contemporary logic failed to meet the demands of set theory.[1] Russell asserted further that a solution to the paradoxes would result only from a reappraisal of the assumptions used in logic, rather than from technical mathematics [1906, 37]. Unperturbed by the paradoxes, Zermelo concentrated on axiomatizing set theory within mathematics rather than on revising the underlying logical assumptions.

During the debate over Zermelo's proof of the Well-Ordering Theorem, there appeared a *potpourri* of set-theoretic principles which some accepted and others denied: Every set has a cardinal number; every ordered set has an order-type; every ordinal number has a successor; the existence of "each" of Cantor's number-classes; the existence of the "set" W of all ordinals; the Well-Ordering Principle; the Axiom of Choice; the Trichotomy of Cardinals; and the Principle of Comprehension, embodied in Cantor's definition of a set. There was much confusion as to which of these principles to retain, since one could not consistently retain all of them. Nevertheless, a majority of those mathematicians who expressed an opinion about Zermelo's proof

[1] Letter of 7 November 1903 to Frege, in Frege 1980, 51–52.

rejected it unequivocally, and usually dismissed the Axiom of Choice in particular.

Thus, in order to protect his proof of the Well-Ordering Theorem, Zermelo needed to specify his assumptions in detail. Making a choice among the possible set-theoretic principles consonant with that proof, he axiomatized set theory. However, when he published this axiomatization, a new controversy arose. Rather than securing his proof, the axiomatization itself was viewed as questionable for the next decade. Furthermore, though Hartogs clarified the status of the Trichotomy of Cardinals, the Axiom was subjected to new attacks—which Hausdorff's paradox seemed to justify. Despite Steinitz' productive use of the Axiom in field theory, the resistance did not disappear.

3.1 Zermelo's Reply to His Critics

During the summer of 1907 Zermelo took stock of the criticisms directed against both his Axiom and his proof of the Well-Ordering Theorem. Recognizing that both had been widely misunderstood, he resolved to modify them in order to avoid subjectivity and further misinterpretation. From this period there emerged two articles, completed within sixteen days of each other and interrelated in numerous ways. One [1908] was a reply to his critics, and the other [1908a] contained the first axiomatization of set theory. Although they appeared separately, they were essentially a single paper. In fact, both articles emerged from a common source—the controversy surrounding Zermelo's proof. Their fundamental unity will be of particular importance in the next section where the role of this controversy and of the paradoxes in motivating Zermelo's axiomatization will be analyzed.

Zermelo's first article [1908] began with a new demonstration of the Well-Ordering Theorem. Although he still upheld his earlier proof, he recognized that certain mathematicians had conflated the notion of well-ordering with a spatio-temporal arrangement. Hence he gave a new definition, purely formal in character, that had nothing to do with such an arrangement: A set M is well-ordered if there is a family $\mathscr{M}$ of subsets of M such that for each a in M there is a unique member $\mathfrak{R}(a)$ of $\mathscr{M}$, called the remainder of a, and every non-empty subset P of M contains a unique element p such that P is a subset of $\mathfrak{R}(p)$. In this way, he hoped to avoid psychologistic interpretations of the Well-Ordering Theorem. In addition, he explicitly stated the postulates that his new proof used, postulates which appeared in his next paper [1908a] as part of his axiomatization for set theory: the Axiom of Separation and the Power Set Axiom. From them he developed the properties of his θ-chains, which generalized Dedekind's earlier concept of chain (see 1.6).[1] These properties supplied the basis for his new proof.

Separating the Axiom of Choice from his theorem, Zermelo reformulated the latter: If every non-empty subset of a given set M contains a distinguished element, then the intersection of all the θ-chains of M well-orders M. Only after establishing this result did he introduce his Axiom, and here the subtlety of his thought became apparent:

> Now in order to apply our theorem to every set, we need merely the assumption that *in principle the simultaneous choice of the distinguished elements is possible for any set of sets*, or more precisely, that the same consequences are valid as if this choice were possible. [1908, 110]

Here he proposed a formalistic interpretation of simultaneous arbitrary choices.

Since Zermelo wished to avoid the subjectivity that supposedly pervaded the notion of simultaneous choices, he revised his Axiom to read as follows:

> If a set S is partitioned into a disjoint family A of non-empty sets, then there exists at least one subset T of S which has exactly one element in common with each member of A.

In effect, this is Russell's Multiplicative Axiom (2.7.1), but Zermelo did not refer to Russell's paper [1906], where it was published. As was discussed in 2.2 and 2.5, Zermelo had independently formulated the Multiplicative Axiom by 1906 and perhaps as early as 1904. Like Russell, Zermelo came to recognize that the Multiplicative Axiom was more intuitively evident than the Axiom of Choice couched in terms of choice functions. But while Russell remained sceptical of the Multiplicative Axiom, Zermelo asserted that its "purely objective character is immediately evident" [1908, 110]. To Zermelo's critics the Multiplicative Axiom may have been objective, but was no more self-evident than its predecessor, the Axiom of Choice.

By and large, in 1907 Zermelo knew of his critics, but there were exceptions. Evidently he remained unaware of the English debate over the Axiom by Hardy, Hobson, Jourdain, and Russell a year earlier.[2] Likewise he was unfamiliar with Brouwer's thesis [1907] and the Youngs' book [1906]. On the other hand, Zermelo had learned very early of certain objections to his proof. At the meeting of the Göttingen Academy of Sciences held on 15 November 1904, he spoke on his proof, and "shed light on the objections to it that E. Borel, J. König, F. Bernstein, and A. Schoenflies have made."[3]

---

[1] Zermelo 1908, 108, 117. For a given set M and a given choice function f on the family of all non-empty subsets of M, T is a *θ-chain* if (1) T is a set of subsets of M; (2) M is in T; (3) if A is in T, then $A - \{f(A)\}$ is in T; (4) if S is a subset of T, then $\bigcap S$ is in T.

[2] Zermelo corresponded with Jourdain in 1907, but apparently their letters focused on a generalization of König's theorem (2.1.2); see Jourdain 1908, 512.

[3] **DMV 14** (1905), 61.

Replying to his critics in detail, Zermelo [1908] divided their objections into four categories. The first included attacks on his Axiom, particularly by Borel and Peano, while the second consisted of Poincaré's comments on impredicative definitions. The third and fourth were interrelated and provided the bulk of Zermelo's reply, perhaps because they were specifically set-theoretical: objections concerning the "set" W of all ordinals (Jourdain, Bernstein, and J. König) and those based as well on the generation of ordinals (Schoenflies).

To those who rejected his Axiom, Zermelo responded at length. Although he had read Borel's article [1905] and the published correspondence between Baire, Borel, Hadamard, and Lebesgue, he concentrated on refuting Peano with whom he had previously feuded over the Equivalence Theorem. He stressed that, by noting the unprovability of his Axiom, Borel and Peano agreed with his own position. Had they also shown the Axiom independent from other postulates, he added, they would have put him in their debt. Here Zermelo turned the arguments of his opponents against them, as he was to do many times in his paper.

In the course of rebutting Peano's arguments, Zermelo presented his own conception of mathematics. Peano had remarked that the Axiom of Choice could not be deduced in his *Formulaire de mathématiques*, and consequently he had objected to the Axiom [1906]. Zermelo countered that the *Formulaire* was incomplete precisely at this point:

> First of all, how does Peano arrive at his own fundamental principles and how does he justify admitting them into the *Formulaire*, since he cannot prove them either? Obviously, through analyzing the rules of inference that have historically been recognized as valid and by referring both to the intuitive evidence for the rules and to their necessity for [mathematical] science—considerations which may be argued just as well for the disputed Axiom .... This Axiom, without being formulated in a scholastic manner,[4] has been applied successfully, and very frequently, in the most diverse mathematical fields, particularly set theory, by R. Dedekind, G. Cantor, F. Bernstein, A. Schoenflies, and J. König among others .... Such extensive usage of a principle can only be explained through its *self-evidence*, which, naturally, must not be confused with its provability. While this self-evidence may be subjective to a certain degree, it is in any case an essential source of mathematical principles [axioms], though not a basis for mathematical proofs. Thus Peano's statement, that self-evidence has nothing to do with mathematics, does not do justice to obvious facts. However, what can be objectively decided, the question of *necessity for [mathematical] science*, I would like now to submit to judgment by presenting a series of elementary and fundamental theorems and problems, which, in my opinion, could not be settled without the Axiom of Choice. [1908, 112–113]

Zermelo conceived of mathematics as "a productive science founded ultimately on intuition ..."—a view in agreement with Poincaré, though not

[4] This was a jibe at Peano's admiration for the Scholastics.

with Poincaré's Kantianism (see 2.4).[5] Since Zermelo also stressed the objectivity of mathematics and its scientific content, he might best be characterized philosophically as a pragmatist and a realist. The extent to which his realism was Platonistic remains uncertain.

To substantiate the mathematical need for his Axiom, Zermelo next cited seven of its applications. Russell had noted three of these in 1906, but Zermelo discovered them independently: Every Dedekind-finite set is finite; the respective unions of two families of disjoint equipollent sets are again equipollent; and the product (multiplicative class) of a disjoint family of sets is empty only if one of the sets is empty. The second of these was required for every sum of cardinals to be well-defined, while the third merely restated the new form of his Axiom. Zermelo credited the fourth and fifth applications to Hamel [1905]: There exists a Hamel basis for all real numbers, and there are discontinuous real functions f such that $f(x + y) = f(x) + f(y)$ for all x and y. A sixth, essential to Cantor's second number-class, was the Countable Union Theorem. As a final application Zermelo attributed the Partition Principle to Beppo Levi [1908, 113–115].

Yet Zermelo was too generous in crediting this principle to Levi. As was discussed in 1.8, Levi avoided the Partition Principle except in those cases where he could prove it *without* resorting to arbitrary choices. Here Zermelo distorted the prehistory of his Axiom, probably to give it a lengthier and more distinguished ancestry. Engaged in an acrimonious debate, Zermelo gave a "Whig interpretation"[6] of his Axiom's history in order to refute his opponents.

With relish Zermelo turned the arguments of one critic against another. Using Russell's paradox, he showed that Peano's *Formulaire* was inconsistent. To rebut Poincaré's critique about impredicative definitions, Zermelo borrowed an argument from Peano, namely, that different definitions may yield notions which are not identical but which possess the same extension. In the eyes of Poincaré, Zermelo's [1904] proof contained an impredicative definition of L_γ and hence a vicious circle. But Zermelo countered that impredicative definitions occurred frequently in analysis. They arose whenever one defined the minimum of a closed set, as in Cauchy's proof of the

[5] Zermelo 1908, 116. During 1906 he corresponded with Poincaré regarding his proof and his axiomatization of set theory. One of Poincaré's letters, unfortunately undated, mentioned that in his article [1906, 315] he had intended to include a passage expressing reservations about deducing the Equivalence Theorem from the Well-Ordering Theorem. Then he added: "This does not mean that I have sided with the critics of your proof. I would be rather inclined to dismiss their objections, but I ask you to allow me the time to reflect further on this question before I adopt a definitive solution." This letter, as well as three others from Poincaré, is kept in Zermelo's *Nachlass* at the University of Freiburg im Breisgau.

[6] In other words, Zermelo interpreted past events so that they naturally culminated in his Axiom. For a general discussion of this style of historical writing, see Butterfield 1965.

Fundamental Theorem of Algebra.[7] Zermelo insisted that whether or not an object A fell under a definition of a concept must be decided by an objective criterion independently of the concept defined; a special relation might then exist between the concept and the object A, as occurred when A was the minimum or intersection of all the objects falling under the concept. By such a special relation, the object A was not created, but only described [1908, 117–118]. Here again Zermelo emerged as a realist, perhaps a Platonist.

Ultimately, Zermelo was concerned with this question: Which propositions should be allowed into mathematics? For him it was vital to preserve as many classical theorems as possible. Thus he reproached Peano by arguing that the *Formulaire* did not even yield proofs for the seven simple applications, mentioned above, of the Axiom. Indeed he described Peano's mathematics as "a science artificially mutilated to some extent ..." [1908, 115]. In a similar vein, he responded to Poincaré's critique by noting that it would invalidate many important proofs, not only Zermelo's. Such remarks were part of Zermelo's emphasis on those axioms and techniques needed to keep mathematics productive.

Zermelo devoted the rest of his reply to those mathematicians who had actively contributed to set theory but for whom his proof was tainted by Burali-Forti's paradox: J. König, Jourdain, Bernstein, and Schoenflies. Though König did not quite fall into this category, Zermelo chided him for implicitly using the Axiom in his Heidelberg lecture of 1904, while, a year later, treating the problem of well-ordering $\mathbb{R}$ as unresolved. Zermelo granted that König had published no specific reservations to his proof. On the other hand, he pointed out that König's version of Richard's paradox did not involve a well-ordering of the continuum, as König asserted.[8]

When Zermelo claimed that Jourdain, Bernstein, and Schoenflies recognized the need for his Axiom, he inflated the number of its supporters. Neither Bernstein nor Schoenflies regarded the Axiom as necessary, although Jourdain did so after an initial hesitation. Zermelo censured all three of them for intruding the "set" W of all ordinals into his proof. There existed no such set W, he insisted, since every well-ordered set A could be extended by a single element to obtain a well-ordered set B of which A was then a segment [1908, 124]. What Zermelo did was to reverse the arguments of Bernstein and Schoenflies; assuming the negation of their conclusion that some well-ordered set cannot be extended, he arrived at the negation of their hypothesis: the non-existence of W. By abandoning W as a set,

---

[7] Zermelo had previously sent this example to Poincaré, who replied in a letter of 16 June 1906 that the impredicative definition in Cauchy's proof could be replaced by "an entirely determinate *process* not admitting any ambiguity or any vicious circle .... It is *not the same*, on the contrary, for your Well-Ordering Theorem. There my doubts remain because I cannot conceive of an analogous *process*."

[8] Zermelo 1908, 118–120; *cf.* 2.6.

Zermelo also dispensed with the Principle of Comprehension, an essential step in axiomatizing set theory (see 3.2).

Against Jourdain's claim [1905a] to priority in proving the Well-Ordering Theorem, Zermelo stressed an assertion that Jourdain had made but left undemonstrated: If A is an infinite set not equipollent to an aleph, then A has a subset equipollent to the totality of all alephs. Zermelo sensed an inherent vagueness when Jourdain made an infinity of successive choices from a set in order to well-order it. Indeed, Zermelo doubted that Cantor, who had invented this process, considered it to be a proof.[9] More telling was Zermelo's observation that Jourdain's proof failed, as he himself admitted, even to ensure that $\mathbb{R}$ can be well-ordered. Lastly, Zermelo dismissed as verbal gymnastics Jourdain's denial of an order-type to some well-ordered sets, a response akin to Russell's upon learning of this proposal.[10]

Toward Bernstein, Zermelo reacted still more negatively. He insisted that Bernstein's requirement, that no order-relation be stipulated between W and an element not in W, was ad hoc:

> As if it only concerned the *word* "order-relation" . . . , and as if by avoiding a *word* an objective mathematical *fact* could be disregarded! Mathematics would not be an international science if its theorems did not possess an objective content independent of the language in which we express them. When a contradiction is examined, it is not at all a question of whether a dubious consequence is *actually derived* and officially *recognized*, but solely, whether it is formally *possible* at all; to deny this possibility just *because* it leads to a contradiction would obviously be begging the question . . . . [1908, 123]

Bernstein had indeed begged the question, Zermelo insisted, since W was self-contradictory. From his perspective, the way to resolve Burali-Forti's paradox was to reject W as a set. As for Bernstein's notion of many-valued equivalence, he pointed out that such a definition failed to determine whether a choice function actually existed, a question which only an axiom could settle [1908, 113]. In this regard his judgment was excellent.

With equal force Zermelo applied to Schoenflies his critique of Bernstein's use of W. Furthermore, Zermelo rejected Schoenflies' distinction, between the general and special parts of Cantor's theory of well-ordering, by demanding that the entire theory be based on the definition of well-ordering. Contrary to Schoenflies, Zermelo denied that his proof of 1904 rested on Cantor's principles for generating ordinals and that a new axiom was needed for each of Cantor's number-classes. Rather, it sufficed to introduce an axiom guaranteeing the existence of a set of type ω.[11] Here Zermelo was

---

[9] As Zermelo later discovered when he edited Cantor's works [1932], Cantor did regard it as a proof, but retained reservations which kept him from publishing it; see 1.6.

[10] Zermelo 1908, 120–121; *cf.* Russell's letter of 12 April 1904 to Jourdain in Grattan-Guinness 1977, 29.

[11] Zermelo 1908, 124–125.

referring to the Axiom of Infinity, an important part of his axiomatization [1908a] (see 3.2).

Schoenflies [1905] had believed the Well-Ordering Theorem to be equivalent to the proposition (2.5.1) that there exists no infinite decreasing sequence of cardinals. Zermelo, who readily admitted that the Well-Ordering Theorem implies (2.5.1), adamantly denied the converse. One must also assume the Trichotomy of Cardinals, he insisted, since it could not be deduced from (2.5.1).[12]

In sum, Zermelo divided all of his opponents except Poincaré into two groups, those who rejected an unprovable principle (the Axiom of Choice) and those who believed Burali-Forti's paradox to lie hidden in his proof of the Well-Ordering Theorem. The first group of critics had failed to recognize that unprovable principles form the basis of every mathematical theory and that this particular principle was essential for the growth of mathematics. On the other hand, the second group did not realize that Zermelo had intended to exclude W as a set. He concluded:

> The relatively large number of critiques directed against my brief note [1904] shows that powerful prejudices bar the way to the theorem that every set can be well-ordered. Still, the fact remains that despite detailed examination, for which I am indebted to all my critics, no one could detect a *mathematical error* in my proof, and further, that the objections raised against my *postulates* contradict each other and, so to speak, cancel out. These facts permit me to hope that in time all of this resistance may yet be overcome through adequate clarification. [1908, 128]

Thus Zermelo responded to his critics by asserting that his proof did not involve Burali-Forti's paradox and that his Axiom was both self-evident and necessary for mathematics. The Axiom's self-evidence, he believed, was corroborated by the fact that, independently of each other, many mathematicians had used it implicitly. As grounds for its necessity, he cited its applications in set theory and analysis. Finally, by rejecting the collection W of all ordinals as a set, Zermelo successfully circumvented Burali-Forti's paradox, and unveiled part of his axiomatization for set theory.

Yet Zermelo's response did not persuade his opponents any more than his new proof, and their later critiques generally failed to refer to his article. Nor are there reasons for believing that his article caused the controversy over the Well-Ordering Theorem to diminish, as van Heijenoort [1967, 183] seems to imply. The peak of the debate was reached during 1905–1906, and by 1907 only Lebesgue, Hobson, Jourdain, and Richard remained actively involved. In the decade following 1908, Borel, Jourdain, Hadamard, Lebesgue, and Russell argued sporadically about the Axiom's validity (see 3.6). This decline in the intensity of the debate was only natural, since opinions had crystallized.

---

[12] *Ibid.* It is not known if (2.5.1) is weaker than the Well-Ordering Theorem.

As the controversy gradually drew new mathematicians into it, it became more technical. Little by little, the number of mathematicians who exploited the Axiom as a legitimate mathematical tool increased. Such supporters tended to adopt the cautious but useful position of stating when a proof depended on the Axiom—an attitude which Zermelo himself was to propose in his axiomatization.

3.2 *Zermelo's Axiomatization of Set Theory*

During the last quarter of the nineteenth century, the use of the axiomatic method in geometry grew considerably more sophisticated as mathematicians attempted to eliminate the role of intuition in proofs. Moritz Pasch [1882] insisted that it must be possible, once the axioms of geometry have been selected, to execute all proofs without referring to the physical meaning of the concepts. Pasch's work on axiomatization was continued by Peano and his colleagues at Turin such as Pieri. In Germany the axiomatic movement culminated in Hilbert's *Grundlagen der Geometrie*, which contained a system of axioms for Euclidean geometry that had no gaps to be bridged by implicit assumptions. As the basis for his system, he introduced three sets of objects (which he called points, lines, and planes) and certain relations on these sets. By using non-standard interpretations, he investigated the independence of his axioms and established their consistency relative to that of the real numbers.[1] Each of these aspects of Hilbert's *Grundlagen*—the use of a domain of objects with a primitive relation, the explicit statement of *all* assumptions as axioms, and the concern with their independence and consistency—was to influence Zermelo's axiomatization of set theory.

Before examining Zermelo's system in detail, we must see to what extent other mathematicians had recognized that set theory should be axiomatized and what postulates they had proposed. Apparently, the first mathematician to suggest the desirability of such an axiomatization was Peano's colleague Burali-Forti. While systematizing the theory of Dedekind-finite sets within Peano's symbolic logic, Burali-Forti wrote:

> We ought to consider the concepts of *class* and of *correspondence* as *primitive* (or *irreducible*) and to assign to them a system of properties (*postulates*) from which it is possible to *deduce* logically all the properties that are usually attributed to these concepts. At present, such a system of postulates is unknown. [1896, 36]

In this spirit he introduced the Partition Principle (lacking the condition of disjointness) as a postulate. Later the same year, in order to deduce the Trichotomy of Cardinals, he suggested a second postulate: the proposition

[1] Hilbert 1899, 1–50.

(1.5.6) that for any uncountable classes S and T there exists a function $f: S \to T$ which is either one-one or onto.[2] It is significant that Burali-Forti expressed his views on axiomatizing set theory, and proposed the first postulates, before he knew of any set-theoretic paradoxes. For it shows that the recognition that set theory needed to be axiomatized arose independently of and prior to those paradoxes.

In two other cases, this need was acknowledged before Zermelo's system appeared in print. Grappling with the Continuum Problem, J. König introduced two axioms, the first of which was psychological and the second metamathematical [1905, 158]:

(1) There are mental processes satisfying the formal laws of logic.
(2) The continuum $\mathbb{R}$, treated as the totality of all sequences of natural numbers, does not lead to a contradiction.

Schoenflies, on the other hand, argued that the Trichotomy of Cardinals should be adopted as a postulate in preference to the "stronger" Axiom of Choice. Furthermore, he regarded the Principle of Comprehension as essential to any future axiomatization for set theory. In the final analysis, he believed, the foundations of set theory must be psychological [1908, 37]. Intriguingly, in all three cases, these postulates were introduced to treat questions closely related to the Axiom of Choice.

Unlike the mathematicians discussed above, Cantor did not conceive of set theory in terms of an axiomatic system and rarely in terms of individual axioms. Indeed, though he had originally proposed the Well-Ordering Principle as a law of thought within logic, by 1895 he believed it to require a proof. The only occasion on which he stated a specifically set-theoretic axiom (in fact, a metamathematical one) was his letter of 28 August 1899 to Dedekind. In it Cantor designated the consistency of each aleph as "the Axiom of Extended Transfinite Arithmetic."[3] On the other hand, there is no evidence that Cantor regarded as *postulates* certain propositions which now seem similar to parts of Zermelo's later axiomatization. Those propositions, which occurred in a letter sent to Dedekind on 3 August 1899, resemble Zermelo's Axioms of Union and Separation as well as Fraenkel's Axiom of Replacement.[4] Likewise, A. E. Harward published in 1905 an article which contained statements analogous to many of Zermelo's axioms, but, as the author has discussed elsewhere, Harward believed himself to possess a formal definition of set "which obviates the necessity for assuming any axiom."[5] Thus he considered his approach to be fundamentally anti-axiomatic. It remained for Zermelo, influenced by Hilbert, to develop a full-scale axiomatization for set theory.

---

[2] Burali-Forti 1896, 46; 1896a, 236.

[3] Cantor 1932, 447–448.

[4] *Ibid.*, 444.

[5] Harward 1905, 459; see Moore 1976. This definition was never published.

In order to grasp Zermelo's system and its relation to the Axiom of Choice, it will be useful to re-examine Russell's article of 1906. There Russell suggested three ways of avoiding the set-theoretic and logical paradoxes then known: zigzag theories, no-classes theories, and theories of limitation of size. A zigzag theory would admit as sets the extension of those propositional functions that were sufficiently simple. There would be a set of all sets and a largest cardinal, but no largest ordinal; the complement of a set would be a set. More radically, a no-classes theory would not assume the existence of sets but would operate instead with propositional functions and substitutions. Such an approach required one to abandon much of Cantor's theory of the transfinite and might not yield even the existence of aleph-one. But a no-classes theory gave considerable security from paradoxes, and for Russell such security was a *summum bonum.*[6] Indeed, he felt most at home with a no-classes theory—until September 1906 when he began elaborating his earlier theory of types.[7]

A theory of limitation of size would, Russell granted, preserve more of the transfinite than would a zigzag or no-classes theory. The only theory of limitation of size that Russell mentioned was due to Jourdain, who had used the class W of all ordinals to distinguish between consistent and inconsistent sets (see 1.6). Despite the apparent plausibility and simplicity of such a theory, Russell regarded it with scepticism. Above all, he objected to the fact that a theory like Jourdain's failed to indicate how far the series of ordinals extended. To decide the question, Russell believed that new axioms would be needed:

> It is no doubt intended by those who advocate this theory that all ordinals should be admitted which can be defined, so to speak, from below, *i.e.*, without introducing the notion of the whole series of ordinals. Thus they would admit all Cantor's ordinals, and they would only avoid admitting the maximum ordinal. But it is not easy to see how to state such a limitation precisely . . . . The merits of this theory, therefore, would seem to be less than they at first appear . . . . [1906, 44]

Nevertheless, the no-classes theory then favored by Russell did not specify precisely which ordinals existed either, and so Russell's criticism remained inconclusive.

During July 1907, unaware of Russell's paper [1906], Zermelo completed the article [1908a] containing his axiomatization, which in Russell's terminology was a theory of limitation of size. In a manner analogous to Hilbert's *Grundlagen*, Zermelo began with a domain (*Bereich*) $\mathfrak{B}$ of objects, among which were the sets, and with the relation $\in$ of membership which held between certain objects in $\mathfrak{B}$. While he used Peano's symbol $\in$ for the membership relation, he adopted Schröder's notation $\Subset$ for the subset relation in preference to Peano's $\supset$. Zermelo stated seven axioms, which he

[6] Russell 1906, 37–40, 45–47.

[7] Russell's letter of 10 September 1906 to Jourdain in Grattan-Guinness 1977, 91.

believed to be independent of each other, but which he confessed himself unable to prove consistent:

I. (Axiom of Extensionality) If, for the sets S and T, $S \underline{\in} T$ and $T \underline{\in} S$, then $S = T$; that is, every set is determined by its members.
II. (Axiom of Elementary Sets) There is a set with no elements, called the empty set, and for any objects a and b of $\mathfrak{B}$, there exist sets $\{a\}$ and $\{a, b\}$.
III. (Axiom of Separation) If a propositional function $P(x)$ is definite (*definit*) for a set S, then there is a set T containing precisely those elements x of S for which $P(x)$ is true. (For Zermelo, a propositional function $P(x)$ was *definit* for a set S if the membership relation on $\mathfrak{B}$ and the laws of logic determined whether $P(x)$ held for each x in S.)
IV. (Power Set Axiom) If S is a set, then the power set of S is a set.
V. (Axiom of Union) If S is a set, then the union of S is a set.
VI. (Axiom of Choice) If S is a disjoint set of non-empty sets, then there is a subset T of the union of S which has exactly one element in common with each member of S.
VII. (Axiom of Infinity) There is a set Z containing the empty set and such that for any object a, if $a \in Z$, then $\{a\} \in Z$.[8]

Zermelo devoted the remainder of his article to developing the theory of cardinals within his system or, more precisely, the theory of equipollence, since he avoided the use of cardinals. As he recognized, the Frege-Russell definition of a cardinal, as an equivalence class under the relation of equipollence, was incompatible with his system. Yet any other definition would have required an additional axiom such as his later Axiom of Foundation (see 4.3). He endowed his Axioms of Choice and of Infinity with a special status by always noting when one of them was used to prove a theorem. From Axiom VI, which was actually the Multiplicative Axiom, he derived what he termed the General Principle of Choice (*Allgemeines Auswahlprinzip*): For each family T of non-empty sets, there is a function f such that $f(S) \in S$ for every S in T. This principle was his original form of the Axiom, given in 1904. Using the Axiom, he deduced that, in effect, the sum and product of any infinite number of cardinals are well-defined.

Zermelo's final application of the Axiom was a new result which is often, though inaccurately, called König's Theorem. Here $A \prec B$ means that the set A is equipollent to a subset of the set B but not conversely:

(3.2.1) Suppose that T and T' are families of disjoint sets, that $f: T \to T'$ for some bijection f, and that for each M in T, $M \prec f(M)$. Then $\bigcup T \prec P$, where P is the product (multiplicative class) of T'.

[8] Zermelo 1908a, 262–267. These names for the axioms were his own.

In terms of cardinals, (3.2.1) stated that if for every t in T, $\mathfrak{m}_t < \mathfrak{n}_t$, then $\sum_{t\in T} \mathfrak{m}_t < \prod_{t\in T} \mathfrak{n}_t$. From (3.2.1) Zermelo deduced the denumerable special case (2.1.2) which Julius König had found in 1904. Many years later it was established that (3.2.1) and the Axiom are equivalent.[9]

Zermelo's particular choice of axioms merits analysis. Since he intentionally refrained from discussing the origin of his axioms, any reconstruction is necessarily tentative [1908a, 261]. Certainly these axioms were consonant with set theory as propounded by Cantor and Dedekind, neither of whom took an axiomatic approach to set theory. Dedekind had stated, as a fact rather than as an axiom, that two sets with the same elements are equal [1888, 2–4]. By proposing his Axiom of Extensionality, Zermelo adopted an extensional view of sets similar to Dedekind's and Peano's.[10] In an 1899 letter to Dedekind, Cantor mentioned a proposition which closely resembled Zermelo's Axiom of Union, but there is no reason to believe that Zermelo had access to this letter by 1908.[11] The Axiom of Elementary Sets may have been motivated by a desire to recognize the empty set to be a legitimate set, as did Schröder but not Frege, and to distinguish a set A from the singleton $\{A\}$, as did Peano but not Dedekind or Schröder. While criticizing Zermelo's proof, Baire had argued that if an infinite set is given it does not follow that all its subsets are given.[12] Baire's claim may have prompted Zermelo to assert the opposite in his Power Set Axiom. Lastly, his Axiom of Separation, which acted in good part as a substitute for Cantor's definition of set (in effect, the Principle of Comprehension), soon generated intense criticism.[13]

The prehistory of Zermelo's Axiom of Infinity can be given in more detail. In 1851 Bolzano had argued that there exists an infinite set, as follows: Consider the proposition that there are truths, and call it A. Then the proposition B, that "A is true," is different from A and also different from the proposition "B is true"; continuing in this way, one obtains an infinite set of propositions [1851, 13–14]. Three decades later, Dedekind transformed this argument into a more psychological one: The totality of all objects of one's thought is Dedekind-infinite, since for any thought A there is also the thought of A, which is distinct from A [1888, 17].

---

[9] Rubin and Rubin 1963, 75–77. Zermelo discovered (3.2.1) between August 1904 and early in 1906, for it occurs on a left-hand page in the notebook discussed in footnote 15 below.

[10] Whitehead, writing to Russell on 25 April 1904, proposed for inclusion in *Principia Mathematica* a postulate that is essentially the Axiom of Extensionality. This letter is kept in the Russell Archives.

[11] Cantor 1932, 444. There is some reason to believe, however, that Zermelo may have read the similar letter of 1896 or 1897 from Cantor to Hilbert; see also footnote 1 of 1.6.

[12] Baire *et alii* 1905, 264. On the relationship between Schröder's and Frege's views of the empty set, see Frege 1895, 437.

[13] See 3.3. In the letter to Dedekind mentioned above, Cantor had stated non-axiomatically that any submultitude (*Teilvielheit*) of a set is a set, but did not specify how to determine a submultitude; see Cantor 1932, 444. In 1905 Lebesgue stated a rudimentary version of Separation, viewing it (incorrectly) as Cantor's definition of a set; see Lebesgue in Baire *et alii* 1905, 265.

Dedekind's argument met with disapproval from several mathematicians. While investigating Dedekind-finite classes, Burali-Forti expressed his dissatisfaction with it, and instead took the existence of an infinite class as a hypothesis wherever necessary [1896, 38]. The following year he again stated that the existence of some infinite class must be assumed, although he did not elevate this presupposition to the status of a postulate, as he had done to the Partition Principle.[14] Independently of Burali-Forti, Keyser criticized Dedekind's argument, together with a similar one that Russell had offered in the *Principles*. Indeed, Keyser expressly assumed what he called the axiom of infinity: The set of possible repetitions of an act, mentally performable once, is infinite [1904, 551]. Russell, who then believed that mathematics did not require any such non-logical assumption, argued against adopting Keyer's axiom and especially against its non-logical character [1904]. When Keyser [1905a] remained unpersuaded, their discussion came to an end. Zermelo, it should be noted, was unacquainted with the articles of Burali-Forti and Keyser. Rather, Zermelo responded directly to Dedekind's argument and rejected it since the domain $\mathfrak{B}$ was not a set in his system [1908a, 265]. As an alternative, he proposed his Axiom of Infinity, which was more precise than Burali-Forti's assumption and lacked the psychologism of Keyser's postulate.

Thanks to material in Zermelo's *Nachlass*, we can see how his system of axioms developed into the published version [1908a]. Zermelo remarked that he had already intended to restrict the concept of set when he gave his 1904 proof [1908, 119]. In the earliest surviving draft of his axiom system, written in a notebook around 1905, such a restriction took the general form of his Axiom of Separation.[15] His second axiom asserted that $M \notin M$, and his third that "if M is any well-defined set of elements m, then there is an element m' which does not belong to the set M." It appears that this third axiom was directed against objections by Bernstein and Schoenflies to Zermelo's extension of a set in his 1904 proof.

Later in the same notebook there occurred a second draft of his system, entitled "Axioms for *Definit* Sets":

I. A well-defined set never contains itself as an element. $M \notin M$.

II. A single element m_0 defines a set $\{m_0\} \neq m_0$. If M is a well-defined set and if m' is any further element that does not occur in M (*e.g.*, M itself!), then $M \cup \{m'\}$ is a well-defined set.

III. If M is a well-defined set and E is any property, which an element m of M can possess or not possess, without any arbitrariness being

---

[14] Burali-Forti 1897, 162n. Pieri [1906, 207] cited Burali-Forti [1896] when adopting the existence of an infinite set as a postulate.

[15] This and the following axioms occur in the notebook, part of Zermelo's *Nachlass*, where he outlined the lectures that he gave on set theory at Göttingen in 1901. The lectures are in shorthand on the right-hand pages, while the axioms and other notes on the left-hand pages were written later. They appear to date from the period 1904–1906.

possible, then the elements m' which have the property E form a well-defined subset M' of M, as does the complementary set M''.

IV. Likewise, the totality of *all* subsets of M forms a well-defined set, as does the set of all those subsets M' which have a given well-defined property.[16]

Here the first postulate eliminated a "set" of all sets, while the second approximated both the Axiom of Elementary Sets and a restricted form of the Axiom of Union. His earlier axiom, stating that every set can be extended, no longer appeared, but it followed easily from postulate II. His Axiom of Separation, now postulate III, became more clearly an attempt to make Cantor's notion of "well-defined set" precise. Finally, Zermelo introduced a version of the Power Set Axiom.

In 1906 his system achieved a stable form. Writing to Poincaré, Zermelo offered to send a summary of the axiomatization. In his reply of 16 June, Poincaré expressed interest but declined the offer for lack of time. Meanwhile, Johannes Mollerup, an assistant at the Polytechnical Institute in Copenhagen, had spoken on 12 June to the Göttingen Academy of Sciences about his proposed substitute for Cantor's definition of a set. A week later Zermelo lectured at the same institution about his own

> attempt to found set theory axiomatically. As a basis he takes a domain of distinct "objects," some of which must be sets and some, indivisible objects. The property that an object a is an "element" of a set M is treated as a primitive fundamental relation. Concerning sets and their elements, he proposes eight postulates, from which all the principal theorems of set theory, including those about equipollence and well-ordering, can be rigorously deduced. However, in order to obtain the order-type of the natural numbers, we must postulate the existence of some "infinite" set. Arithmetic and the theory of functions require this last postulate, while "general set theory," which treats finite and infinite sets in the same fashion, can dispense with it.[17]

Here he referred for the first time to urelements (also known as individuals or atoms), *i.e.*, non-sets which have no members but which can belong to a set.

During the summer of 1906, Zermelo gave a course entitled "Set Theory and the Number Concept" at Göttingen, and began with lectures on his system of nine axioms. Of these, the last seven were essentially the ones given

---

[16] In German, Zermelo's four axioms read: "I. Eine wohldefinirte Menge enthält niemals sich selbst als Element. $M \notin M$. II. Ein einziges Element m_0 definirt eine Menge $\{m_0\} \neq m_0$; ist M eine wohldefinirte Menge und m' ein beliebiges weiteres Element, das in M nicht vorkommt (z.B. M selbst!), so bildet auch (M, m') eine wohldefinirte Menge. III. Ist M eine wohldefinirte Menge und E irgend eine Eigenschaft, die einem Element m von M zukommen oder nicht zukommen kann, ohne das noch eine Willkür möglich ist, so bilden die Elemente m' welche die Eigenschaft E haben eine wohldefinirte Menge, eine Teilmenge M' von M, sowie der Komplementive M''. IV. Auch die Gesamtheit *aller* Teilmengen von M bildet selbst eine wohldef. Menge, ebenso alle diejenige Teilmengen M' welche eine wohldefinirte Eigenschaft besitzen."

[17] **DMV 15** (1906), 406–407.

later in his article [1908a]. The sole exception was the Axiom of Infinity, which in 1906 stated only that there exists a Dedekind-infinite set. On the other hand, the first of the nine axioms, his "Axiom of Definiteness," asserted that for any two objects a and b it is *definit* whether $a = b$ or $a \neq b$, while the second, his "Axiom of Sets," stated that some objects are sets.[18] In his article [1908a] the first axiom was absorbed into his definition of *definit* property, and the second became his notion of "domain." Finally, in 1906 he discarded his original axiom I ensuring that $M \notin M$, presumably because he was no longer concerned about Russell's paradox.

What motivated Zermelo to formulate his axiomatization? Traditionally, it has been regarded as a response to the set-theoretic paradoxes—an attempt to place set theory on a secure foundation.[19] Such a view contains a morsel of truth but ignores essential differences between Zermelo and his contemporaries. The spectrum of views can best be seen by examining three publications that appeared in the same year: Russell [1908], Hausdorff [1908], and Zermelo [1908a].

Preoccupied with the paradoxes whose solution he had long sought, Russell [1908] began with a lengthy analysis of all those known at the time. From this analysis he concluded that each such paradox violated his Vicious Circle Principle (which prohibited impredicative definitions) and could be avoided by his theory of types. Privately in a letter of 1908 to Jourdain[20] and publicly two years later in the preface to *Principia Mathematica*, Russell held that *only* some version of the theory of types could circumvent the paradoxes. Even though his article of 1908 explicitly set aside the philosophical questions raised by the theory of types, questions that he intended to discuss on some other occasion, Russell's philosophical background permeated his approach to the foundations of set theory. Among the seven antinomies that he analyzed at length in his article was the ancient paradox of the Cretan Liar, to which Zermelo [1908a] did not even allude. Russell believed that his own theory had "a certain consonance with common sense which makes it inherently credible" [1908, 222]. Two years earlier, he had contended that all the paradoxes of the transfinite belonged to logic and would be solved only by modifying the current logical assumptions [1906, 37]. Similarly, in 1908 he began by referring to the theory of types as a theory of the underlying symbolic logic, rather than as a specifically mathematical system.

At the other end of the spectrum was Hausdorff, whose article of 1908 laid the foundations for the theory of ordered sets—within mathematics, not

---

[18] Zermelo's lecture notebook of 1906 is kept in his *Nachlass* at the University of Freiburg im Breisgau.

[19] See, for example, Beth 1959, 494; Bourbaki 1969, 47–48; Kline 1972, 1185; Quine 1966, 17; and van Heijenoort 1967, 199.

[20] Russell's letter of 15 March 1908 in Grattan-Guinness 1977, 109.

logic. Fundamentally pragmatic in approach, Hausdorff avoided such self-contradictory sets as those containing all the ordinals or all the cardinals. Otherwise, he ignored the paradoxes. As a working mathematician, he had a certain contempt for the lengthy discussions of his contemporaries on the foundations of set theory. Rather, he argued, one should develop the young discipline of set theory itself.[21] Admirably he fulfilled his own directive. His anti-foundational attitude was evident even in his famous textbook *Grundzüge der Mengenlehre*, where he devoted only two sentences to Zermelo's axiomatization [1914, 2].

Zermelo's position lay between Russell's and Hausdorff's. Unlike Russell, in 1908 Zermelo was not preoccupied with the paradoxes (popular misconception to the contrary), but with the reception of his 1904 proof. Yet, like Russell, he realized that Cantor's theory of the transfinite needed some thoughtful pruning and, specifically, that Cantor's definition of a set—as any totality of well-distinguished objects of our thought—was untenable. Both Hausdorff and Zermelo considered set theory to be a part of *mathematics*, rather than of philosophy or logic.[22] However, in contrast to Hausdorff, Zermelo regarded an axiomatization as essential to the sound future development of set theory.

What role did the paradoxes play in Zermelo's axiomatization? To answer this question, we must bear in mind the close relationship between Zermelo's two papers [1908] and [1908a]. Both were completed at Chesières, France, during 1907—the former on 14 July, the latter on 30 July. Moreover, there are numerous internal links connecting them. Zermelo's first paper, while devoted to a new proof of the Well-Ordering Theorem and to refuting critics of his 1904 proof, explicitly used three of the seven postulates from his [1908a] axiomatization: the Axiom of Separation, the Power Set Axiom, and the Axiom of Choice. In addition he alluded to a fourth, his Axiom of Infinity.[23] Three times he mentioned his forthcoming article [1908a], where he referred in turn to the earlier paper [1908] for a justification of the Axiom of Choice.[24] Thus his second article was at least partially written before the first one was completed. But the link between the two papers is even stronger: In conception and motivation they were a unit.

Both articles corroborate the fact that Zermelo saw the threat of the paradoxes, from a mathematical standpoint, as more apparent than real. In his first article, the paradoxes served merely as a club with which to bludgeon critics of his 1904 proof. He used Russell's paradox to show that Peano's *Formulaire* was inconsistent, and he employed the Burali-Forti paradox to refute objections by Bernstein, Jourdain, and Schoenflies.[25]

---

[21] Hausdorff 1908, 435–437.

[22] Zermelo 1908a, 261; *cf.* 1908, 115–116.

[23] Zermelo 1908, 107, 110, 125.

[24] *Ibid.*, 110, 115–116, 124; *cf.* 1908a, 266.

[25] Zermelo 1908, 115, 118–127.

Yet in Zermelo's second article the situation was more subtle. In order for his axiomatization to be satisfactory, he knew that it must circumvent all known paradoxes of the transfinite. He felt certain that one must abandon Cantor's definition of set; instead, one must postulate certain initial sets and then state principles for producing new sets from given ones, as he proceeded to do.[26]

In his second paper, Zermelo devoted minimal space to discussing the paradoxes—less than a paragraph in three separate places. Although he mentioned in passing that his Axiom of Separation did not permit the usual derivation of the set of all sets, the set of all ordinals, or Richard's paradox, he did not bother to show why the usual arguments for such paradoxes could not be carried out in his system—a sharp contrast with Russell [1908] and a sign that Zermelo considered the paradoxes of limited concern. Instead of refuting Russell's paradox, Zermelo compelled it to do mathematical work by transforming it into the theorem that every set M has a subset S which is not a member of M [1908a, 264]. In fact, Zermelo had independently discovered Russell's paradox, several years prior to its publication by Russell in 1903, and had communicated it to Hilbert.[27] But Zermelo did not publish it. This too suggests strongly that the paradox was much less compelling to him than it was to Russell, perhaps because Zermelo was more mathematically and less philosophically inclined.

If the paradoxes did not motivate Zermelo to axiomatize set theory, then what did? As we have seen, between 1904 and 1907 Zermelo's proof had been subjected to numerous detailed criticisms by many eminent mathematicians. Throughout his career, controversy spurred Zermelo to his greatest efforts, and this polemical aspect of his personality is essential to understanding what motivated his axiomatization.[28] True, he wanted to create a firm foundation for set theory, which in turn would serve as the basis for all of mathematics. But the evidence shows that he regarded the paradoxes as only an apparent threat.

What *was* threatened, however, was his proof that every set can be well-ordered. How could he secure this proof? His answer was two-fold. First, he replied to his critics at length and gave a new demonstration, which still depended as heavily as the first on the Axiom of Choice. Second, he created a rigorous system of axioms for set theory and embedded his proof within it. He knew that in order to preserve the entire proof, such a system had to include his Axiom of Choice. Thus his axiomatization was primarily motivated by a desire to secure his demonstration of the Well-Ordering Theorem and, in particular, to save his Axiom of Choice. For Zermelo, in profound

---

[26] Zermelo 1908a, 261; *cf.* 1908, 124. He elaborated this idea two decades later in his cumulative type theory; see 4.9.

[27] Zermelo 1908, 118–119. Rang and Thomas [1981] have analyzed what is known about Zermelo's discovery.

[28] See, for example, the description of Zermelo's character in Fraenkel 1967, 149.

contrast to Russell, the paradoxes were an inessential obstacle to be circumvented with as little fuss as possible.

3.3 The Ambivalent Response to the Axiomatization

Zermelo's system was not well received at first. In view of the diversity of set-theoretic principles which had been advocated during the debate over the Axiom, it was reasonable that mathematicians did not immediately acknowledge Zermelo's choice among such principles as a natural and correct one. However, in contrast to the Axiom, almost no one applied Zermelo's system. Two mathematicians used it briefly in the theory of well-ordering, but no one obtained an important new theorem within his system until 1915, when Hartogs established that the Trichotomy of Cardinals is equivalent to the Axiom (see 3.4). Furthermore, the fundamental questions of the consistency and independence of Zermelo's axioms remained uninvestigated. Instead, several of Zermelo's previous critics objected to the Axiom of Separation and especially to the notion of *definit* property. Both Russell and Poincaré doubted the consistency of Zermelo's system, but did not succeed in exhibiting a contradiction within it. At the time no one noticed, as Fraenkel and Skolem were to do later, that the system failed to guarantee the existence of sets even as large as $\aleph_\omega$ (see 4.9). Nevertheless, a modified version of Zermelo's axiomatization would eventually become synonymous with set theory.

Jourdain and Russell exchanged their objections to Zermelo's system early in 1908. Russell regarded the system as a theory of limitation of size since it disallowed collections, such as that of all ordinals, which were in some sense too large. While Jourdain's letter to Russell on this subject has not survived, Russell's reply of 15 March illuminates the views of both of them:

> I have only read Zermelo's article [1908a] once as yet, and not carefully, except his new proof of [the] Schröder–Bernstein [Theorem], which delighted me. I agree with your criticisms of him entirely. I thought his axiom [of separation] for avoiding illegitimate classes so vague as to be useless; also, since he does not recognise the theory of types, I suspect that his axioms will not really avoid contradictions, *i.e.*, I suspect new contradictions could be manufactured specially designed to be consistent with his axioms. For I feel more and more certain that the solution [to the paradoxes] lies in types as generated by the vicious circle principle, *i.e.*, the principle "No totality can contain members defined in terms of itself". . . .
>
> But I think Zermelo a very able man, and on the whole I admire his work greatly. I am much obliged to you for writing to him about me and referring him to my articles.
>
> I am going to Rome for the mathematical congress in April; I wonder whether he will be there.[1]

[1] Russell in Grattan-Guinness 1977, 109.

Eight days later Russell wrote Zermelo a cordial letter to thank him for sending his two papers [1908, 1908a], and enclosed "two articles that I have written about the contradictions in set theory."[2] Russell added that in a forthcoming paper [1908] he had shown the Axiom of Choice to be equivalent to the Multiplicative Axiom. However, his letter did not discuss Zermelo's system.

Zermelo presented a paper [1909a] on the theory of finite sets to the Rome Congress, where, as he wrote Cantor on 24 July 1908, he spoke at length with both Russell and Peano.[3] We may only wonder what aspects of Cantor's set theory, Zermelo's axiomatization, Russell's theory of types, and the Axiom of Choice they discussed. In any case, neither Russell nor Zermelo influenced the other's outlook substantially. Although Russell failed to manufacture a contradiction in Zermelo's system, he remained convinced that only some version of his theory of types would eliminate the paradoxes of logic and set theory. This faith found expression in his masterwork, *Principia Mathematica*, written jointly with Whitehead.[4]

In 1909, after objecting to Russell's theory of types on the grounds that it presupposed the theory of ordinals, Poincaré criticized Zermelo's axiomatization. According to Poincaré [1909a], Zermelo had proposed a system of *a priori* axioms as the basis from which to derive all mathematical truths. If Zermelo considered his axioms to be an implicit definition of the term "set," then he needed to prove them consistent—as Hilbert had done for geometry. Yet Zermelo did not and could not do so, Poincaré added, since he would have to rely on previously established truths. This was impossible because he wished his axioms to stand entirely by themselves as a foundation for mathematics [1909a, 473]. On the other hand, Poincaré overlooked the fact that Zermelo had regarded a consistency proof as a deep problem for future investigation.[5]

Since Zermelo's postulates did not suffice as a definition of set, Poincaré continued, then they must be self-evident, a question which he considered

[2] This letter can be found in Zermelo's *Nachlass* at the University of Freiburg; the two articles are probably Russell 1906 and 1906a. The *Nachlass* also contains a brief card from Dedekind, dated 18 December 1907, and a similar card from Frege, dated 29 December 1907, both thanking Zermelo for sending his articles on set theory—presumably [1908] and [1908a]. Dedekind and Frege made no substantive comments, and no other correspondence from them exists in the *Nachlass*. It appears that these two articles were in print by December 1907, unless Zermelo sent page proofs.

[3] Zermelo in Meschkowski 1967, 267.

[4] Russell and Whitehead 1910, vii. Zermelo's *Nachlass* contains notes by Kurt Grelling of the course on mathematical logic that Zermelo gave at Göttingen during the summer of 1908. In the first lecture Zermelo asserted that the chief question was to what extent mathematics is "a logical science." He carefully took the middle ground between those, from Leibniz to Peano and Russell, who affirmed that arithmetic is part of logic, and those, from Kant to Poincaré, who claimed that it is not. In conclusion, Zermelo argued that logic should be developed in terms of Hilbert's axiomatic method.

[5] Zermelo 1908a, 262.

to be primarily psychological. Nevertheless, Zermelo accepted certain postulates but rejected others that Poincaré found just as intuitively evident as those retained. Since Zermelo had abandoned Cantor's definition of set, from where, Poincaré inquired, would the intuition of set arise?[6]

In the course of listing Zermelo's seven axioms, Poincaré left the word for set as *Menge* rather than translating it into French. He did so because he doubted that these axioms preserved the intuitive meaning of the word "set." To be sure, he noted, the first six axioms were self-evident—if one understood *Menge* intuitively but considered only a finite number of objects. Under this intepretation he found another axiom to be equally self-evident: (8) Every collection of objects forms a *Menge*. But Zermelo explicitly excluded the latter axiom. Why, Poincaré asked, did this eighth axiom (essentially the Principle of Comprehension) cease to be self-evident for infinite collections whereas Zermelo's first six axioms did not?[7]

In point of fact, Poincaré denied the existence of the actual infinite [1909a, 463], an attitude essential for understanding his criticism of Zermelo's system: "If a *Menge* M has infinitely many elements, it does not mean that these elements can all be conceived as existing together in advance, but that new ones can endlessly come into being from it" [1909a, 477]. It was precisely because Poincaré was thinking in terms of the potential infinite that he discarded axiom (8). For, although any objects whatever formed a collection, he believed that such a collection need not be well-defined and might be altered by the adjunction of unexpected elements.

On the other hand, Poincaré objected to the Axiom of Separation because of the term *definit* occurring in it. A proposition that was *definit*, he insisted, might not be well-defined. As an example, he supposed that a certain proposition $P(x)$ held for x if x satisfied a given relation with respect to all members of a set M. If M was infinite, one could always introduce into M new elements which would alter whether x satisfied $P(x)$ or not. Here again, the potential infinite was vital to his argument, but he probably had in mind Richard's paradox as well. Furthermore, he criticized Zermelo's term *definit* as vague,[8] and so unwittingly echoed the same objection advanced by Russell and Jourdain.

By way of conclusion, Poincaré dismissed Zermelo's axioms because they were not self-evident and could not be shown to be consistent:

> The attempts made to escape these difficulties [the paradoxes of set theory] are interesting in more than one respect, but they are not entirely satisfactory. Zermelo wished to construct an impeccable system of axioms. But these axioms cannot be regarded as arbitrary decrees, since one would need to prove that the decrees are not contradictory. And since Zermelo made a *tabula rasa*, there is nothing on which one could base such a proof. Thus these axioms must be

[6] Poincaré 1909a, 473.

[7] *Ibid.*, 474–475.

[8] *Ibid.*, 476–477.

> self-evident. But what is the mechanism by which they were constructed? He took axioms which are true for finite collections; he could not extend all of them to infinite collections, and so he made this extension only for a certain number of them, chosen more or less arbitrarily. Besides, as I have said previously, I believe that any proposition concerning infinite collections cannot be intuitively self-evident. [1909a, 481–482]

Later the same year Zermelo answered Poincaré in an article which deduced the principle of mathematical induction on $\mathbb{N}$ from his axiomatization:

> If these axioms [for set theory] ... are nothing more than purely logical principles, then the principle of mathematical induction is too; if, on the contrary, they are intuitions of a special sort, then one can continue to regard the principle of mathematical induction as a result of intuition or as a "synthetic *a priori* judgment." As for me, I would not dare at present to decide this purely philosophical question.[9]

After attempting to catch Poincaré on the horns of a dilemma, Zermelo presented himself as an agnostic vis-à-vis both Poincaré's Kantianism and the philosophical foundations of his own axiomatization. He preferred to discuss only the mathematical aspects of his system.

Influenced by Zermelo, two German mathematicians soon used his axiomatization to investigate the theory of well-ordering. In a paper on the foundations of transfinite ordinal arithmetic Ernst Jacobsthal, then at Berlin, took Zermelo's system as his starting point [1909, 146]. On the other hand, Jacobsthal did not refer explicitly to any of Zermelo's axioms while deducing results.

By contrast, Hessenberg relied on Zermelo's system explicitly and in detail to develop the theory of well-ordering.[10] At the same time, Hessenberg accepted the Axiom of Choice, toward which he had been ambivalent three years earlier. He based his new paper [1909] on the theory of θ-chains, which had originated in Zermelo's second proof of the Well-Ordering Theorem. Generalizing the concept of θ-chain, which Zermelo had derived from Dedekind's notion of chain, Hessenberg studied the relationship between this generalization and the Well-Ordering Theorem. On the other hand, he avoided the Axiom of Choice while establishing a result about the images of functions.[11] Apparently he felt, as Zermelo did, that one should use the Axiom only when it was necessary for a given theorem.

At Göttingen, Zermelo's system was received positively. Under the direction of Hilbert, Kurt Grelling completed his doctoral thesis on founding arithmetic upon set theory [1910]. Influenced by Russell's paradox, Grelling

---

[9] Zermelo 1909, 192. During 1907 he had corresponded with Poincaré about submitting this article to the *Revue de Métaphysique et de Morale*. In his reply of 19 June 1907, Poincaré found the paper too mathematical for readers of the *Revue* but added that he had proposed it to Mittag-Leffler for publication in *Acta Mathematica*. This is where the paper appeared two years later.

[10] Hessenberg 1909. He had already spoken approvingly of Zermelo's system in his previous paper [1908, 147] which criticized Schoenflies' 1908 report on set theory.

[11] Hessenberg 1909, 82, 86–90, 130–133, 103.

believed that the insecure foundations of logic also affected set theory and arithmetic. On the other hand, he believed that Zermelo had removed this danger from set theory: "Zermelo has set up a system of axioms from which all the essential theorems of classical set theory can be deduced, but not Russell's contradiction. It will not be investigated here whether this system is definitive" [1910, 6–7]. After accepting both the Axiom and Zermelo's proof of the Well-Ordering Theorem, Grelling offered a modest extension of the notion of *definit* property.[12]

That same year Hermann Weyl, then a young *Privatdozent* at Göttingen, made a more fundamental contribution to the development of Zermelo's system. Convinced that the system was essentially correct, Weyl attempted to modify the Axiom of Separation, but failed to obtain a definition of *definit* property that fully satisfied him. Nevertheless, he characterized a property as *definit* if it was built up from the relations $\in$ and $=$ by a finite number of unspecified "definition-principles." Until these principles had been precisely formulated, he believed, one could not hope to solve Cantor's Continuum Problem. Finally, he regretted that his revision of *definit* property presupposed the natural numbers, which, following Dedekind and Zermelo, he wished to construct from set theory itself [1910, 112–113].

In 1911, aware of Hessenberg's article but unaware of those by Poincaré and Weyl, Schoenflies objected to Zermelo's system. In contrast to Poincaré and Russell, who had criticized the Axiom of Separation on the grounds that the notion of *definit* property was vague, Schoenflies considered this notion to be merely unnecessary. Thus he argued that Zermelo's reasoning lacked meaning in the following case: $a = b$ is *definit* because it is equivalent to $a \in \{b\}$.[13] Here he failed to understand that Zermelo used the word *definit* in a technical sense to mean decidable on the basis of the membership relation which exists between the sets of the given universe. Apparently Schoenflies considered *definit* to mean simply "well-defined." In addition, he thought it a weakness that the Axiom of Separation permitted sets to be obtained only from those sets already known to exist, rather than independently, thereby restricting the domain of sets unnecessarily. Set theory, he insisted, should include *every* set definable independently as a mathematical object in a consistent way.[14] Like other set-theorists of the time, he conceived of set theory as a closed system rather than as a system which could be extended indefinitely.

Schoenflies' other objections often missed the point as well. Thus Schoenflies claimed that Zermelo had made a logical error when he deduced the theorem, that his universe (*Bereich*) of sets was not itself a set, without

[12] Grelling 1910, 9, 21.

[13] Schoenflies 1911, 227, 241.

[14] *Ibid.*, 231–232, 251. A serious difficulty with Schoenflies' dictum, but one which he did not realize, was the following: Two sets A and B might each be consistent with a given system S of set theory; but if both A and B were adjoined to S, the resulting system could be inconsistent.

referring to the universe in his axioms [1911, 242]. In point of fact, Zermelo had obtained this theorem by reformulating Russell's paradox. More generally, Schoenflies criticized what he regarded as philosophical undercurrents, due to Peano and Russell, in Zermelo's system. In particular, Schoenflies blamed Russell for the fact that Zermelo specifically allowed the possibility that $A \in A$ for some set A in his system. After chiding Zermelo for mentioning Russell's paradox at all, Schoenflies emphasized that such paradoxes need not concern mathematicians, only philosophers. He saw Hilbert's axiomatic method as a refuge from such philosophical speculation.[15]

What Schoenflies did not realize was that Zermelo had initially taken as a postulate that $A \notin A$, but had abandoned it. Much later, Zermelo introduced the Axiom of Foundation, which disallowed not only sets such that $A \in A$ but all sets that were not well-founded (see 4.9). Schoenflies proceeded only a little way in this direction when he proposed, as part of an alternative axiomatization for set theory, the postulate that there exist no sets such that $A \in B$ and $B \in A$. His five other postulates included the Axioms of Union and Extensionality, but he modeled the remainder too closely on axioms of incidence in geometry [1911, 245–249]. As an axiomatization for set theory, Schoenflies' system was quite inadequate, and attracted little interest.

In 1914 Hausdorff objected to Zermelo's axiomatization on grounds that were more equivocal than Schoenflies'. Although Hausdorff readily acknowledged set theory to be the proper foundation for all of mathematics, he remarked that no foundation had been generally accepted for set theory itself. Nevertheless, the collection of all cardinal numbers and that of all ordinal numbers could not be sets. He concluded that it was essential to restrict the process of unlimited set construction and that, in fact, Zermelo's system did so. Yet Hausdorff excluded this system from his textbook for pedagogical and other reasons:

> At present, these extremely ingenious investigations [by Zermelo] cannot be regarded as completed, and introducing a beginner to set theory by this [axiomatic] approach would cause great difficulties. Thus we wish to permit the use of naive set theory here, while still observing the strictures which bar the way to the paradoxes. [1914, 2]

Despite the fact that Hausdorff had greatly advanced the development of set theory, Zermelo would find no support for his axiomatization there.

As late as 1919, German views of Zermelo's system remained ambivalent. Defending set theory against finitism, the doctrine that mathematics should be founded solely on the natural numbers, Bernstein at first appeared to support Zermelo—as Grelling had done a decade earlier:

> The works of Zermelo, although they have given rise to the most debate, are those which, from the main point of view, offer the least reason for doing so. Zermelo sets forth a system of axioms, from which everything positive in set theory can

[15] *Ibid.*, 244, 229, 254, 222.

certainly be deduced, but none of the paradoxes which we wish to avoid. [1919, 71–72]

However, Bernstein added, this system allowed one to prove too much, since it yielded a well-ordering of the continuum. In this way the Well-Ordering Theorem, against which he had inveighed in 1905, kept him from accepting Zermelo's axiomatization.

When Abraham Fraenkel, a young *Privatdozent* at Marburg, published his textbook on set theory [1919], it differed from Hausdorff's earlier book in two respects. First of all, Fraenkel had written a truly introductory text, which contrasted with Hausdorff's compendium of recent research. Secondly, Fraenkel accepted Zermelo's system *in toto* and gave a lengthy exposition of it. Fraenkel believed the axiomatic method to be an appropriate foundation for set theory, since in this way Zermelo had excluded the paradoxes while preserving Cantor's principal results.[16] Not until two years later did Fraenkel become aware of an important set which could not be obtained within Zermelo's system (see 4.9).

What, Fraenkel inquired, could be determined about the consistency of this system? A proof for its consistency could be sought only outside mathematics, he insisted, and would be fraught with difficulties. Even a proof of relative consistency must come from logic and not from mathematics. As a mathematician, he concluded, one must be content with an axiomatic construction of set theory, such as Zermelo's, excluding the known paradoxes. For Fraenkel, the need for a final appeal to the subjective was by no means unique to set theory. Rather, it was shared by all scientific and mathematical disciplines, which were founded ultimately on confidence in the validity of certain assumptions or axioms, rooted in faith or experience.[17] Later Fraenkel's philosophical position became considerably more Platonistic.

In sum, during the first decade after Zermelo's system appeared in print, it made few converts and attracted much criticism. Not only did he fail to secure his Axiom of Choice against further objections, but his new Axiom of Separation also came under attack. The notion of *definit* property, central to the latter axiom, remained obscure to many mathematicians and urgently required clarification, which was not forthcoming. As a result, the prospects for his axiomatization looked dim, and outside Germany no one employed it at all.

In part, the deficiencies of Zermelo's notion of *definit* property can be viewed as a failure to specify and develop an acceptable underlying logic. This situation reflected the fact that no conception of mathematical logic had gained the confidence of mathematicians. Yet in 1917, Weyl had already seen how logic could express the notion of *definit* property, and five years

[16] Fraenkel 1919, 134–135.

[17] *Ibid.*, 136–137, 125–128.

later Thoralf Skolem would independently formulate this notion within first-order logic (see 4.9).

Finally, Zermelo had given no explicit rationale for his choice of axioms, except that they yielded the main theorems of Cantorian set theory. Such a rationale had to await the cumulative type hierarchy which Zermelo introduced two decades later. During the transitional decade 1909–1919, the disputed Axiom of Choice proved to be more secure than the axiomatic system that he introduced to serve as its foundation.

3.4 The Trichotomy of Cardinals and Other Equivalents

Although over one hundred propositions have been proven equivalent to the Axiom of Choice,[1] only two were publicly recognized as such by 1908: the Well-Ordering Theorem and the Multiplicative Axiom. Before the 1920s three more propositions, including the Trichotomy of Cardinals, were shown to be equivalents. This fact reflects the limited interest at the time in determining the deductive strength of various propositions relative to the Axiom of Choice, an interest which later became quite intense (see 4.3–4.5 and the Epilogue).

A different type of proposition, not proven equivalent to the Axiom until the 1930s, slowly acquired mathematical significance after 1907: the maximal principles. In effect, they asserted that under certain conditions there exists a maximal element in a given partial order, that is, an element such that no other element is strictly greater in the partial order.[2] If the partial order was that of inclusion, then the element was later termed an $\subseteq$-maximal element. In 1902, when Whitehead found that under certain conditions a set can be extended to a member of a given multiplicative class, he used the Assumption implicitly to obtain an $\subseteq$-maximal element.[3] Likewise, Zermelo's proof of 1904 involved the construction of such an $\subseteq$-maximal element. Nevertheless, the concept of maximality, though not the name, emerged as a distinct notion only with Hausdorff's researches of 1907 on ordered sets of real

---

[1] For a list of equivalents, see Rubin and Rubin [1963, 111–124] as well as Appendix 2.

[2] Although the term *maximal principles* has become standard in the recent mathematical literature, a more appropriate designation would be extremal principles. Indeed, for many maximal principles there is a corresponding minimal principle asserting the existence of a minimal element, *i.e.*, an element such that no other element is strictly smaller in the given partial order. For a history of maximal principles whose interpretation differs somewhat from the present book, see Campbell 1978.

[3] Whitehead 1902, 383; see (1.7.14).

functions. In particular, he applied the Well-Ordering Theorem to establish the existence of a pantachie, *i.e.*, a maximal set of real-valued sequences ordered according to which sequence was eventually greater.[4]

When Hausdorff gave a second proof for this result two years later, he based it on a new set-theoretic theorem which he deduced from the Well-Ordering Theorem. To state his theorem, he first supposed that S was a family of sets such that, for every limit ordinal α, if $\{B_\beta: \beta < \alpha\} \subseteq S$ and if $B_\gamma \subseteq B_\delta$ whenever $\gamma < \delta < \alpha$, then there existed some set B_α in S with $B_\beta \subseteq B_\alpha$ for every $\beta < \alpha$. In this case he termed S "extendable by limits," and also discussed the particular case when $B_\alpha = \bigcup_{\beta<\alpha} B_\beta$. Then he expressed his theorem as follows [1909, 300]:

(3.4.1) Let M be an infinite set and let S be a non-empty family of subsets of M. If S is extendable by limits, then S contains a set that is not a proper subset of any member of S.

This result, the first which clearly expressed a maximal principle, was essentially what later came to be known as Zorn's Lemma (see 4.4).

Hausdorff's theorem (3.4.1) did not receive the attention that it deserved. When he returned to it in his textbook on set theory [1914], he phrased it in two variant forms, each of which was later termed Hausdorff's Maximal Principle. However, it is important to realize that, despite the name later given to them, he did not propose either of these variants as an axiom or a principle:

(3.4.2) Every partially ordered set M has an ordered subset A that is greatest [$\subseteq$-maximal] among ordered subsets of M.

(3.4.3) Every partially ordered set M has an ordered subset A that is greatest [$\subseteq$-maximal] among ordered subsets and also includes B, a given ordered subset of M. [1914, 140–141]

Hausdorff, who deduced both variants from the Axiom of Choice, made no attempt to show their equivalence to it.[5] Nor did he apply them to obtain further results. Like their predecessor (3.4.1), these two theorems attracted little attention from Hausdorff's contemporaries. As a consequence, maximal principles were rediscovered independently several times, the most important instances occurring in articles by Kazimierz Kuratowski [1922] and Max Zorn [1935].

[4] Hausdorff 1907, 117–118; see 2.5.

[5] Contrary to the claim of Grattan-Guinness [1977, 61, 159], Hausdorff did not use a maximal principle in order to avoid the Axiom of Choice and transfinite induction.

While these rediscoveries are the subject of 4.4, Kuratowski's is of interest here because it was influenced by the earlier work of Zygmunt Janiszewski [1910] as well as by that of Brouwer [1911]. Janiszewski, a young Polish mathematician who wrote his doctoral thesis under Lebesgue at Paris, investigated the topology of Euclidean n-space. Both Janiszewski and Brouwer were particularly concerned with the properties of continua, and both applied the notion of irreducible continuum that Ludovic Zoretti had suggested: A continuum S containing points p and q is irreducible between p and q if every proper subset of S containing p and q is not a continuum.[6]

Janiszewski's principal result, which he deduced from the Well-Ordering Theorem, was that in n-space every continuum containing the given points p and q includes a continuum irreducible between p and q [1910, 198]. Later the same year another Pole, Stefan Mazurkiewicz, published a demonstration of this result that did not use the Axiom of Choice.[7] Thus, Janiszewski's result was only suggestive of a maximal principle; deductively it did not constitute a special case of one. The same is true as well of Brouwer's generalization of the result in 1911.[8] Indeed, both results could be deduced in Zermelo's system without the Axiom by a Bolzano–Weierstrass type of argument. On the other hand, Kuratowski, who was motivated like Brouwer and Mazurkiewicz by the desire to avoid transfinite ordinals, would use the Axiom in 1922 to establish a minimal principle which consciously generalized Janiszewski's and Brouwer's results but which was much stronger deductively.[9] In fact, Kuratowski's principle was later shown equivalent to the Axiom. What Janiszewski contributed was, not a special case of a minimal principle, but the terms "irreducible" and "saturated," which he expressed exactly as mathematicians now define $\subseteq$-minimal and $\subseteq$-maximal [1912, 85–86].

Unlike Janiszewski and Brouwer, who did not seek to discover equivalents to the Axiom, Bertrand Russell remained interested in the Axiom's deductive strength at the same time that he doubted its truth. In the first volume of *Principia Mathematica*, which appeared in 1910, Russell showed the Axiom equivalent to each of the following:

(3.4.4) For every class K and every relation S with $K \subseteq \text{dom}(S)$, there is a function f with $f \subseteq S$ and $\text{dom}(f) = K$.

---

[6] Zoretti 1909, 487. In a topological space, a *continuum* is a perfect set which is connected, *i.e.*, not the union of two disjoint closed sets.

[7] Mazurkiewicz 1910, 296–298. Nevertheless, his stated aim was not to eliminate the Axiom from the proof but rather to avoid the use of transfinite ordinals.

[8] Brouwer 1911, 138. Consequently we must reject Campbell's claim [1978, 78] that these two results were special cases of a maximal principle, as well as his claim [1978, 80] that Brouwer used the Well-Ordering Theorem in his proof. Such a use would have been illegitimate from Brouwer's intuitionistic standpoint.

[9] Kuratowski 1922, 88–89.

(3.4.5) For every class K and every function f with $K \subseteq \text{rng}(f)$, there is a one-one function g with $g \subseteq f$ and $\text{rng}(g) = K$.[10]

In other words, every relation includes a function with the same domain, and every function includes a one–one function with the same range. These two propositions remained largely unknown at the time, perhaps because of the general lack of interest in the Axiom's equivalents or because of the rigors of reading *Principia Mathematica.* Three decades later Paul Bernays independently stated (3.4.4) as a form of the Axiom [1941, 1–2], and later authors credited him alone with this result.[11]

In contrast to the maximal principles (which were still in their infancy until the 1930s) and to Russell's two equivalents, the Trichotomy of Cardinals had been a proposition of note for two decades when Friedrich Hartogs, an analyst at Munich, used it in 1915 to give a new proof of the Well-Ordering Theorem. Hartogs established that, in Zermelo's system without the Axiom, the Trichotomy of Cardinals implies that every set can be well-ordered. Since the 1890s mathematicians had recognized that if every set can be well-ordered, then Trichotomy holds. But the converse was not at all obvious to them. As late as 1908 Schoenflies had insisted that Trichotomy is essentially weaker than the Axiom.[12] Similarly, before he learned of the Axiom, Jourdain had claimed that Trichotomy is weaker than the Well-Ordering Theorem [1904, 68]. Yet three years later he considered Trichotomy to be equivalent to the Axiom and hence to the Well-Ordering Theorem as well. Nevertheless Jourdain gave no argument, not even a heuristic one, for this equivalence [1907, 366]. It remained for Hartogs [1915] to provide a convincing proof.

Hartogs began by establishing, in Zermelo's system without the Axiom, the lemma that no set has a power greater than the power of every well-ordered set. Since Zermelo had not developed the theory of ordered and well-ordered sets within his system, Hartogs showed that the concepts and results needed for his own proof did not involve the Axiom. Then Hartogs argued as follows, explicitly using the Axioms of Power Set and of Separation: The set M_w, consisting of all well-ordered subsets of a given non-empty set M, exists and is non-empty. Then M_w can be partitioned into a set E of equivalence classes, where two sets are equivalent, and hence in the same class, if they are order-isomorphic. It follows that E is ordered; namely, let $G < H$ for equivalence classes G and H if for every $g \in G$ and every $h \in H$, g is order-isomorphic to a segment of h. In fact, this ordering well-orders E,

[10] Russell and Whitehead 1910, 561–565. In an unpublished manuscript of 1906, Russell had conjectured that (3.4.4) and (3.4.5) were equivalent to the Axiom of Choice and the Multiplicative Axiom respectively; *cf.* (2.7.4) and (2.7.5). At that time he thought the latter to be weaker than the former.

[11] See, for example, Rubin and Rubin 1963, 6.

[12] Schoenflies 1908, 36.

since every segment $\{A \in E\colon A < G\}$ of E is order-isomorphic to the set of all segments of any g in G and so every segment of E is well-ordered. Furthermore, every well-ordered subset of M is order-isomorphic to a segment of E. Consequently, since a well-ordered set cannot be order-isomorphic to a segment of itself, E is not order-isomorphic to a subset of M. So $\overline{\overline{E}} \leq \overline{\overline{M}}$ does not hold, and the lemma follows. This lemma eliminated the possibility, entertained earlier by both Hardy and J. König, that some set could have a power greater than every aleph, and in particular that $\mathbb{R}$ might be such a set (see 1.6 and 2.1).

At this point Hartogs easily derived his principal result that the Trichotomy of Cardinals implies the Well-Ordering Theorem. For, by the lemma and Trichotomy, it followed that $\overline{\overline{M}} < \overline{\overline{E}}$ and hence that M can be well-ordered. Therefore Trichotomy, the Axiom, and the Well-Ordering Theorem were all equivalent [1915, 438].

In this fashion Hartogs established that if the infinite cardinals were to satisfy the minimum requirements for numbers, then the Axiom of Choice held. This argument in favor of the Axiom had little influence on mathematicians at the time, probably because Cantor's infinite cardinals were still under attack on other grounds. The fruitfulness of the Axiom was to prove convincing not so much in set theory, which at the time was peripheral to the interests of most mathematicians, but in algebra, which was much closer to the center.

3.5 *Steinitz and Algebraic Applications*

During the decade after 1908 abstract algebra made rapid advances, especially in field theory, by building on the classical algebra of the nineteenth century. Until 1910 the Axiom of Choice played a very minor role in these advances. The only significant exception was Hamel's proof of 1905 that there exists a Hamel basis, *i.e.*, a basis for the real numbers as a vector space over the rationals (see 2.5). Yet in his seminal study of 1910 on abstract fields, Ernst Steinitz recognized that the Axiom was necessary for many important results in field theory. In other parts of algebra the Axiom penetrated more slowly, although both Zermelo and Emmy Noether used it to investigate integral domains. Gradually the Axiom was proving fruitful in modern algebra, a trend that accelerated over the following decades as algebra grew more abstract and general.

In the introduction to his article of 1910, Steinitz, a professor at Breslau, discussed the Axiom's role vis-à-vis algebraically closed fields.[1] The first

[1] An *algebraically closed field* is a field in which every polynomial in one indeterminate can be decomposed into linear factors. The *algebraic closure* of a field F is the least algebraically closed field which includes F.

proposition where he pointed out the need for the Axiom fell within the classical theory of fields:

(3.5.1) Up to isomorphism, there is a unique algebraic closure for the field of rational numbers.

In the case of the rationals, the existence of an algebraic closure did not require the Axiom. Yet for an arbitrary field F, he insisted, the Axiom was essential in order to demonstrate not only the uniqueness of an algebraic closure for F but the existence of one as well [1910, 170]. Later research bore out his contention.[2]

These theorems led Steinitz to stress the importance of the Axiom in algebra and, indeed, in mathematics generally:

> Many mathematicians still stand opposed to the Axiom of Choice. With the increasing recognition *that there are questions in mathematics which cannot be decided without this axiom*, the resistance to it must increasingly disappear. On the other hand, in the interest of purity of method it seems expedient to avoid the above-named axiom in so far as the nature of the question does not require its use. I have endeavored to make these limits [on its use] conspicuous. [1910, 170–171]

Here he adopted a position which closely resembled the one taken by Zermelo a year earlier. In order to follow his own precept, Steinitz organized his article by first developing the theory of fields as far as possible without the Axiom and then deducing those additional theorems explicitly obtained from it.

Significantly, the most far-reaching of Steinitz' results all depended on the Axiom. In addition to those on algebraic closure mentioned above, they concerned the decomposition and extension of fields. One of them involved absolutely algebraic fields:

(3.5.2) Any two absolutely algebraic fields, which have a finite prime subfield and the same proper finite subfields, are isomorphic.[3]

On the other hand, the Axiom enabled him to give a canonical decomposition of any extension field:

(3.5.3) If a field F_2 is an extension of a field F, then F_2 can be obtained from F by a purely transcendental extension F_1 followed by an extension algebraic with respect to F_1.[4]

---

[2] Läuchli [1962, 5–7] used the Fraenkel–Mostowski method to provide a model of **ZFU** in which a certain field had no algebraic closure, while Pincus [1972, 722–723] obtained a model of **ZF** containing such a field.

[3] Steinitz 1910, 251. A field is *absolutely algebraic* if all its elements are algebraic relative to its prime subfield. A *prime field* is a field which has no proper subfield.

[4] *Ibid.*, 171, 292–293. A field F_1 is a *purely transcendental extension* of a field F if F_1 is isomorphic to a field of quotients obtained by adjoining some number of indeterminates to F. F_2 is an *algebraic extension* of F_1 if every element of F_2 is algebraic with respect to F_1.

Lastly, the Axiom allowed Steinitz to define the "transcendence degree" of a field K which extended a field F. Loosely speaking, the transcendence degree of K relative to F is the cardinality of the indeterminates that one must adjoin to F in order to obtain K. The Axiom ensured that this cardinality was invariant for given fields F and K.[5]

In all of these results Steinitz relied on the Axiom to justify the Well-Ordering Theorem, by means of which he then obtained a well-ordering appropriate to the particular result. Consequently, he was able to use transfinite induction on the elements of the given field, the essential step in his results. As Hadamard had done earlier and Fraenkel would later, Steinitz emphasized that Zermelo's proof guaranteed only the existence of a well-ordering for any set; it did not provide a means whereby such a well-ordering could actually be constructed.

Among the propositions whose dependence on the Axiom Steinitz first noticed was the following:

(3.5.4) The family of all finite subsets of an infinite set M has the same power as M.

Steinitz considered this proposition important because it implied that the cardinal number of the polynomials of one indeterminate in an infinite field was the same as the power of the field. Thus an algebraic extension did not increase the cardinality of an infinite field [1910, 301]. In this instance the Axiom was less important than he realized, since (3.5.4) can be proved without it if M is denumerable or has the power of the continuum (see 4.1).

While Steinitz used the Axiom as a tool to study fields, Zermelo applied it in a paper, written during December 1913 and published the following year, to a problem that combined integral domains with linear algebra. As a first step, Zermelo modified Hamel's proof of 1905, in which the Axiom guaranteed that a Hamel basis exists, so as to obtain a basis for the vector space of complex numbers over the rational numbers. Such a basis then permitted Zermelo to solve the problem of transcendental integers (*ganze transzendente Zahlen*), due to Edmond Maillet and others. This problem, which extended the notion of algebraic integer, asked for a set B of complex numbers satisfying the following four conditions:

(i) The sum, difference, and product of any two members of B is again in B.
(ii) Every complex number is a quotient of two members of B.
(iii) Every complex algebraic integer is in B.
(iv) No other complex algebraic number is in B.

Since no one had successfully constructed such an integral domain B, Zermelo inquired whether the four conditions were consistent. In fact

[5] *Ibid.*, 288–289, 299.

they were, he replied, since the existence of such an integral domain followed from the existence of a well-ordering for the complex numbers and hence from the Axiom [1914, 434–435].

Emmy Noether, who elaborated on Zermelo's ideas during March 1915, while at Erlangen after finishing her doctorate, classified all possible integral domains B that satisfied the conditions (i)–(iv). Noether too began her investigations with a modified Hamel basis. Accepting Zermelo's proof of the Well-Ordering Theorem in a matter-of-fact way, she obtained from it a representative set C of integral domains such that any integral domain B satisfying (i)–(iv) was isomorphic to a unique member of C.

Noether believed Zermelo to have "constructed" (*konstruiert*) an integral domain B satisfying (i)–(iv),[6] whereas Zermelo considered himself to have shown only the consistency of these conditions. For, he stressed, whether or not a complex number was a transcendental integer in B depended on the Hamel basis employed, and hence on a particular well-ordering of the complex numbers; yet no such well-ordering had been specified uniquely.[7] Clearly Zermelo's concern with the distinction between a constructive proof and an abstract proof of existence was not shared by all of those who applied his results. Moreover, as Noether's comment illustrates, the notion of a construction remained broad and rather ill-defined for many mathematicians.

By 1916 Dénes König, who had been sceptical of the Axiom eight years earlier (see 2.6), applied it without hesitation in the theory of graphs—a discipline then emerging from set theory on the one hand and from algebra on the other. Apparently his acceptance of the Axiom stemmed from his father's decision to accept it as well (see 3.6). From the Axiom, Dénes König deduced that if an infinite graph of second degree is a pair graph, then it is the disjoint union of two graphs of first degree.[8] In a similar fashion he established that the vertices of an infinite pair graph can be partitioned into two sets such that only vertices from different sets are connected by an edge [1916, 464]. As he emphasized, both results required the Axiom. In contrast to Steinitz and Zermelo, König did not attempt to restrict the Axiom to those proofs where it was indispensable.[9]

All in all, the Axiom had only begun to contribute to abstract algebra, a fact which partly reflected the nascent state of the discipline. As abstract algebra developed further during the 1920s and 1930s, the Axiom gradually came to be regarded as an essential tool—either through the Well-Ordering Theorem or indirectly through maximal principles such as Zorn's Lemma.

[6] Noether 1916, 103–106, 123.

[7] Zermelo 1914, 442.

[8] D. König 1916, 461. A graph is of *nth degree* if each vertex is on exactly n edges. A *pair graph* is a graph such that any closed path contains an even number of vertices.

[9] *Ibid.*, 461–463.

Here the work of Steinitz pointed the way. If abstract algebra was to progress, it required powerful new methods, and for the deepest results the Axiom would necessarily be among those methods.

3.6 A Smoldering Controversy

Although the debate over the Axiom of Choice was most intense during 1905 and 1906, the controversy continued sporadically throughout the decade after 1908. Except for a few supporters in England and France, the Axiom still encountered sharp criticism outside Germany. On the other hand, Wacław Sierpiński, who later became a forceful Polish advocate for the Axiom, published his first studies of it near the end of this decade. Within France, Germany, and Hungary, several mathematicians changed their minds about the Axiom—some coming to accept it and others abandoning it. These changes in attitude had diverse causes, but doubts about the Axiom's role in logic, as well as constructivistic beliefs, remained prominent among the grounds for rejecting it. All in all, the Axiom appeared quite suspect to many mathematicians during this transitional decade. Despite the efforts of Steinitz and Zermelo, the ordinary working mathematician had not yet understood how the Axiom was needed in his own research.

Logic was the key to understanding the objections raised in 1908 by Edwin Wilson, a professor at the Massachusetts Institute of Technology. His logical views had been shaped by the American postulate theorists Edward Huntington and Oswald Veblen as well as, to a lesser extent, by Peano. Although Wilson regarded Zermelo's system as potentially valuable, the logical status of the Axiom of Choice disturbed him. Since Zermelo, like most mathematicians, did not state postulates for the underlying logic, Wilson found it impossible to determine what deducible meant in this context. Nevertheless, he believed that what Zermelo had actually done was to introduce the Axiom of Choice as a new postulate of logic:

> In view of the facts that it may be doubted whether our logic is yet complete and that Zermelo's postulate is apparently not in contradiction with the other logical postulates, it is difficult to see how anyone can deny him the right to proceed as he does. The only question appears to be whether his method is valuable. [1908, 438]

Apparently Wilson interpreted the word valuable to mean constructive, since he insisted that a well-ordering remained all but meaningless without an algorithm to execute it. On such grounds, which owed much to Baire, Borel, and Lebesgue, he was quite sceptical of the Axiom.[1]

Across the Atlantic, Bertrand Russell also inquired, in a lecture delivered during March 1911 to the *Société Mathématique de France*, whether the

---

[1] Wilson 1908, 440–442.

Axiom belonged among the postulates of logic. There he summarized the known applications of the Axiom in set theory but contributed little beyond the results found in his papers of 1906 and 1908. An exception was his remark that much in Hausdorff's seminal articles of 1906 and 1907 on order-types depended on the Axiom and especially on Cantor's theorem (1.4.4), previously unrecognized as a consequence of the Axiom: The limit of a sequence of denumerable ordinals is denumerable.[2]

As his lecture revealed, Russell's scepticism toward the Axiom stemmed largely from his conception of logic. Though his logic owed much to Peano, in this instance he took pains to distinguish his position from his mentor's:

> Peano says [regarding the Axiom] . . . that the question of self-evidence is a psychological question which does not concern logic. But logic depends on the axioms of logic, and these axioms are accepted because they are self-evident; at least, the self-evidence either of the axioms or of their consequences is the sole reason which makes us accept the axioms. Thus we may ask if the multiplicative axiom is true, although not deducible from the other axioms [of logic] .... When we speak of an infinite class, it must be given by means of a property possessed by all its members and by nothing else .... Therefore the multiplicative axiom must assert that for a set of classes, there is always some property possessed by one element, and only one, in each class of the set. But, in my opinion, this is not at all self-evident. Thus I find myself led to the conclusion that the axiom ceases to be self-evident as soon as we grasp what it means. [1911, 32–33]

Here Russell's Platonic realism was wedded to a constructivism that had no room for entities not uniquely definable by a finite number of symbols. To this was added an intensional view of classes that increased his scepticism toward the Axiom of Choice. Since he found it exceedingly implausible that every class, and especially the continuum, could be well-ordered, he proposed to his French colleagues that they employ the Axiom only in those arguments likely to lead to a contradiction and thereby to show the Axiom false [1911, 33]. One can almost hear Borel and Lebesgue applauding.

In *Principia Mathematica* [1910–1913], Russell and Whitehead treated the Axiom with almost as much scepticism as Russell evinced in his 1911 lecture. These authors refused to adopt the Axiom as a postulate of logic, and they confined it to statements of the form, "If the Multiplicative Axiom holds, then P holds," where P was some proposition. Moreover, they were the first to consider seriously what mathematics would be like if the Axiom were false. One possibility was the existence of mediate cardinals (later called Dedekind cardinals) which were too large to be finite but too small to be Dedekind-infinite. As they demonstrated, even the Denumerable Axiom was sufficient to ensure that no mediate cardinal existed. Without using the Axiom, they established that if a set A is infinite, then the power set of the power set of A is Dedekind-infinite—a result which restricted the possibilities for mediate cardinals.[3]

---

[2] Russell 1911, 31; this lecture is translated in Grattan-Guinness 1977, 161–174.

[3] Russell and Whitehead 1910, 503, and 1912, 190, 228, 278.

Meanwhile, the French critics of Zermelo's Axiom did not remain silent. Their number was increased by Poincaré, who, at an earlier date, had tentatively accepted the Axiom of Choice [1906a, 313]. When Poincaré objected to Zermelo's axiomatization in 1909, he did not single out the Axiom for censure.[4] Rather, in another article of the same year, he reiterated his approval: "I call attention, in passing, to a very interesting proof of Zermelo's Axiom which this researcher [Richard] has just published in *Enseignement mathématique*, where he expresses himself on this subject with the greatest clarity."[5] Yet Poincaré's approval was limited by the fact that Richard had merely deduced the Denumerable Axiom from the assumption that every infinite set contains an element definable in a finite number of words (see 2.4). For Richard, this sufficed to demonstrate the Axiom itself, since he insisted that in practice one only considered countable sets which contained definable objects [1907, 97].

In the last of his articles on the foundations of mathematics, Poincaré abandoned his previous stance toward the Axiom and seemed to reject it altogether [1912]. That is, while he did not repudiate the Axiom directly, such a repudiation followed from his assertions. In particular, he referred to two schools of mathematicians, which he named Pragmatists and Cantorians. This distinction paralleled the one, which Lebesgue had borrowed from du Bois-Reymond, between Empiricists and Idealists.[6] The Pragmatists, Poincaré remarked, were suspicious of Zermelo's proof of the Well-Ordering Theorem, while the Cantorians tended to accept it. What Pragmatists distrusted about the proof was Zermelo's failure to provide a rule, expressible in a finite number of words, that showed how to well-order even Euclidean space. Since by his own definition Poincaré was a Pragmatist and since he asserted that Pragmatists rejected Zermelo's proof because it gave no constructive rule to determine the well-ordering, by 1912 Poincaré had probably come to reject the Axiom. Yet, whether he rejected the Axiom or not, he had already divested it of most of its scope by accepting Richard's truncated version.

As Poincaré emphasized, the Pragmatists made a principle out of the concept of definition by a finite number of words. In 1912, when one philosopher questioned the legitimacy of including words such as "indefinitely" and "infinitely" in such definitions, Borel quickly rose to the defense. What mattered to him was not such a "philosophical distinction," as he termed it, but a mathematical one: the unique determination of the mathematical object defined [1912, 219].

---

[4] Poincaré 1909a, 472–478.

[5] Poincaré 1909, 196.

[6] Poincaré 1912, 2–4; *cf.* 2.3. One notable difference between the two distinctions was that Poincaré regarded the Pragmatists as philosophical idealists, whereas Lebesgue branded the Empiricists' opponents with that term.

Borel, who reiterated that such a determination was not provided by infinitely many arbitrary choices, made a concession of sorts to Zermelo and Hadamard:

> Of course it is possible to reason about a class of mathematical objects, for example all the real numbers or all the continuous functions; this class is defined by means of a finite number of words, although not *all* the members can be defined in that way. Thus one obtains the general properties of the *class* . . . . [1912, 224]

This concession soon prompted Hadamard to write a letter to Borel in reply:

> I doubt that I have ever said otherwise. For indeed, all the members must exist in some way in order to form the class . . . . Zermelo would thus have demonstrated, if not that there exists *one* way of well-ordering the continuum, at least that there exists a [non-empty] *class* of such orderings. . . . This means, in sum, that (if Zermelo's argument is not ultimately perfected) we shall only be able to reason about the properties common to *all* these orderings. I willingly believe it. There are so many other things that we shall never know.[7]

While granting Zermelo the right to endow any abstract entities with any non-contradictory properties that he wished, Borel stressed that such formal logic led to nothing but purely verbal conclusions, unrelated to reality [1914a, 75–76]. Despite Borel's scepticism, Hadamard's distinction between an object which can be defined uniquely and a non-empty class of objects, no one of which can be defined uniquely, would later assume considerable importance in mathematical logic.

Hadamard continued to argue for the Axiom with Lebesgue as well. On the occasion of a Congress of Philosophy, held at Paris in 1914, they exchanged what Lebesgue later described as an impassioned correspondence over the Axiom. The outbreak of the First World War prevented its publication.[8] When he next wrote about the Axiom, Lebesgue reiterated his contention that any existence proof must provide a uniquely defined object rather than just a non-empty class of objects [1917, 137]. However, the following year he disputed the Axiom on new grounds: the Axiom was unacceptable because a choice function could not be characterized by a finite number of premises. Unable to comprehend how one could reason with infinitely many premises, Lebesgue concluded that the inadequacy of Zermelo's proof was essentially logical in nature [1918, 238]. Here his critique resembled the one that Peano had put forward in 1906 (see 2.8).

During the winter of 1915 Charles de la Vallée-Poussin, forced to stop teaching at the University of Louvain by the German invasion of Belgium, gave a course of lectures at the *Collège de France* in Paris. His lectures analyzed the relationship between Baire functions, Borel sets, and Lebesgue measure.

---

[7] Hadamard in Borel 1914a, 72–73.

[8] Lebesgue 1922, 61. Correspondence with his heirs has failed to uncover any surviving manuscript of this exchange.

In the book that emerged from these lectures, de la Vallée-Poussin used the Axiom implicitly without any awareness of the fact.[9] One such use occurred in a theorem which he credited to the American mathematician Dunham Jackson and which was based on Lindelöf's concept of condensation point:

(3.6.1) Every uncountable subset A of $\mathbb{R}^n$ contains a condensation point. [1916, 3]

De la Vallée-Poussin easily deduced from (3.6.1) that every uncountable subset A of $\mathbb{R}^n$ can be partitioned into a set B, consisting of all the condensation points of A that belong to A, and a countable set C. Two years later, Sierpiński recognized that (3.6.1), which can be deduced from the Countable Union Theorem, implies a special case of the Denumerable Axiom.[10]

Sierpiński published his first two articles on the Axiom in the *Comptes Rendus* of the Paris Academy of Sciences [1916; 1917a], where the second of them attracted Lebesgue's attention. It contained a theorem that Lebesgue found disagreeable. Sierpiński had established that, even without the Axiom, the existence of a non-measurable set can be deduced from the following classical theorem of set theory:

(3.6.2) The family F of all denumerable subsets of $\mathbb{R}$ has the power of the continuum.

Lebesgue, who denied that a non-measurable set had ever been exhibited, disputed Sierpiński's reasoning and claimed that Sierpiński had used the Axiom without recognizing his slip.[11] Yet Lebesgue was mistaken. All that Sierpiński had done was to note that (3.6.2) implies the existence of a one–one function with domain F and range $\mathbb{R}$; he did not need to choose an element from each set in F. Originating with this exchange, the debate between Lebesgue and Sierpiński over the Axiom continued for more than two decades (see 4.1 and 4.11).

What was at issue here was how far analysis could be developed without the Axiom. This question had already been raised in 1913 by the Italian analyst Michele Cipolla, particularly vis-à-vis the theory of limits of real functions. He had also claimed that, since the Axiom implied that every set can be well-ordered, it limited the general notion of set. More importantly, he added, the Axiom should be avoided in applications so as to escape a possible contradiction [1913, 1–2].

---

[9] In the second edition [1934] of his book, de la Vallée-Poussin attempted to eliminate all his earlier uses of the Axiom.

[10] Sierpiński 1918, 124–125; see (4.1.13) and (4.1.20).

[11] Lebesgue 1918, 238–239.

Cipolla believed that the theory of limits of real functions possessed greater simplicity and elegance when it was united with the theory of real sequences. Thus far, such a connection had relied on the Axiom, especially to obtain the principal result desired:

(3.6.3) If p is a limit point of a subset M of $\mathbb{R}$, then p is a sequential limit point of M.

He also noted that (1.2.4), the Bolzano–Weierstrass Theorem for sequential limit points, required the Axiom. Finally, he observed that the Axiom was needed to prove Cantor's theorem (1.2.2): A real function is continuous at a point if and only if it is sequentially continuous there.[12]

The aim of Cipolla's paper was to extend the notion of limit point to families of sets and thereby to establish a form of (1.2.2), among other results, without the Axiom. In this way he hoped to preserve the elegance and simplicity given by the Axiom while avoiding such a dubious assumption. To carry out his program, Cipolla defined a sequence $A_1, A_2, \ldots$ of non-empty sets in $\mathbb{R}$ to converge to a number p if for every positive ε there was a positive integer m such that for every $n > m$, A_n was a subset of the open interval $(p - \varepsilon, p + \varepsilon)$. Consequently, he was able to deduce a weakened version of (3.6.3) without the Axiom: If p is a limit point of a set M, then one can construct a non-constant sequence of sets included in M and converging to p [1913, 5]. Similarly, he demonstrated that the Axiom was not required to obtain the following result which connected continuity and sequential continuity: A real function f is continuous at a point p of the open interval (a, b) if and only if, for every sequence $A_1, A_2, \ldots$ of subsets of (a, b) converging to p, the sequence $f''A_1, f''A_2, \ldots$ converges to $f(p)$.[13] Thus, by collecting all the real numbers satisfying a given condition into a set, he was able to form a denumerable sequence of sets instead of having to choose a single object from each such set. In this fashion Cipolla successfully avoided the Axiom.

On 16 November 1913, Peano submitted to the Turin Academy of Sciences a paper from Leonida Tonelli which was strongly influenced by [Cipolla 1913].[14] A few years earlier, Constantin Carathéodory had pointed out how the Axiom was used implicitly in the calculus of variations to show that certain integrals have a minimum value [1906, 493]. But while Carathéodory, a friend of Zermelo's at Göttingen, found such a procedure unthinkable without the Axiom, Tonelli hoped to eliminate the need for the Axiom in as many special cases as possible. Thus Tonelli wished to specify, for each integral, a unique sequence of curves converging to a curve along which the value of the integral was minimized. He succeeded

[12] Cipolla 1913, 1–2.

[13] *Ibid.*, 7.

[14] Kennedy 1980, 191. Tonelli had just obtained his doctorate from the University of Bologna.

in doing so when the integral was a lower semi-continuous function of the curves along which it was evaluated.[15] Likewise he used sequences of sets, after the fashion of Cipolla, in order to avoid the Axiom in a weakened version of Arzelà's Theorem that an infinite set of functions, equicontinuous and uniformly bounded, has a limit-function.[16]

A few years later two mathematicians from Moscow, Nikolai Luzin and his student Mikhail Suslin, undertook to extend what could be shown in real analysis without the Axiom of Choice. In this vein Suslin sought to obtain a definition of the Borel sets in $\mathbb{R}$ that did not use either the Axiom or infinite ordinals. On 8 January 1917, Hadamard presented to the Paris Academy of Sciences an article in which Suslin proposed such a definition. As a first step, he termed a function H from finite sequences of positive integers to closed intervals as a defining system. Then a real number p was said to be associated with H if there was some sequence $m_1, m_2, \ldots, m_n, \ldots$ of positive integers such that p belonged to all of the sets

$$H(m_1), H(m_1, m_2), \ldots, H(m_1, m_2, \ldots, m_n), \ldots.$$

Finally, a subset E of $\mathbb{R}$ was called analytic if there existed some defining system H such that E contained all and only those real numbers associated with H. Suslin's first result determined the cardinality of the family A of analytic sets [1917, 89]:

(3.6.4) There are exactly $2^{\aleph_0}$ analytic subsets of $\mathbb{R}$.

Since A included all the closed intervals and was itself closed under countable unions and countable intersections, he concluded that:

(3.6.5) Every Borel set in $\mathbb{R}$ is analytic.

Indeed, he characterized the Borel sets by the following theorem [1917, 90]:

(3.6.6) A subset E of $\mathbb{R}$ is a Borel set if and only if both E and its complement are analytic.

This theorem (3.6.6) was linked with Suslin's attempt to establish the existence of an analytic set which was not a Borel set. For unknown reasons, Suslin insisted that any example of such a set must be obtained without using the Axiom; there were only $2^{\aleph_0}$ analytic sets, whereas, so he asserted, every class of examples given by the Axiom would have power $2^{2^{\aleph_0}}$ [1917, 90].

---

[15] Tonelli 1913, 6. A function f is *lower semi-continuous* at a point x if $f(x)$ equals the limit, as r approaches zero, of the greatest lower bound of f within an open ball of radius r.

[16] *Ibid.*, 7–13; *cf.* 1.8. It seems that the only Italian mathematician to accept the Axiom explicitly at this time was Giuseppe Bagnera of Palermo. In a calculus text Bagnera adopted the Denumerable Axiom for sets of real numbers [1915, iii].

When he claimed to define such an example explicitly, he erred. For, in the model of **ZF** introduced by Feferman and Levy [1963], every subset of $\mathbb{R}$ is a Borel set. Since Suslin did not publish the demonstrations of his results, it is difficult to ascertain precisely how he used the Axiom implicitly. However, the Feferman–Levy model reveals that (3.6.4) and (3.6.5) cannot both be deduced in **ZF**. Whether (3.6.6) can be shown, without using the Axiom, remains uncertain.

Luzin, in his article [1917] accompanying Suslin's in the Paris *Comptes Rendus*, asserted that the additional theorems which he had discovered about analytic sets did not use the Axiom. There he stated two results connecting these sets to the concepts of measure and category:

(3.6.7) Every analytic set in $\mathbb{R}$ is measurable.[17]

(3.6.8) Every analytic set in $\mathbb{R}$ has the Baire property.

Furthermore, Luzin mentioned that Suslin had obtained a related result:

(3.6.9) Every uncountable analytic set in $\mathbb{R}$ includes a perfect subset.[18]

However, it must be observed, if (3.6.5) can be proved in **ZF**, then (3.6.7), (3.6.8), and (3.6.9) all require the Axiom. For there is an uncountable Borel set in the Feferman–Levy model that is non-measurable, lacks the Baire property, and has no perfect subset.[19]

In Germany and Hungary, where the Axiom was slowly gaining support, Schoenflies and J. König abandoned their previous opposition to Zermelo's 1904 proof. When criticizing the proof, Schoenflies [1905] had objected not to the Axiom but to certain procedures concerning well-ordered sets. Three years later he regarded the proof as establishing only the equivalence of the Axiom and the Well-Ordering Theorem. The Axiom, he believed, merely shifted the Well-Ordering Problem to a different footing, but did not resolve it [1908, 36]. Yet in 1913 he devoted an entire chapter of his third report on set theory to the Well-Ordering Theorem—and accepted both of Zermelo's proofs.[20] Unfortunately, the report provided few clues as to what motivated this change. At moments he even appeared dubious of Zermelo's proofs. For, although he understood that the proofs showed the cardinal of every infinite set to be an aleph, he still entertained the possibility that

[17] While Luzin did not publish a proof of (3.6.7) in his article [1917], Luzin and Sierpiński did so in their joint article [1918, 44–47] on analytic sets. There they relied indirectly on the Axiom when they used the proposition that a countable union of measurable sets is measurable.

[18] Luzin 1917, 93–94. Hausdorff [1916] had established that every uncountable Borel set in $\mathbb{R}$ has a perfect subset, but here too the Axiom was used in an unavoidable way. Paul Alexandroff [1916], a student of Luzin, independently arrived at the same result as Hausdorff.

[19] See the forthcoming doctoral thesis of Juris Steprans at the University of Toronto.

[20] Schoenflies and Hahn 1913, 170.

$2^{\aleph_0}$ might be greater than every aleph. On the other hand, he was the first to recognize that the following theorem (4.3.4) depended on the Axiom; they were shown to be equivalent a decade later: For all infinite cardinals $\mathfrak{m}_1, \mathfrak{n}_1, \mathfrak{m}_2, \mathfrak{n}_2$, if $\mathfrak{m}_1 < \mathfrak{m}_2$ and $\mathfrak{n}_1 < \mathfrak{n}_2$, then $\mathfrak{m}_1 + \mathfrak{n}_1 < \mathfrak{m}_2 + \mathfrak{n}_2$.[21]

Julius König, who had originally doubted Zermelo's proof, came to accept it more unreservedly than Schoenflies. In 1904 and 1905 König had argued that the continuum cannot be well-ordered, but had not criticized Zermelo's proof directly (see 2.6). Several years prior to 1913, according to his son Dénes, Julius König came to accept the validity of Zermelo's proof for what he termed "Cantorian sets" and in particular for the continuum.[22]

Julius König's public acceptance of the Axiom occurred in a book published posthumously in 1914, *New Foundations for Logic, Arithmetic, and Set Theory*, which was largely philosophical in tone and heavily indebted to Poincaré. By psychological means, König tried to redeem set theory from the paradoxes which beset it. He began with certain observations on experience and consciousness, quoting with approval Kant's dictum that all human knowledge has its origin in intuitions. In a rather vague fashion, König designated a certain class of sets which were "possible" and "consistent" as the Cantorian sets. Belonging to this class, he claimed, were all the sets considered by Cantor.[23]

In his efforts to redeem Cantorian set theory, König wished especially to preserve the Well-Ordering Theorem and the Axiom of Choice. While attempting to do so, he revealed his philosophical position [1914, iv]. Like Cantor, Hilbert, and Poincaré, he subscribed to the formalist view that the consistency of a mathematical axiom system implies that some mathematical objects satisfy the system. Furthermore, intuitive self-evidence remained just as important to him as it did to Poincaré. On the other hand, he took a much stronger position concerning the self-evidence of the Axiom than Poincaré had ever done. To König the Axiom was just as evident "as any other fundamental intuition of the logico-mathematical sciences" [1914, 163].

As part of his program, König proposed to demonstrate the consistency of the Axiom. To begin, he stressed the distinction between the logical existence of a choice function and a rule of choice. Next, from the intuition of an exactly specified "domain of thought" D, it followed that D was possible and logically consistent. Under the influence of Poincaré's Kantianism, König regarded the existence of a choice function as demonstrated not by

---

[21] *Ibid.*, 180, 224. Perhaps the acceptance of Zermelo's proofs in this report on set theory reflected the views of its coauthor Hans Hahn more than of Schoenflies. Certainly, in his later textbook on real functions, Hahn [1921, 25] embraced the proof without much ado. At that time, however, Schoenflies also accepted it within the context of Zermelo's axiomatization [1922, 102].

[22] Dénes König in Julius Konig 1914, v.

[23] J. König 1914, 1–2, 148.

logical deduction but by a new synthetic judgment. However, he felt certain that in general a rule of choice did not exist [1914, 169]. Not surprisingly, his amorphous philosophical approach to the problem of the Axiom's consistency found little favor among other mathematicians.

During the period that Schoenflies and König abandoned their opposition to Zermelo's proof, a new critic arose within Germany, the philosopher Hugo Dingler.[24] In 1911 Dingler criticized Zermelo's two proofs on grounds similar to those which Bernstein and Schoenflies had put forward in 1905: the role of Burali-Forti's paradox, and in particular of the collection W of all ordinals, in the proofs. Specifically, Dingler argued that if there were a set M with power greater than $\overline{\overline{S}}$ for every well-ordered set S, then M would have a subcollection order-isomorphic to W. From the Axiom of Separation it would then follow that W was a set, although Zermelo had shown the contrary.[25] Four years after Dingler's article appeared, Hartogs proved without the Axiom of Choice that no such set M could exist (see 3.4). Thereby Dingler's objection lost its force.

Among the mathematicians who discussed the Axiom during the period 1908–1919, those who eventually proved most influential were Abraham Fraenkel and Wacław Sierpiński (see Chapter 4). Yet when Fraenkel first expressed his views on the Axiom in 1919, his opinion was simply one among many. He considered Cantor's process of well-ordering an arbitrary set, by successively selecting elements from it, to be inadmissible because it appeared to provide a method for actually constructing a suitable well-ordering. On the other hand, while accepting Zermelo's Axiom and proof, he analyzed certain criticisms of them dispassionately. In Fraenkel's opinion, the most reasonable criticism was the one that rejected the Axiom and regarded the Well-Ordering Theorem as unproven. Indeed, some mathematicians mistrusted both principles because neither helped to solve the Continuum Problem. They need not feel such mistrust, he added, if they would distinguish between the possibility of a well-ordering (which the Well-Ordering Theorem provided) and its actual execution (which it did not). To Fraenkel, the Axiom of Choice was no less evident than other mathematical axioms, all of which were plausible but unproven assumptions.[26]

In sum, the arguments for and against the Axiom of Choice and Zermelo's first proof of the Well-Ordering Theorem did not change substantially during the decade after he axiomatized set theory. Those who supported the Axiom and proof—such as Hadamard, Steinitz, J. König, and Fraenkel—distinguished strongly between the existence of a well-ordering and its actual construction. They characterized Zermelo's demonstration as an

---

[24] Likewise, in his first study of large cardinals, the German mathematician Paul Mahlo [1911, 187] briefly expressed reservations about the Axiom.

[25] Dingler 1911, 13–14, 8.

[26] Fraenkel 1919, 125–128, 146–148, 143.

existence proof and doubted that a general procedure for constructively well-ordering an arbitrary set would ever be discovered. At the same time, they rejected Cantor's procedure for well-ordering a set by successively selecting arbitrary elements from it. They remained unperturbed that the Axiom had not been shown consistent.

The mathematicians who opposed the proof—such as Borel, Lebesgue, Russell, and Wilson—did so largely because of the non-constructive character of the Axiom which served as its base. Most of them emphasized the concept of definition by a finite number of words. In this vein Lebesgue regarded the Axiom as an attempt to reason with an infinite number of premises, for him an utter impossibility. Yet, as Richard's paradox should have made clear to Zermelo's opponents, definition by a finite number of words was a rather nebulous concept. Working within a formal system, Russell was less vulnerable to such pitfalls than the other critics. But when Borel permitted the words infinitely and indefinitely to appear in his definitions, the concept of finite definition became very hazy indeed.

Moreover, few of the opponents other than Russell suspected how much mathematics they would have to abandon because they disowned the Axiom. Only Cipolla understood that analysis might be seriously restricted and, in particular, that the use of sequences would have to be seriously re-examined. The sole consequences of the Axiom to appear improbable were the Well-Ordering Theorem and the existence of a non-measurable set, while there was an abundance of useful consequences which various opponents had applied inadvertently. Nevertheless, a radical change occurred in 1914 when Hausdorff discovered a consequence of the Axiom that other mathematicians found truly implausible: Hausdorff's paradox.

3.7 Hausdorff's Paradox

Although many mathematicians criticized the Axiom of Choice because of its non-constructive character and although some suspected that it could lead to a contradiction, prior to 1914 no one had substantive evidence of any such inconsistency. Even Lebesgue, who labored to refute the Axiom, had established only that certain choice functions are non-measurable. In 1914 there finally appeared what seemed to be irrefutable evidence against the Axiom: Hausdorff's paradox that one half of a sphere was congruent to one third of the same sphere. Nevertheless, Hausdorff did not regard his "paradox" in this way, for he did not consider his result to be at all paradoxical. In order to understand his point of view, we must first turn to his book, *Grundzüge der Mengenlehre*, published the same year.

In the *Grundzüge* Hausdorff discussed the Well-Ordering Problem at length and expressed his faith in the Axiom's validity and significance. After noting Cantor's belief that the Well-Ordering Principle was a law of thought, Hausdorff stated the intuitive argument for it: To well-order a set B, one picked an element b_0, a next element b_1, and so on; this process would exhaust B since the set of all ordinals could not be emptied and still leave members of B. While Hausdorff did not share most of the objections raised against this argument, he rejected the existence of a "set" of all ordinals as well as the undesired reference to the passage of time. In fact, if each act of choice took some minimum length of time, no one could ever select the element b_ω. Thus, he concluded, one must abandon such practical and psychological considerations. The element b_ω should be conceived in the sense of transfinite induction, and the series $b_0, b_1, \ldots$ should be regarded as completely timeless. He believed that Zermelo had acted in harmony with such timelessness by postulating the simultaneous choice of an element from each non-empty subset of B. In this way a system of successive acts of choice was replaced by a system of simultaneous ones—which could not be executed in practice either. Finally, Hausdorff set forth in detail both of Zermelo's proofs for the Well-Ordering Theorem, a proposition which, he felt, endowed set theory with appropriate simplicity.[1]

Shortly before his book appeared, Hausdorff published an article [1914a] containing the theorem that later became known as Hausdorff's paradox. He did not consider his theorem to cast any aspersions on the Axiom but rather to resolve Lebesgue's Measure Problem (see 1.7). This problem required that one associate with each bounded set A in a Euclidean space $\mathbb{R}^n$ a non-negative number $m(A)$, called the measure of A, satisfying four conditions:

(a) The n-dimensional unit cube has measure one.
(b) Congruent sets have the same measure.
(c) $m(A \cup B) = m(A) + m(B)$ if A and B are disjoint and bounded.
(d) $m(\bigcup_{i=1}^{\infty} A_i) = \sum_{i=1}^{\infty} m(A_i)$ if all A_i are disjoint and their union is bounded.

Hausdorff remarked that Lebesgue's constructive definition of measure did not associate a measure with all bounded sets, only with the Lebesgue-measurable ones, and thus did not provide a solution to the Measure Problem.[2] Lebesgue, on the other hand, had repeatedly doubted that non-measurable sets actually exist (see 2.3 and 3.6).

At this point Hausdorff deduced from the Axiom that the Measure Problem had no solution. The existence of a solution for $\mathbb{R}^{n+1}$, he remarked, would provide a solution for $\mathbb{R}^n$. But there existed no solution even for the Euclidean line $\mathbb{R}$ since there was none for a circle of unit radius. Hausdorff's demonstration for the case of the circle was strongly reminiscent of Vitali's

[1] Hausdorff 1914, 133–138.
[2] Hausdorff 1914a, 428.

[1905], Lebesgue's [1907], and Van Vleck's [1908] examples of sets that were not Lebesgue-measurable, and, as they had done, he applied the Axiom to an uncountable family of sets. For Hausdorff, the essential point of his proof was that it used only conditions (a)–(d) and not Lebesgue's particular definition of measure.[3]

Would a solution be possible, Hausdorff inquired, if condition (d) were dropped? While this question remained open for one- and two-dimensional Euclidean space, he established that there exists no solution satisfying (a), (b), and (c) for three or more dimensions.[4] In particular, he demonstrated that there is no solution for the sphere; for, if such a solution existed, then half of a sphere would have the same measure as a third of a sphere.

In essence Hausdorff's proof, which again extended earlier arguments yielding a non-measurable set, was as follows: He wished to decompose a sphere S into four sets A, B, C, D such that A, B, C, and $B \cup C$ were all congruent, while D had measure zero. To do so, he let ϕ and ψ be rotations of S, through 180 and 120 degrees respectively, on distinct axes passing through the center of S. With ϕ, ψ, and ψ^2 all treated as different formal factors, a group G of motions was generated. By avoiding a certain denumerable set of angles between the initial axes, all members of G other than the identity had the representation ϕ or

$$\phi^{i_1}\psi^{j_1}\phi^{i_2}\psi^{j_2} \ldots \phi^{i_{n-1}}\psi^{j_{n-1}}\phi^{i_n}$$

for some n, where i_1 and i_n equaled zero or one, every other i_k with $1 < k < n$ equaled one, and every j_k equaled one or two. Moreover, all members of G whose formal representation differed were distinct. In consequence, each member of G other than the identity had two fixed points; so the set D of those fixed points was denumerable and hence of measure zero.

Next he determined A, B, and C. For each point x of $S - D$, he defined $P_x = \{g(x) : g \in G\}$, and thus for every x and y in $S - D$, P_x and P_y were either equal or disjoint. Using the Axiom to select a point from each distinct P_x, Hausdorff called the resulting choice set M. Thus $S - D$ was the union of the images of M under all g in G. At this stage he specified a partition of the images $g''M$ into three disjoint sets A, B, and C such that for every g in G,

(i) one of the two images $g''M$ and $(g\phi)''M$ was included in A and the other in $B \cup C$, and
(ii) exactly one of the three images $g''M$, $(g\psi)''M$, and $(g\psi^2)''M$ was included in A, one in B, and one in C.

It followed that A, B, C, and $B \cup C$ were all congruent to each other since $\psi''A = B$, $(\psi^2)''A = C$, and $\phi''A = B \cup C$. But then A had the measure of

---

[3] *Ibid.*, 429.

[4] Stefan Banach [1923] established that there is a solution for the line and the plane; see 4.11.

one half and of one third of a sphere, a contradiction. Consequently, Lebesgue's Measure Problem had no solution for three or more dimensions.[5]

The first to respond to Hausdorff's paradox—indeed, the first to characterize it as a paradox—was Borel, who felt quite certain that the culprit was the Axiom of Choice. In the second edition of his *Leçons sur la théorie des fonctions* [1914], Borel concluded his exposition of the paradox with a polemic against the Axiom:

> If, then, we designate by a, b, c the probability that a point in S belongs to A, B, or C respectively and if we grant that the probability of a point belonging to a set E is not changed by a rotation around a diameter (this is what Lebesgue expresses by saying that two congruent sets have the same measure), one obtains the contradictory equalities: $a + b + c = 1$, $a = b$, $a = c$, $a = b + c$.
>
> The contradiction has its origin in the application . . . of Zermelo's *Axiom of Choice*. The set A is homogeneous on the sphere; but it is at the same time a half and a third of it . . . . The paradox results from the fact that A *is not defined*, in the logical and precise sense of the word *defined*. If one scorns precision and logic, one arrives at contradictions.[6]

At last, Borel believed, his decade of opposition to the Axiom had been fully justified. The Axiom generated a contradiction; hence its illogical and, above all, imprecise character could not help but be obvious to mathematicians. Yet Hausdorff, who had discovered the paradox, did not agree. Nor did Stefan Banach and Alfred Tarski, who analyzed the paradox in depth a decade later (see 4.11). As the source of the paradox, one could cite either the Axiom of Choice or the existence of a solution to Lebesgue's Measure Problem, and hence one could reject either of them. In fact, many mathematicians came to regard Hausdorff's paradox and its successor, the Banach–Tarski paradox, as sufficient reason to be wary of the Axiom. Nevertheless, the struggle over the Axiom had not been decided by Hausdorff's paradox. It was only intensified anew.

3.8 An Abortive Attempt to Prove the Axiom of Choice

Philip Jourdain's attempts to resolve the Well-Ordering Problem spanned the first two decades of this century. While in one sense the history of these attempts is highly personal and idiosyncratic, in a larger sense it illuminates the difficulties which mathematicians of the time experienced vis-à-vis the Axiom. These difficulties frequently concerned to what extent the Axiom was

[5] Hausdorff 1914a, 430–433.

[6] Borel 1914, 255–256. Since Gödel's work on constructible sets (see 4.10), it is known to be consistent with **ZF** that the non-measurable sets A, B, C given by Hausdorff are definable (by means of ordinals). In particular, they are definable if we assume the Axiom of Constructibility.

essential to their researches, where it was used, and how it could be avoided. In this regard Jourdain's changing attitude reflects the ambivalence of a large number of mathematicians toward the Axiom.

Jourdain's attitude was molded as much by his contact with Cantor and Russell as by his own discoveries. In his argument for the Well-Ordering Theorem, published early in 1904, Jourdain did not remark his use of dependent arbitrary choices any more than Cantor had done in his related argument. Even in 1905, after learning of Zermelo's proof, Jourdain did not consider his own argument to require the Axiom. Yet by early in 1906 Russell had persuaded Jourdain that the Axiom was inescapable in any proof of the Well-Ordering Theorem (see 1.6 and 2.7).

Furthermore, it was Russell's interpretation of what the Axiom meant that influenced both Jourdain's future research and his view of the Axiom: The Axiom required the existence of a propositional function which determined a particular choice function.[1] This interpretation, which in effect demanded a rule by which to execute arbitrary choices, had led Russell to doubt the Axiom's validity. On the other hand, it caused Jourdain to seek a new proof for the Well-Ordering Theorem, one that would dispense with arbitrary choices and thereby yield the Axiom as well. The desired proof would not require a new mathematical principle but must be based on the actual construction of a well-ordering from a definite rule, albeit a transfinite one. While seeking such a rule, Jourdain wrote several articles to determine those theorems which required the Axiom and those which could do without it.[2]

The breakthrough came early in 1918. On 12 March, Jourdain sent the editor of *Nature* a letter which sketched his purported proof of the Well-Ordering Theorem, and consequently of the Axiom itself, from accepted mathematical principles [1918]. Within a few days he mailed a more detailed argument to the *Comptes Rendus* of the Paris Academy of Sciences, followed by a similar account to *Mind* [1918a,b].

Jourdain's sketch in *Nature* was strongly influenced by Hartogs' proof of 1915 (which did not use the Axiom) that no set has a power greater than the power of every well-ordered set. Yet Jourdain endeavored to show that an arbitrary set M can be well-ordered, not by relying on the Trichotomy of Cardinals as Hartogs had done, but by arranging the well-orderings of subsets of M into classes such that for any two members of a given class, one was a segment of the other. Then he used a variant of Burali-Forti's paradox to deduce that one of these classes contained a well-ordering of M [1918, 84]. However, this new argument did little more than rework the one which Jourdain had published in 1904. Despite his claim to the contrary, the Axiom lay hidden in the new argument as well.

---

[1] Jourdain 1906a, 16.

[2] The last of these was Jourdain 1910. See also Jourdain 1906a, 1906b, 1907, 1908, 1908a, discussed briefly in 2.7.

The articles in *Mind* and the *Comptes Rendus* made Jourdain's reasoning more detailed and precise. His basic tool was what he termed a chain of M of type α, *i.e.*, a bijection from some subset of M onto the set of all ordinals less than α. To well-order M, he first placed all the chains of M of type α into a set K_α for each ordinal $\alpha > 0$. Then the members of each K_α were assigned to sets V_α defined by transfinite induction as follows: V_1 was the set of all $\{f\}$ for each f in K_1, and V_2 was the set of all $\{f, g\}$ for each f in K_1 and each g in K_2 such that $f \subseteq g$. More generally, V_α was the family of all sets S, ordered by inclusion, that consisted of exactly one member of K_β for each non-zero $\beta \leq \alpha$. His principal lemma was the following:

(3.8.1) For any limit ordinal α, V_α is non-empty if V_β is non-empty for each non-zero $\beta < \alpha$.[3]

Using Burali-Forti's paradox, he concluded that there existed some ordinal α, and hence a least one, such that $V_{\alpha+1}$ was empty. Consequently M was well-ordered by some chain of M of type α.

The error in Jourdain's reasoning occurred when he claimed to establish (3.8.1) by transfinite induction. While his inductive hypothesis asserted only that V_β was non-empty for each non-zero $\beta < \alpha$, he inadvertently and unjustifiably assumed that it gave him a class C containing sets A, each of which contained a member of V_β for every non-zero $\beta < \alpha$ and such that $f \subseteq g$ or $g \subseteq f$ for all f, g in A. Then it would indeed have followed that V_α was non-empty, since $V_\alpha = \{A \cup \{\bigcup A\} : A \in C\}$. In actuality, Jourdain needed either the Axiom or an equivalent non-constructive assumption to obtain C. What apparently induced him to believe that he had proved (3.8.1) without it was his confusion between possessing sets like A and simply having V_β non-empty for all non-zero $\beta < \alpha$.

At first Jourdain was unconcerned with this step in his proof. Instead he concentrated on how to obtain a maximal set V_α, which he now termed a complete class of direct continuations. He then published a revision of his proof in another letter to *Nature* [1918c], as well as in articles for the *Comptes Rendus* and for *Science Progress* [1918d,e].

By early in 1919 several English mathematicians had objected privately to Jourdain's argument, but he remained fully convinced of its validity. On 11 March he replied by letter to Russell, one of his principal critics:

> Neither your nor Whitehead's remarks seem to touch the point of the thing because the very essence of it is *actual construction* by induction of a class of direct continuations which determines a chain that is of the sort that cannot be continued. The construction avoids Zermelo's principle because *many* classes of direct continuations are constructed *simultaneously*.[4]

---

[3] Jourdain 1918b, 388. The same lemma occurs in his articles 1918d, 985; 1918e, 303; and 1921, 240.

[4] Jourdain in Grattan-Guinness 1977, 146.

That same day Jourdain sent *Nature* a letter [1919] answering his critics. There he stated that all of them wished to see (3.8.1) proved in detail when $\alpha = \omega$. While demonstrating this case, he made the same error that he had committed earlier for the general case of (3.8.1). In fact, for the case that $\alpha = \omega$, Jourdain's assertion was closely related to König's Infinity Lemma, which had not yet been formulated (see 4.5).

By this time Jourdain was becoming anxious, while his critics grew increasingly impatient with analyzing the endless variants of his proof. On 24 May he sent Russell a frustrated letter:

> When I wrote out my suggestions more than a year ago I imagined that Hartogs [1915] had provided a certain necessary step and did not repeat what I thought he had done. Whitehead, you, and Hardy all pointed out the gap in my proof: I examined Hartogs again, found he had not done what I thought he had, and supplied the link. Whitehead and Hardy simply repeated what they said at first, thereby proving that they did not read my new paper with understanding. When I pointed this out Hardy left my letter unanswered and Whitehead said he had no time to read any more of my stuff.[5]

Soon afterward, Jourdain explained his proof again in *Mind*, this time attacking his critics with venomous sarcasm [1919a].

Jourdain's health, which had never been good, was deteriorating rapidly. As he lay dying, he grew more and more obsessed with his proof and with Russell's inability to acknowledge its validity. Many years later J. E. Littlewood recounted what then occurred:

> When he was dying he sent a request to a party staying in Lulworth, including Russell, Dorothy Wrinch, and myself, asking someone to come to listen to his final proof. Dorothy Wrinch and I went, and he gave his "proof", fallacious as usual. (The subject is a great trap, many eminent mathematicians have assumed the axiom [of choice] when they thought they were not.) . . . I said that a new point was involved, and I would have to consider it closely and at length. Then he burst out: "My dear fellow, you know perfectly well that you can see whether a proof is right or wrong in five minutes." Strange outbreak of complete rationality![6]

Belatedly, Russell decided to visit Jourdain and sent him a telegram to that effect. In anguish, Jourdain's wife replied to Russell on 26 September:

> Your telegram came just too late to arrange for Philip to see you. He is now quite unable to talk or see anyone, but just lies in a semi-unconscious state. Why didn't you make an effort to come a little sooner? You have made him so unhappy by your inability to see his well-ordering. You are the only person he wanted to see and talk with months ago.[7]

Five days later Jourdain died.

Jourdain's final paper, a detailed version of his "proof," was published posthumously in *Acta Mathematica* by its editor Mittag-Leffler, who

---

[5] Jourdain in Grattan-Guinness 1977, 149.

[6] Littlewood in Grattan-Guinness 1977, 152.

[7] Laura Jourdain in Grattan-Guinness 1977, 153.

disagreed with the proof but described the paper as valuable for its new points of view.[8] On the other hand, Mittag-Leffler requested that no one send him any further arguments of this type, and with good reason. As he well understood, the Well-Ordering Theorem could not be deduced from the usual mathematical assumptions; some non-constructive principle such as the Axiom of Choice was required. Unfortunately, to the end Jourdain repeated the same fundamental error in his "proof" of the Axiom. To the end he remained blind to his error.

3.9 Retrospect and Prospect

For the Axiom of Choice, the decade from 1908 to 1918 was a transitional period. Zermelo's axiomatization, motivated primarily by a desire to secure both his Axiom and his proof of the Well-Ordering Theorem, did not initially fulfill its purpose. During the debate over this proof between 1904 and 1908, many set-theoretic principles had been stated, but it was impossible to accept all of them without generating a contradiction. When Zermelo formulated his system, few mathematicians were content with the particular axioms that he proposed. Indeed, several mathematicians insisted that his axiomatization was seriously flawed. They believed the most glaring defect to reside in his Axiom of Separation and especially in his vague notion of *definit* property. Only Weyl attempted to make this notion precise and, by his own admission, he did not succeed. A further uncertainty was the system's consistency, which had to be established if the system was to provide a foundation for all of mathematics, as Zermelo had intended.

Why did Zermelo overlook these difficulties and thereby remain convinced that his axiomatization was satisfactory? Why, moreover, did he offer an axiomatization as the foundation for set theory and for his proof of the Well-Ordering Theorem in particular? Clearly, he relied on an axiomatization because he was influenced while at Göttingen by Hilbert's axiomatic method, embodied in the *Grundlagen der Geometrie*. Yet Hilbert had based the consistency of geometry on the consistency of the real numbers, and so we may inquire how Zermelo intended to secure the consistency of his own system. The answer appears to be: by logic. Nevertheless, Zermelo did not specify an underlying logic for his system, though he referred in passing to Schröder's researches on the algebra of logic and adopted Schröder's symbol for inclusion. In fact, Zermelo's failure to provide a satisfactory foundation for the logic required by his system lay at the center of Poincaré's

[8] Mittag-Leffler in Jourdain 1921, 239. It may well be that Mittag-Leffler, who did not specify these new points of view, published the paper mainly out of respect for Jourdain.

and Russell's objections. This failure reflected the larger fact that mathematical logic remained in a fluid and unsatisfactory state. A few years later Skolem would employ a first-order logic, adapted from Schröder's system, as a way to render Zermelo's Axiom of Separation precise (see 4.9).

Although Zermelo acknowledged that he had not yet proved his system consistent, he insisted that from it all of Cantorian set theory could be deduced. Likewise, he believed that his notion of *definit* property was sufficiently clear. It must be remembered that, except for a few scattered axioms, no one had attempted to axiomatize set theory before Zermelo. The distinction between consistent and inconsistent sets, proposed by Cantor and Jourdain, had been published by the latter in 1904 only as a part of his argument for the Well-Ordering Theorem, an argument rejected by other mathematicians. Developed independently along similar but much more precise lines, John von Neumann's axiomatization did not appear until 1925. However, the criticisms directed against the notion of *definit* property eventually led Zermelo to clarify it.

In contrast to Zermelo's axiomatization, which had not gained widespread acceptance by 1918, the Axiom of Choice found more varied applications. Of these, the most important occurred in the theory of fields. Steinitz' work on algebraic and transcendental extensions of fields, fundamental to future research in algebra, depended explicitly on the Axiom and included a plea for its acceptance. On the other hand he adopted the convention, as Zermelo had earlier and many others would later, of employing the Axiom only when no means was known to avoid it. Steinitz remained convinced that in general this did not constitute a failure of present knowledge, but reflected a genuine need for the Axiom in order to obtain deep results in algebra. As algebra grew increasingly abstract, events bore him out. Yet there was still no method for establishing that a given theorem required the Axiom, other than by proving their equivalence.

Except in algebra, the Axiom was rarely applied consciously to obtain new results between 1908 and 1918, but mathematicians increasingly recognized its previous implicit uses. In 1915 Hartogs showed that the Axiom was indispensable to any adequate theory of infinite cardinals since the Axiom and the Trichotomy of Cardinals are equivalent. The first mathematician to grasp how deeply analysis depended on the Axiom was Cipolla, who recognized that it implied the equivalence of limit point and sequential limit point in $\mathbb{R}$, as well as of continuity and sequential continuity. Cipolla attempted to circumvent the Axiom, as far as possible, by employing sequences of sets rather than of points, but his influence was slender. The researches which Sierpiński began to publish in 1916 on the Axiom's uses in analysis would affect future developments much more directly.

In 1918 most mathematicians remained ignorant of the Axiom's significance for several reasons. Firstly, mathematicians still did not understand the extent to which they depended on the Axiom, even to prove theorems which they considered well-established without it. Such was the case for the classical theorem (3.6.1) that the family of all denumerable subsets of $\mathbb{R}$ has the same power as $\mathbb{R}$, a result from which Sierpiński deduced the existence of a non-measurable set. Lebesgue, who accepted (3.6.1), believed that Sierpiński had unknowingly used the Axiom to obtain this set. Thus Lebesgue erroneously thought it possible to preserve the theorem (3.6.1) while avoiding both the Axiom and its undesirable consequences such as the existence of a non-measurable set. This exchange between Lebesgue and Sierpiński exemplifies the subtlety of the Axiom, as does Jourdain's attempt in 1918 to prove the Well-Ordering Theorem constructively.

Secondly, it remained uncertain whether the Axiom would lead to a contradiction, as Cipolla and Russell suspected it might. When Hausdorff's paradox appeared in 1914, Borel trumpeted it as confirmation of the Axiom's inconsistency—ignoring the fact that one could reach this conclusion only if one assumed Lebesgue's Measure Problem to have a solution. In fact, the question of the Axiom's consistency lacked a clear formulation, since both its supporters and its opponents other than Russell were not then thinking in terms of a formal system with an underlying logic. Working within such a formal system, his axiomatization for set theory in first-order logic, Gödel established two decades later that the Axiom could not lead to a contradiction unless the rest of Zermelo's axioms were already inconsistent (see 4.10).

Thirdly, the controversy surrounding the Axiom was emblematic of that between constructive mathematics and abstract mathematics. Euclid had formulated his postulates for geometry as constructions, and even within the increasingly abstract mathematics of the early twentieth century there remained a strong tendency to view mathematics as a sequence of constructions. Yet the question remained: What constituted a construction? At the very least, a construction required the unique determination of the object constructed. Even the Axiom's supporters granted that it could not provide an actual construction by which to well-order a given set; instead it implied the abstract existence of such a well-ordering. Repeatedly Zermelo's critics stressed that this well-ordering was not uniquely defined and that a mathematical object did not exist unless it could be uniquely defined. Thereby they identified mathematical existence with actual construction—without giving a precise statement of what was permitted in such a construction. In this fashion, as the ongoing debate between Hadamard and Borel illustrates, the Axiom served as a focus for the discussions on the nature of mathematical existence between Pragmatists and Cantorians, Empiricists and Idealists, Constructivists and Platonists.

Fourthly, the Axiom had not yet found an advocate, a mathematician who would thoroughly investigate its consequences and who would repeatedly defend its fundamental importance in mathematics. To some extent Zermelo had been such an advocate, since it was through his researches that the Axiom first attained recognition. Yet after 1908 no one published a series of articles which revealed important new consequences of the Axiom.

Generally the mathematicians who contributed to the subject, such as Hamel and Hartogs, did so only once. Russell, who had uncovered many of the Axiom's consequences and who might have become such an advocate, remained dubious of the Axiom.

Then the situation changed, and an advocate appeared—the young Polish mathematician Wacław Sierpiński. Although he had written about the Axiom as early as 1916, his influence became pronounced two years later when he published a detailed survey of the Axiom's uses (see 4.1). Soon the Warsaw school of mathematics, which he helped to found in 1918, accepted the Axiom and explored its interrelations vigorously. The fact that he not only supported the Axiom and clarified its role but also motivated his students to investigate it in depth proved to be crucial to the Axiom's development.

Chapter 4

The Warsaw School, Widening Applications, Models of Set Theory (1918–1940)

> The essential idea on which the Axiom of Choice is based constitutes a general logical principle which, even for the first elements of mathematical inference, is necessary and indispensable.
>
> David Hilbert [1923, 152]

When, shortly after 1918, a school of mathematicians emerged at Warsaw under Sierpiński's tutelage, the Axiom finally gained the attention that it deserved, and thereby became the object of much careful research. Within a few years Polish mathematicians discovered interconnections between the Axiom and many other propositions in various branches of mathematics. Sierpiński's survey of the Axiom's uses was soon followed by Tarski's research on definitions of finite set whose equivalence required the Axiom. Furthermore, Tarski discovered that each of several propositions in cardinal arithmetic implies the Axiom and hence is equivalent to it. On the other hand, Banach and Tarski extended Hausdorff's paradox by demonstrating via the Axiom that any sphere S can be decomposed into a finite number of pieces and reassembled into two spheres with the same radius as S. Neither author regarded this result as any slight on the Axiom, but later mathematicians were to call it the Banach–Tarski paradox.

While these theorems were largely set-theoretic, outside Poland the Axiom assumed an increasingly vital role in algebra, logic, and topology. Among the algebraic applications were Artin's and Schreier's work on real fields, Stone's Representation Theorem for Boolean algebras, and Zorn's Lemma. In mathematical logic Hilbert did what Peano, Russell, and Wilson had entertained as a possibility but had rejected: He formulated the Axiom of Choice as a postulate of logic, which he called the ε-axiom. When the first researches began in model theory, the Axiom helped to establish the Löwenheim–Skolem Theorem. As for topology, a particularly fruitful application occurred in Tychonoff's Compactness Theorem, later shown to be equivalent to the Axiom. But most important of all was the gradual realization that abstract algebra, mathematical logic, and general topology would all be severely limited without some form of the Axiom.

During the same period, the place of the Axiom within Zermelo's system began to be investigated. Fraenkel succeeded in proving that even the Denumerable Axiom is independent from the rest of the postulates in the system, although he required denumerably many urelements to do so. After Fraenkel applied his method of independence proofs to several other questions involving the Axiom, Adolf Lindenbaum and Andrzej Mostowski refined and extended it. Still more important eventually, though lacking immediate influence, was Gödel's definition of the constructible sets,[1] by which he established the relative consistency of the Axiom of Choice. This consistency result resolved the question of whether the Axiom could generate a contradiction, and did so in a way quite different from the answer suggested earlier to some mathematicians by Hausdorff's paradox. From the researches of Fraenkel, Mostowski, and Gödel there emerged a new field of mathematics: models of set theory.

By 1940 the Axiom's significance and relative consistency were evident to all informed mathematicians, but the Axiom itself remained an unpleasant assumption to many of them, one best avoided whenever possible. Even supporters of the Axiom such as Tarski preferred to deduce their theorems without it when they could do so. Nevertheless, it had become clear that, while one could phrase the Axiom in more intuitively pleasing forms, modern mathematics would be decimated without it. As Zermelo had foreseen, the Axiom's usefulness finally won it a secure place within the mainstream of mathematical research.

4.1 *A Survey by Sierpiński*

Prior to the First World War only four mathematicians taught at universities in partitioned Poland, which remained under the control of Russia, Germany, and Austria–Hungary. During 1911 Sierpiński, the three other professors, and a historian of mathematics attended the mathematical section of a congress for Polish scientists, but they could not discuss mathematics in depth since each of them did research in a different specialty. This incident led Sierpiński to the conclusion that several Polish mathematicians would have to undertake research in a single field if mathematics were to flourish in Poland [1963, 27].

With the reunification of Poland in 1918, Sierpiński and his associates began to fulfill this hope. Janiszewski, whom in 1913 Sierpiński had invited to

[1] These sets were constructible in a technical sense far too broad to render them acceptable to constructivist mathematicians such as Borel and Lebesgue. In particular, the definition of constructible set presupposed all the transfinite ordinals (see 4.10).

join the mathematics faculty at the University of Lwów, published a seminal article, "On the Needs of Mathematics in Poland," in the journal *Polish Science* [1918]. Above all, Janiszewski believed that Polish mathematics stood in need of co-researchers working together in a suitable mathematical environment. To compel students with the same specialty to concentrate at a single location, he proposed granting different Polish universities exclusive power to confer doctorates in particular fields of mathematics. Equally, he stressed the importance of founding a new journal which would specialize in a single area—unlike other mathematical journals of the time. Accepting papers only in the major mathematical languages of English, French, German, and Italian, such a journal would attract foreign co-workers and become indispensable to specialists. Above all, it would help to create a close-knit group of Polish mathematicians.[1]

In 1919 Janiszewski's ideas bore fruit. He, Sierpiński, and Stefan Mazurkiewicz (who had written his doctoral thesis under Sierpiński at Lwów) were the first professors of mathematics at the newly reopened University of Warsaw. Despite Janiszewski's untimely death soon afterward, research in set theory and topology continued to flourish there. "We decided," Sierpiński later wrote, "to carry out Janiszewski's idea of publishing a foreign-language periodical devoted to set theory, topology, theory of real functions, and mathematical logic. This was the origin of *Fundamenta Mathematicae* ...."[2] Its pages were to include many articles bearing on the Axiom of Choice.

Sierpiński, who apparently became intrigued by the Axiom through reading the Paris lecture of 1911 that Russell gave on this subject, extended Russell's work in uncovering the Axiom's previous implicit uses. While Russell had pointed out many such uses in cardinal arithmetic, Sierpiński concentrated on real analysis where he emphasized the central role of the Denumerable Axiom. In 1916, independently of Cipolla, Sierpiński discovered its role in proving, for real functions, that continuity at a point and sequential continuity at a point are equivalent. Sierpiński established not merely that all known proofs of equivalence relied on the Axiom but that this equivalence was itself equivalent to the following non-constructive assumption, a new form of the Axiom weaker than the Denumerable Axiom:

(4.1.1) If $A_1, A_2, \ldots$ is a disjoint family of non-empty subsets of $\mathbb{R}$, then there is a sequence $a_1, a_2, \ldots$ of real numbers such that different terms belong to different sets A_n. [1916, 689]

---

[1] Janiszewski's article is translated in Kuzawa 1968, 112–118. For a detailed account of the rise of the Warsaw school, see especially Kuratowski 1980 and Kuzawa 1968, but also Kuratowski 1975 and Sierpiński 1959.

[2] Sierpiński 1963, 27. On the origin of this journal, see also Kuzawa 1970. Lebesgue encouraged the journal but feared that its scope was too narrow [1922a, 36]. He used its pages to publish his own research as well as to reiterate his arguments against the Axiom [1921, 259–260].

Except for a few cases, such as the Trichotomy of Cardinals, where a proposition had been shown to be equivalent to the Axiom, this was the first attempt to establish conclusively that a proposition's use of the Axiom was unavoidable. Yet, like all other mathematicians at the time, Sierpiński lacked a method for proving that (4.1.1) required the Axiom.[3] Furthermore, he did not work within an axiomatic system for set theory, and thus in many instances he could not make precise claims about the dependence of theorems on the Axiom.

All the same, Sierpiński [1916] emphasized the fact that every known proof of various theorems discovered by Baire and Lebesgue relied on the Denumerable Axiom. Sierpiński's first example was Baire's result (1.7.4) that each function in Baire class two is the iterated limit of some double sequence of continuous functions $f_{m,n}$. Even more important was Sierpiński's second example:

(4.1.2) Lebesgue measure is countably additive.

Here Sierpiński recognized that (4.1.2), the most characteristic feature of Lebesgue measure, depended on the Denumerable Axiom, which Lebesgue opposed. Sierpiński left the reader to draw his own conclusions as to how these two French critics of the Axiom should respond to its appearance so deep within their own research. In fact, they ignored the challenge.

On the other hand, Lebesgue did respond to Sierpiński's article [1917a] concerning non-measurable sets. There Sierpiński had shown that each of three propositions implies the existence of a non-measurable set, even without the Axiom. The first was the theorem (3.6.2) that the family of all denumerable subsets of $\mathbb{R}$ has the power of the continuum, while the others concerned real functions:

(4.1.3) Baire class two has the power of the continuum.

(4.1.4) The set of all real functions can be ordered.

Lebesgue [1918, 238] mistakenly believed that, in order to obtain a non-measurable set from (3.6.2), Sierpiński had inadvertently used the Axiom. Probably Lebesgue did not respond to the other cases because Sierpiński had obtained a non-measurable set from (4.1.3) by showing, without the Axiom, that (4.1.3) implies (3.6.2); as for (4.1.4), Sierpiński had provided no details. Nevertheless, the error in Lebesgue's argument remained in force.

In addition to determining where the Axiom had been used implicitly, Sierpiński wished to ascertain where it could be avoided. The same year he discovered how to establish without the Axiom certain theorems previously

---

[3] This was shown by Jaegermann [1965], using Cohen's method.

shown only through its implicit use. Such was the case for the Cantor–Bendixson Theorem,[4] as well as for Julius König's result that the continuum does not have power $\aleph_\omega$ (see 2.1). In the paper containing the latter proof, written jointly with Luzin, there were mentioned two more propositions whose known demonstrations depended on the Axiom:

(4.1.5) If $\mathbb{R}$ is partitioned into two sets A and B, then either A or B has the power of the continuum.

(4.1.6) If B is the projection of a subset A of $\mathbb{R}^2$ into $\mathbb{R}$, then $\bar{\bar{B}} \leq \bar{\bar{A}}$.[5]

In particular, they noted that (4.1.6) sufficed to prove (4.1.5) without the Axiom. At the time Sierpiński did not inquire about the relation between the Axiom and the following proposition which generalized (4.1.6):

(4.1.7) For every set A and function f, $\overline{\overline{f''A}} \leq \bar{\bar{A}}$.

A year later, however, he remarked that (4.1.6) was a special case of the Partition Principle [1918, 109]. Indeed, (4.1.7) is equivalent to this principle.

Of all Sierpiński's investigations concerning the Axiom, the most impressive remains his lengthy survey [1918], in which he detailed the Axiom's contributions to set theory and analysis. There he assumed an attitude of studied neutrality toward the Axiom. At no point did he assert its truth or falsehood. Nevertheless, he provided three types of substantive evidence for the Axiom:

(i) Many particular cases of the Axiom had been verified without using it.
(ii) Mathematicians had deduced numerous consequences from the Axiom, none of which had led to a contradiction.
(iii) The Axiom was essential to the proofs of many important theorems in set theory and analysis.

For these reasons he believed that it should be determined precisely which proofs relied on the Axiom—even if, like many mathematicians, one rejected it [1918, 97–98].

Much of the controversy surrounding the Axiom, Sierpiński contended, arose from divergent interpretations of its meaning. In this vein he considered the name "Axiom of Choice" to be inappropriate. For the Axiom did not enable a mathematician to choose an element *effectively* even from a single non-empty set. In fact there existed sets, proven non-empty by the Axiom, in which it was not known how to determine a single element uniquely, that is, to give a characteristic property distinguishing this element

[4] Sierpiński 1917.
[5] Luzin and Sierpiński 1917.

from all others in the set. One such set was that of all non-measurable subsets of $\mathbb{R}$. On the other hand he knew of no set, shown to be non-empty without the Axiom, in which a particular element could not be determined.

For Sierpiński, the Axiom had to be interpreted as merely asserting the abstract existence of a choice function, an attitude shared by many supporters of the Axiom but not by its critics. Concomitantly, he raised the question of what it signified when a proposition was deduced from the Axiom. According to Sierpiński, Luzin viewed such a deduction as establishing that the proposition in question could not be proved to be false, but certainly did not prove it to be true. Indeed Luzin believed that, although the Axiom would never lead to a contradiction, it was useful *solely* as a heuristic device.[6]

At this point Sierpiński considered particular cases of the Axiom, in the guise of the Multiplicative Axiom (2.7.1), by varying the cardinality of the infinite family M of disjoint sets involved. He credited Luzin with a notation that facilitates the treatment of the Axiom's various cases: $(\mathfrak{m}, \mathfrak{n})$, where $\mathfrak{m}$ is the power of M and $\mathfrak{n}$ is the least upper bound of $\{\bar{\bar{P}}: P \in M\}$. It will be useful to revise this notation in accord with later usage by employing $\mathbf{C}^{\mathfrak{m}}_{\leq \mathfrak{n}}$ for Luzin's $(\mathfrak{m}, \mathfrak{n})$. Then $\mathbf{C}^{\mathfrak{m}}_{\mathfrak{n}}$ is $(\mathfrak{m}, \mathfrak{n})$ if every set in M has power $\mathfrak{n}$, and $\mathbf{C}^{\mathfrak{m}}_{<\mathfrak{n}}$ is $(\mathfrak{m}, \mathfrak{n})$ if every set in M has power less than $\mathfrak{n}$. When a subscript or superscript is omitted, there is no limitation on the corresponding cardinality.[7] Thus $\mathbf{C}^{\aleph_0}$ is the Denumerable Axiom, and $\mathbf{C}_2$ is the Axiom restricted to any family of unordered pairs. Eventually, the clear recognition of the Axiom's different cases led to deep investigations of their relative deductive strength. Here Sierpiński already raised the question of determining

(1) those cases of $\mathbf{C}^{\mathfrak{m}}_{\leq \mathfrak{n}}$ which imply a given proposition,
(2) those cases implied by the proposition, and
(3) those cases equivalent to it.

In his survey he provided examples of all three, while acknowledging that it was not always possible to establish an equivalence [1918, 104].

Sierpiński's first examples, taken from set theory, concerned the various definitions of finiteness. All known proofs that a set is finite if and only if it is Dedekind-finite relied on the Denumerable Axiom. Furthermore, he mentioned several other definitions of finite set which had thus far been shown equivalent to Dedekind-finiteness only through using $\mathbf{C}^{\aleph_0}$. All of these equivalences could be deduced without the Axiom from the following proposition, but its known proofs depended on the Axiom:

(4.1.8) If A is uncountable and B is denumerable, then $A - B$ has the same power as A.[8]

---

[6] Sierpiński 1918, 98–99, 103.

[7] For this modern notation, see Azriel Levy 1965, 222.

[8] Sierpiński 1918, 104–108. However, (4.1.8) can be proved in **ZF** when A is the set of all real numbers; see Jech 1978, 32.

Likewise, the equivalence of finiteness and Dedekind-finiteness sufficed to obtain two further propositions, whose demonstrations used the Axiom:

(4.1.9) If A is infinite and B is denumerable, then $A \cup B$ has the same power as A.

(4.1.10) If $\aleph_0 < \bar{\bar{A}}$ and $\aleph_0 = \bar{\bar{B}}$, then $\aleph_0 < \overline{\overline{A-B}}$.

In contrast to (4.1.8), Sierpiński observed, the Axiom was not needed to show that if A is uncountable, B is denumerable, and $A-B$ has a denumerable subset, then $A-B$ has the same power as A. Six years later one of his students, Alfred Tarski, investigated these questions in much greater depth (see 4.2).

Within cardinal arithmetic, Sierpiński remarked, the Axiom could be seen to play a role, sometimes quite subtle, in many propositions. For instance, no one knew how to establish without the Axiom that:

(4.1.11) If an uncountable set A has the same power as each of its uncountable subsets, then A has the power $\aleph_1$.

Nor was it known how to prove $\aleph_1 \leq 2^{\aleph_0}$ without using $\mathbf{C}^{\aleph_1}$ even though the Axiom was not required to partition $\mathbb{R}$ into $\aleph_1$ non-empty sets, as he proceeded to show. Moreover, one could demonstrate without the Axiom the proposition that the family of all infinite sequences of real numbers has the power of the continuum, but not the proposition (3.6.2) that the family of all denumerable subsets of $\mathbb{R}$ has the power of the continuum. Yet, since $\mathbb{R}$ is an ordered set, he could deduce that the family of all finite subsets of $\mathbb{R}$ has the power of the continuum without any use of the Axiom [1918, 111]. This last comment suggests that Sierpiński already possessed the crux of the argument employed by Kuratowski six years later to prove that if every set can be ordered, then $\mathbf{C}_{<\aleph_0}$ holds (see 4.2).

One proposition whose importance Sierpiński underlined concerned unions and their cardinality:

(4.1.12) If every member of an infinite set A has the power of A, then the union of A has the power of A.[9]

[9] The author would like to point out that (4.1.12) and (1.4.12) are jointly equivalent to the Axiom, for they jointly imply proposition (4.3.2) that $\mathfrak{m} = \mathfrak{m}^2$ for every infinite $\mathfrak{m}$, which Tarski [1924] proved equivalent to the Axiom. To establish this assertion, suppose that A is any infinite disjoint family of sets, each equipollent to A, where $\bar{\bar{A}} = \mathfrak{m}$. By (4.1.12), A is equipollent to $\bigcup A$. Since each a in A is equipollent to $A \times \{a\}$, then from (1.4.12) it follows that $\bigcup A$ is equipollent to $\bigcup \{A \times \{a\} : a \in A\}$, *i.e.*, to $A \times A$. Thus $\mathfrak{m} = \mathfrak{m}^2$, as desired.

Even very restricted cases of this proposition, such as the Countable Union Theorem, relied on the Axiom. In fact, Sierpiński succeeded in establishing a stronger connection:

(4.1.13) The Countable Union Theorem implies $\mathbf{C}_{\aleph_0}^{\aleph_0}$.

For if $A_1, A_2, \ldots$ was a sequence of disjoint denumerable sets, then the Countable Union Theorem yielded that $\bigcup_{i=1}^{\infty} A_i$ was denumerable and hence well-ordered. Thus the set consisting of the first element in each A_i was a choice set for $A_1, A_2, \ldots$. On the other hand, he did not know whether the converse of (4.1.13) was true, but only that $\mathbf{C}_{2^{\aleph_0}}^{\aleph_0}$ yields the Countable Union Theorem. The Axiom also appeared to be necessary to demonstrate (4.1.12) when A had the power $2^{\aleph_0}$ (*cf.* 1.7.2), despite the fact that $\mathbb{R}$ could be partitioned into $2^{\aleph_0}$ sets of power $2^{\aleph_0}$ without using the Axiom. A particularly useful case of the Countable Union Theorem was the following proposition:

(4.1.14) $\mathbb{R}$ is not the union of a countable family of countable sets.

Except by employing the Axiom, it was not even known how to ensure that a countable union of countable sets does not have a power greater than that of the continuum.[10] In the same vein, the Axiom appeared necessary to prove that:

(4.1.15) $\mathbb{R}$ cannot be partitioned into a family F of sets such that $\overline{\overline{F}} > 2^{\aleph_0}$.

For the usual demonstration of (4.1.15) relied on the Partition Principle [1918, 112–114].

Various cardinal inequalities, Sierpiński continued, depended unavoidably on the Axiom. Of these the most familiar was the Trichotomy of Cardinals. On the other hand, there existed many other equalities and inequalities which had been proven only via the Axiom but which had not been shown equivalent to it. Among these was proposition (1.6.9) that for every infinite cardinal $\mathfrak{m}$, $2^{\mathfrak{m}} = \mathfrak{m}^{\mathfrak{m}}$.[11] The same was true of the propositions (4.3.4) and (4.3.5): For all cardinals $\mathfrak{m}_1, \mathfrak{m}_2, \mathfrak{n}_1, \mathfrak{n}_2$, if $\mathfrak{m}_1 < \mathfrak{m}_2$ and $\mathfrak{n}_1 < \mathfrak{n}_2$, then $\mathfrak{m}_1 + \mathfrak{n}_1 < \mathfrak{m}_2 + \mathfrak{n}_2$ and $\mathfrak{m}_1\mathfrak{n}_1 < \mathfrak{m}_2\mathfrak{n}_2$. If one replaced $<$ everywhere by $\leq$ in these two propositions, then their proofs did not require the Axiom.

[10] On the other hand, it can be deduced in **ZF** that $\aleph_2$ is not a countable union of countable sets; see Jech 1978, 28. By contrast, Feferman and Levy [1963] gave a model of **ZF** in which (4.1.14) is false.

[11] Sierpiński 1918, 114–116. Schoenflies [1913, 180] had also pointed out the use of the Axiom in proving (1.6.9).

However, Sierpiński continued, the logician Stanisław Leśniewski had discovered that (4.1.16), of which (4.1.5) is a special case, implies the Trichotomy of Cardinals and hence is equivalent to the Axiom:

(4.1.16) The sum of two infinite cardinals $\mathfrak{m}$ and $\mathfrak{n}$ cannot be greater than both $\mathfrak{m}$ and $\mathfrak{n}$.

In other words, if $\mathfrak{m}$ and $\mathfrak{n}$ are infinite cardinals such that $\mathfrak{m} < \mathfrak{p}$ and $\mathfrak{n} < \mathfrak{p}$, then $\mathfrak{m} + \mathfrak{n} \neq \mathfrak{p}$. Finally, Sierpiński pointed out the Axiom's intimate connection with one other proposition involving unions:

(4.1.17) If $S = \bigcup_{i=1}^{\infty} E_i$, where $E_1, E_2, \ldots$ is a sequence of disjoint sets, and if $\overline{\overline{T}} < \overline{\overline{S}}$, then $\overline{\overline{T}} < \overline{\overline{\bigcup_{i=1}^{n} E_i}}$ for some n.

In 1926 Tarski claimed (4.1.17) to be equivalent to the Axiom, but he did not publish a proof of this equivalence until two decades later.[12]

As in cardinal arithmetic, many results in the theory of ordered sets depended on the Axiom. Such was the case, Sierpiński noted, for the proposition (1.6.6) that every ordered set lacking a subset of type $*\omega$ is well-ordered, as well as for the following equivalent result:

(4.1.18) Every ordered set without a last element has a subset of type ω.

Furthermore, the Axiom appeared liberally in Hausdorff's incisive researches on ordered sets, such as the next proposition:

(4.1.19) Every ordered set is cofinal with some well-ordered set.[13]

Proposition (4.1.19) was later named the Cofinality Principle.

The theory of point-sets in $\mathbb{R}^n$ connected the Axiom's uses in set theory with those in real analysis. Here Sierpiński built on his paper of 1916 by distinguishing carefully between a limit point and a sequential limit point.[14] Consequently, the notions of closed set, set dense-in-itself, perfect set, derived set, and isolated point all bifurcated (*cf.* 1.2). Their respective equivalence depended on that between limit point and sequential limit point. While in 1916 Sierpiński had established that his proposition (4.1.1), a weak form of the Axiom, is true if and only if continuity and sequential continuity are equivalent, here he demonstrated that (4.1.1) is equivalent to the inter-

[12] Lindenbaum and Tarski 1926, 312; Tarski 1948, 82–84. See 4.3.

[13] Sierpiński 1918, 117–118; *cf.* Hausdorff 1908, 444, and 1914, 132. Morris [1969] showed that, contrary to the claim of Felgner [1969], the proposition (4.1.19) is weaker than the Axiom. On the other hand, Morris found that (4.1.19) and the Order Extension Principle are jointly equivalent to the Axiom. (See also Harper and Rubin 1976.)

[14] He designated a limit point as a *point d'accumulation* and a sequential limit point as a *point limite*.

changeability of the notions of limit point and sequential limit point. Many theorems—including those of Bolzano–Weierstrass, Cantor–Bendixson, and Heine–Borel—could dispense with the Axiom if phrased in terms of limit points, but required the Axiom if expressed in terms of sequential limit points. Thus one knew how to demonstrate the following theorems without the Axiom if one omitted the words in parentheses, but only with the Axiom if one included these words: Every (sequentially) isolated subset of $\mathbb{R}^n$ is countable; every (sequentially) closed subset of $\mathbb{R}$, bounded above, contains its least upper bound; if A' denotes the (sequentially) derived set of A, then A'' is a subset of A'; every (sequentially) perfect set has the power of the continuum.[15]

Within the theory of point-sets, Sierpiński stressed that the Axiom was particularly intertwined with Lindelöf's notion of condensation point. In the absence of the Axiom, it was not known how to show that every uncountable subset A of $\mathbb{R}$ possesses a condensation point (which might or might not belong to A), but a demonstration was known if it was also required that A be bounded. Yet even then the Axiom was apparently needed to ensure that A possesses more than one such point. While it had been established that the set S of all condensation points of A is closed, all known proofs relied on the Axiom to show that S is dense-in-itself and hence perfect. Of particular interest was a theorem on condensation points which was equivalent to the Countable Union Theorem for subsets of $\mathbb{R}$:

(4.1.20) Every uncountable subset A of $\mathbb{R}$ has more than one condensation point. [1918, 125]

Finally, the Axiom was used to prove Lindelöf's Covering Theorem (1.7.7) that every family of open intervals covering a subset A of $\mathbb{R}$ has a countable subfamily that covers A.

As for real analysis, Sierpiński underlined the Axiom's role in the theory of Lebesgue measure on $\mathbb{R}$. Not only was the Axiom required in order to deduce that the union of countably many measurable sets is measurable, but even to demonstrate a more limited result:

(4.1.21) The union of countably many sets of measure zero has measure zero.

For (4.1.21) implied the proposition (4.1.14) that $\mathbb{R}$ is not the union of a countable family of countable sets. While the Axiom was not needed to prove that every measurable set is the union of some closed set and some set of arbitrarily small positive measure, it was apparently needed to show that:

(4.1.22) Every measurable set E equals $A \cup B$ for some F_σ set A and some set B of measure zero.

[15] Sierpiński 1918, 119–126; *cf.* (1.2.5).

On the other hand, he added, it was possible to establish without the Axiom that $E = (A—C) \cup B$ where E, A, B were as in (4.1.22) and C was some set of measure zero. In the same vein, it was not known how to dispense with the Axiom in the proof of Luzin's Theorem [1912] that every measurable function is continuous on an interval if one neglects a set of arbitrarily small positive measure, for this theorem too implied (4.1.14). Luzin's Theorem followed easily from what Borel [1912b, 414] had called the third fundamental theorem:

(4.1.23) Given a measurable function $f(x)$ that is bounded almost everywhere and given two arbitrarily small positive numbers ε and η, there is a polynomial $P(x)$ such that the set $\{x: |f(x) - P(x)| > \varepsilon\}$ has measure less than η.[16]

Thus Borel's theorem also required the Axiom. Likewise, the Axiom entered all known proofs of Egoroff's Theorem that if a sequence $f_1, f_2, \ldots$ of measurable functions converges almost everywhere on an interval I, then $f_1, f_2, \ldots$ converges uniformly on $I—E$, where E is a set of arbitrarily small positive measure. Sierpiński concluded that measure theory depended heavily on the Axiom even to obtain its most basic results for the real line.[17]

Measure theory also connected the Axiom to Baire's classification of real functions. Sierpiński noted that all known examples of a non-measurable set depended on the Axiom, as did the proof of the proposition (2.3.1) that every function in Baire's classification is analytically representable and hence measurable. Without using the Axiom, Sierpiński had deduced that if Baire class two has the power of the continuum, then there exists a non-measurable set. Such a set also existed if it was even true that:

(4.1.24) Baire class two can be ordered.

Furthermore, the Axiom appeared necessary to show that there exists a real function escaping from Baire's classification, or even a function in Baire class four. On the other hand, in the absence of the Axiom it was not known how to demonstrate that if $f_1(x), f_2(x), \ldots$ is a convergent sequence of functions in Baire's classification, then its limit also belongs to Baire's classification [1918, 133–137].

In a similar fashion Sierpiński discussed how the Axiom intruded on the theory of analytic sets, developed by Suslin and Luzin (see 3.6). Among other results, Suslin had claimed to give an effective example of a real function that did not belong to Baire's classification. More precisely, while this function was uniquely defined, the proof that it escaped from Baire's classification relied on the Axiom by using the proposition (3.6.5) that every Borel

[16] *Almost everywhere* means except on a set of measure zero.

[17] Sierpiński 1918, 126–128, 138–139.

set in $\mathbb{R}$ is an analytic set. But (3.6.5) depended on the Axiom via the following proposition:

(4.1.25) The union of a denumerable family of analytic sets in $\mathbb{R}$ is an analytic set.

In order to establish (4.1.25), one used the Axiom to choose a defining system for each analytic set in the denumerable family [1918, 136].

Finally, Sierpiński examined the Axiom's role in the construction of examples. In addition to such implausible examples in $\mathbb{R}^n$ as a non-measurable set and an uncountable set lacking a perfect subset, he cited several others shown to exist by means of the Axiom. The first of these was due to Mazurkiewicz, the second to Luzin, and the third to both Luzin and Sierpiński:

(4.1.26) There is a subset A of $\mathbb{R}^2$ such that every line has exactly two points in common with A.

(4.1.27) There is an uncountable subset of $\mathbb{R}$ which is of first category on every perfect set.

(4.1.28) There is an uncountable subset A of a given interval I such that every continuous bijection of I transforms A into a set of measure zero.

While the Axiom led to many such instances of sets with counter-intuitive properties, Sierpiński stressed that other improbable sets could be shown to exist even without the Axiom—such as a plane set partitioned into two sets and congruent to each of them. Thus he concluded that results such as Hausdorff's paradoxical decomposition of a sphere did not constitute evidence against the Axiom.[18]

Over the next two decades Sierpiński continued to explore the Axiom and its consequences. Many of these investigations, such as [1920a] and [1920d], concerned the interplay between the Axiom and non-measurable sets. Earlier, Luzin and Sierpiński [1917a] had obtained from a well-ordering of $\mathbb{R}$ the proposition that the closed interval [0, 1] can be partitioned into $2^{\aleph_0}$ non-measurable sets. Twenty years later, Sierpiński deduced from the same assumption that there exists an almost disjoint family F of non-measurable subsets of [0, 1] such that the power of F is greater than $2^{\aleph_0}$.[19]

[18] *Ibid.*, 140–145.

[19] Sierpiński 1937. Two sets are *almost disjoint* if their intersection has a power less than that of each of the sets.

He had established a related theorem from cardinal arithmetic during the intervening period [1928]:

(4.1.29) Every infinite set of power $\mathfrak{m}$ can be partitioned into a family, whose power is greater than $\mathfrak{m}$, of almost disjoint sets.

Also in cardinal arithmetic, he had derived from the Axiom a result that extended early work by Cantor to the case where the Continuum Hypothesis is false (*cf*. 1.4.9):

(4.1.30) Every subset A of $\mathbb{R}^n$ can be partitioned into a nowhere dense set and a countable family of homogeneous sets.[20]

However, even more than the Axiom, Sierpiński studied the Continuum Hypothesis in numerous articles which culminated in his book *Hypothèse du continu* [1934a].

Another matter of continuing interest to Sierpiński was the relationship between constructive mathematics and the Axiom. In one article [1921] on effective examples, he stressed that the Axiom was often essential, not to construct a particular object, but to show that the object constructed had the desired property. This was the case for Lebesgue's example [1905] of a real function escaping from Baire's classification, as well as of Sierpiński's own example [1921] of a well-ordered set having the power of the continuum. While this set was constructed effectively, without the Axiom one could only prove that its power was neither greater nor less than that of the continuum.

Unfortunately, Sierpiński made a serious error as to which sets could be obtained effectively in the Borel hierarchy. This error is particularly surprising since Sierpiński had correctly recognized that, in the absence of the Axiom, $\mathbb{R}$ may be a countable union of countable sets (*cf*. 4.1.14). In this case every set of real numbers would be a Borel set, and even an $F_{\sigma\delta\sigma}$ set, as in the later Feferman–Levy model of **ZF**. Yet Luzin and Sierpiński published a joint paper [1923] purporting to define an effective example of a subset of $\mathbb{R}$ that was not a Borel set. At the same time Sierpiński's student Kuratowski [1923] claimed to give an effective example of a set in each of the ω_1 classes of the Borel hierarchy, and soon Sierpiński [1924] put forward a supposedly effective example different from Kuratowski's. In this instance the nuances of the Axiom misled even such astute and careful mathematicians as Luzin, Kuratowski, and Sierpiński.

In what respect did Sierpiński act as an advocate for the Axiom of Choice when he wrote his survey article of 1918? To be sure, he did not actively defend the Axiom as Zermelo had done a decade earlier. Nor did Sierpiński even claim that the Axiom was true. Like the subject which he discussed, his

---

[20] Sierpiński 1920b. A subset B of $\mathbb{R}^n$ is *homogeneous* if, whenever it intersects an open set, the intersection always has the same cardinality as B.

position remained subtle. On the one hand, he illustrated in detail how thoroughly the Axiom was intertwined, not just with set theory, but with real analysis as well. He argued that a number of the Axiom's uses were unavoidable because they implied some special case of the Axiom. On the other hand, he granted that many mathematicians had opposed the Axiom and that their position might be justified. Nevertheless, he turned this opposition into a compelling reason to investigate the Axiom's role. If one opposed the Axiom, surely it became vital to recognize when it appeared in a proof, when it could be avoided, and when a proposition depended on it in an essential way. The challenge which Sierpiński's survey laid down was taken up primarily by himself and his students, such as Tarski, in numerous articles on set theory, topology, and real functions. By serving as a focus for their research, the Axiom led to the discovery of techniques useful in solving a broad spectrum of mathematical and metamathematical problems.

4.2 Finite, Infinite, and Mediate

Many opponents of the Axiom of Choice used the notion of finiteness as high ground from which to attack the Axiom. Thus it is of interest to determine whether the theory of finite sets can be developed fully without the Axiom. To a large extent the answer depends upon the particular definition of finite set that is adopted. Here the term finite will continue to designate a set which is either empty or equipollent to $\{1, 2, \ldots, n\}$ for some positive integer n.

Another common definition was the one which Dedekind offered in 1888 and which was later called Dedekind-finite: A set A is Dedekind-finite if and only if A is not equipollent to any of its proper subsets. By late in 1905 Russell had recognized that one could prove the equivalence of the notions of finite set and Dedekind-finite set only by using the Axiom. Furthermore, he remarked that the Axiom appeared in all known proofs that the power set of a Dedekind-finite set is Dedekind-finite.[1] Later, in *Principia Mathematica*, Russell and Whitehead added that the Denumerable Axiom sufficed to obtain these results. Thus the boundary between the finite and the infinite was affected by one's acceptance or rejection of weak forms of the Axiom. In the latter case, there might exist mediate cardinals which were greater than every finite cardinal but were not Dedekind-infinite, a possibility that Russell and Whitehead seriously considered.[2]

---

[1] Russell 1906, 49; *cf.* (1.3.4).

[2] Russell and Whitehead 1912, 190, 278–280; see 3.6.

In his survey article on the Axiom, Sierpiński investigated the relationship between definitions of finiteness more deeply. There he mentioned several such definitions, each of which could be shown equivalent to the usual definition, or to Dedekind's, without the Axiom.[3] This was a clue that such definitions split naturally into more than one class. Any two definitions in the same class could be shown equivalent without the Axiom, but any two from different classes required the Denumerable Axiom to establish their equivalence. While Sierpiński was only aware of two such classes, their number was soon to increase.

In 1924, influenced by Sierpiński's article, Tarski published an extensive paper which systematized the theory of finite sets. To do so, he operated explicitly within Zermelo's system but used neither the Axiom of Choice nor the Axiom of Infinity. After he introduced still another definition for finite set, Tarski proved it equivalent to the usual one.[4] From his definition he developed those theorems that were reasonable for finite sets, including a form of mathematical induction and the following propositions, among others: A set is finite if and only if its power set is finite; if A is a family of sets, then the union of A is finite if and only if both A is finite and every member of A is finite.[5] If the term finite was replaced by Dedekind-finite in each of these propositions, the resulting propositions depended on the Denumerable Axiom (see 1.3).

In addition, Tarski considered the role of the Axiom of Choice vis-à-vis propositions about finite sets. It was known how to show that every finite set can be ordered, but not how to deduce the more general proposition that every set can be ordered (later called the Ordering Principle), except by using the Axiom. He remarked that even without the Axiom the following proposition (which could be viewed as the equivalence of the usual definition of finite set with that defined below by the clause after "only if") was itself equivalent to the Ordering Principle:

(4.2.1) A set A is finite if and only if every relation which orders A is a well-ordering.

Furthermore, Kuratowski had supplied Tarski with a proof of a related result:

The Ordering Principle implies $\mathbf{C}_{<\aleph_0}$.[6]

In other words the Axiom, when restricted to families of finite sets, followed from (4.2.1) and hence from the equivalence of two definitions for finite set, a theme which Tarski was to explore in detail a decade later.

---

[3] Sierpinski 1918, 104–108.

[4] Tarski 1924a, 48–49. He defined a set A to be finite in his sense if every non-empty family of subsets of A contains an $\subseteq$-minimal element.

[5] *Ibid.*, 48–63.

[6] *Ibid.*, 78, 82.

At the end of his article, Tarski presented some open problems concerning the equivalence of five definitions of finite set:

(4.2.2) A set A is finite if and only if every non-empty family of subsets of A has an $\subseteq$-maximal element.

(4.2.3) A set A is finite if and only if every non-empty family of subsets of A, ordered by inclusion, has an $\subseteq$-maximal element.

(4.2.4) A set A is finite if and only if no proper subset of the power set of A is equipollent to the power set of A.

(4.2.5) A set A is finite if and only if no proper subset of A is equipollent to A. (Dedekind-finiteness)

(4.2.6) A set A is finite if and only if A is not the union of two disjoint non-empty sets, each equipollent to A.

Tarski proved that a set finite in any of these five senses, say (4.2.m), could be shown to be finite in sense (4.2.n) if m was less than n (where $2 \leq m,n \leq 6$). But if m was greater than n, the proof appeared to require the Axiom. In any case, the definition on which Tarski had based his article, namely (4.2.2), was equivalent to the usual definition without the Axiom. Hence there existed at least five classes of definitions for finite set. Within each class any two definitions could be shown equivalent without the Axiom, but one needed the Axiom to obtain such an equivalence between classes [1924a, 93–95]. In this way there arose a hierarchy of classes of such definitions, a hierarchy determined by the Axiom. Later mathematicians established that all five of Tarski's classes were, in fact, distinct (see 4.9).

Thus it was possible that, if the Denumerable Axiom were false, there might exist sets which were infinite in the usual sense but were finite in one or more of the other senses. Independently of Tarski this possibility was studied by two authors, both influenced by *Principia Mathematica*, for the case of mediate cardinals. The Polish logician Leon Chwistek briefly discussed infinite Dedekind-finite sets but did not contribute significantly to their understanding [1924, 136]. Meanwhile the English mathematician Dorothy Wrinch, an associate of Russell's, generalized the notion of mediate cardinal in a different direction. She defined a mediate cardinal of degree $\aleph_\alpha$ to be a cardinal $\mathfrak{m}$ such that $\mathfrak{m}$ is greater than all cardinals less than $\aleph_\alpha$ but is neither greater nor less than $\aleph_\alpha$ itself. Thus the usual mediate cardinals were the mediate cardinals of degree $\aleph_0$. She pointed out the Axiom's equivalence to the proposition that there do not exist mediate cardinals of any degree. Correctly, she found no reason to believe the stronger assertion that the nonexistence of the usual mediate cardinals implies the Axiom [1923].

Whether such nonexistence yielded the Denumerable Axiom was a question that she did not investigate. In fact, it does not.

The properties of mediate cardinals received little attention for several decades.[7] On the other hand, Tarski's article [1924a] about finite sets motivated Fraenkel [1927, 147] to pose two additional questions concerning their definitions: Can the definitions in Tarski's five classes be shown equivalent to each other by some assumption weaker than the Axiom? Do there exist two definitions of finite set whose equivalence implies the Axiom? The first question received an affirmative answer only recently, when the Israeli mathematician Gershon Sageev showed the proposition (1.7.10), that $\mathfrak{m} + \mathfrak{m} = \mathfrak{m}$ for all infinite $\mathfrak{m}$, to be weaker than the Axiom [1975]. As for the second, in 1938 Tarski himself established that its answer was positive by offering three definitions for finite set. The equivalence of each of them to the usual definition implied the Axiom and hence was equivalent to it:

(4.2.7) A set A is finite if and only if, for each well-ordering S of A, the relation converse to S well-orders A.

(4.2.8) A set A is finite if and only if A has at most one element or can be partitioned into two sets B and C, each of smaller power than A.

(4.2.9) A set A is finite if and only if A has at most one element or is of smaller power than the Cartesian product $A \times A$.

For, he argued, if the definition of finiteness in (4.2.7) was equivalent to the usual definition, then the Well-Ordering Theorem held. In the second instance (4.2.8), it followed that for every infinite $\mathfrak{m}$, if $\mathfrak{m} = \mathfrak{n} + \mathfrak{p}$, then $\mathfrak{m} = \mathfrak{n}$ or $\mathfrak{m} = \mathfrak{p}$, a form of Leśniewski's proposition (4.1.17) which was known to imply the Axiom. And in the third instance, $\mathfrak{m} = \mathfrak{m}^2$ for every infinite $\mathfrak{m}$, a proposition which Tarski had proved equivalent to the Axiom in 1924 (see 4.3). In addition, he now stated an example of a definition of finite set whose equivalence with the usual definition implied an even stronger result, the Generalized Continuum Hypothesis:

(4.2.10) A set A is finite if and only if A has at most one element or there exists a set B whose power is greater than that of A and less than that of the power set of A. [1938, 163]

In sum, by 1938 Tarski had shown conclusively that the usual definition of finite set sufficed to develop the theory of finite sets without the Axiom but that many alternative definitions required the Axiom to do so. Moreover, a number of such alternative definitions implied the Axiom if they were taken

---

[7] For later developments, see Tarski 1965 and Ellentuck 1965.

to be equivalent to the usual definition for finite set. All of this suggested that the boundary between the finite and the infinite was considerably less clear than Zermelo's opponents had supposed. The Hungarian mathematician John von Neumann regarded this fuzziness as grounds for entertaining objections both to constructivist philosophies, such as intuitionism, and to set theory itself [1925, 240]. If the Axiom were rejected, there might exist not only mediate cardinals but many other types which were infinite in some senses but not in others, and thus the Axiom acted as a simplifying influence on definitions of finite set. Its opponents were unimpressed by this contribution, however, since they doubted the Axiom's truth.

4.3 Cardinal Equivalents

When in 1918 Sierpiński proposed that mathematicians actively seek to determine the deductive strength of various propositions relative to the Axiom of Choice, only a few such propositions were known to be equivalent to it. Chief among these were the Well-Ordering Theorem, the Multiplicative Axiom, and the Trichotomy of Cardinals. Gradually, mathematicians increased the number and variety of known equivalents, and by 1963 Herman and Jean Rubin had listed over one hundred of them. These two authors classified equivalents into cardinal forms, maximal principles, and algebraic forms—the subject matter of the next three sections respectively [1963, xi].

Such equivalents did not proliferate all at once. Most of the algebraic forms arose after 1950. The maximal principles, the first of which Hausdorff had formulated in 1909, attracted attention rather slowly, only becoming prominent with the publication of Zorn's Lemma in 1935. Moreover, the equivalents involving relations and functions, equivalents such as those that Russell and Whitehead had stated in *Principia Mathematica*, never stimulated enough interest to evolve into a separate category. The cardinal equivalents, those propositions in cardinal arithmetic that are equivalent to the Axiom, had their origin in Hartogs' paper [1915] on the Trichotomy of Cardinals and in Leśniewski's proposition (4.1.17): If $\mathfrak{m}$ and $\mathfrak{n}$ are infinite, then not both $\mathfrak{m} < \mathfrak{m} + \mathfrak{n}$ and $\mathfrak{n} < \mathfrak{m} + \mathfrak{n}$. But the most significant contribution to cardinal equivalents was made by Tarski during the period 1924–1926.

In 1924 Tarski published an article which demonstrated, within the framework of Zermelo's axiomatization, that each of seven propositions in cardinal arithmetic is equivalent to the Axiom. At that time the only known definition of cardinal number, in terms of sets or classes, was the Frege–Russell definition: The cardinal number of a set A is the class of all sets equipollent to A. Since such a class was not a set in Zermelo's system, this system offered no apparent means for defining cardinals, and so Tarski introduced them via two new postulates:

(1) Every set has a cardinal number.
(2) Two sets have the same cardinal number if and only if the sets are equipollent.

On the other hand, Tarski was well aware that his results on cardinal equivalents could be formulated in terms of equipollence alone, in which case his two postulates would not be necessary. As von Neumann recognized four years later, the Axiom could be used to define the cardinal of a set as the least ordinal equipollent to the set.[1]

The seven cardinal equivalents that Tarski published in 1924 were quite diverse, involving both addition and multiplication as well as inequalities. All but the first two of them generalized theorems known for finite cardinals to the infinite:

(4.3.1) $\mathfrak{m}\mathfrak{n} = \mathfrak{m} + \mathfrak{n}$ for every infinite $\mathfrak{m}$ and $\mathfrak{n}$.

(4.3.2) $\mathfrak{m} = \mathfrak{m}^2$ for every infinite $\mathfrak{m}$.

(4.3.3) If $\mathfrak{m}^2 = \mathfrak{n}^2$, then $\mathfrak{m} = \mathfrak{n}$.

Two of these equivalents stated the monotonicity of $+$ and $\cdot$:

(4.3.4) If $\mathfrak{m} < \mathfrak{n}$ and $\mathfrak{p} < \mathfrak{q}$, then $\mathfrak{m} + \mathfrak{p} < \mathfrak{n} + \mathfrak{q}$.

(4.3.5) If $\mathfrak{m} < \mathfrak{n}$ and $\mathfrak{p} < \mathfrak{q}$, then $\mathfrak{m}\mathfrak{p} < \mathfrak{n}\mathfrak{q}$.

The final equivalents were cancellation laws:

(4.3.6) If $\mathfrak{m} + \mathfrak{p} < \mathfrak{n} + \mathfrak{p}$, then $\mathfrak{m} < \mathfrak{n}$.

(4.3.7) If $\mathfrak{m}\mathfrak{p} < \mathfrak{n}\mathfrak{p}$, then $\mathfrak{m} < \mathfrak{n}$. [1924, 148]

Tarski's results relied heavily on earlier research by Bernstein and Sierpiński. In his thesis of 1901, Bernstein had established that if $\mathfrak{m}\mathfrak{n} = \mathfrak{m} + \mathfrak{n}$, then $\mathfrak{m} \leq \mathfrak{n}$ or $\mathfrak{n} \leq \mathfrak{m}$ (*cf.* 1.6.2). However, as Tarski recognized, Bernstein's proof had used arbitrary choices. To avoid the Axiom here, Tarski weakened Bernstein's proposition to the following lemma:

$$\text{If } \mathfrak{m}\aleph_\alpha = \mathfrak{m} + \aleph_\alpha, \text{ then } \mathfrak{m} \leq \aleph_\alpha \text{ or } \aleph_\alpha \leq \mathfrak{m}.$$

With the lemma in hand, Tarski easily established that (4.3.1) implies the Axiom, and hence is equivalent to it, by using one further theorem. This

[1] Von Neumann 1928, 727–731; see 4.9.

theorem, by which Sierpiński had sharpened a result of Hartogs [1915], was the existence for every infinite $\mathfrak{m}$ of an aleph $\aleph(\mathfrak{m})$ which is neither greater nor less than $\mathfrak{m}$.[2] From (4.3.1) and the lemma, Tarski deduced that $\mathfrak{m} \leq \aleph(\mathfrak{m})$ or $\aleph(\mathfrak{m}) \leq \mathfrak{m}$, and consequently that $\mathfrak{m} = \aleph(\mathfrak{m})$; thus every infinite cardinal was an aleph, and the Axiom followed [1924, 148–150]. To prove the remaining equivalences, he utilized similar techniques.

Before these results appeared in *Fundamenta Mathematicae* in 1924, Tarski sent Lebesgue a note showing that (4.3.2) is equivalent to the Axiom, and asked him to submit it to the *Comptes Rendus* of the Paris Academy of Sciences. Lebesgue returned the note on the grounds that he opposed the Axiom, but suggested sending it to Hadamard. When Tarski did as Lebesgue advised, Hadamard also returned the note—saying that, since the Axiom was true, what was the point of proving it from (4.3.2)?[3]

Two years later Tarski asserted that six more propositions are each equivalent to the Axiom. One of these was Sierpiński's proposition (4.1.18) that if $E_1, E_2, \ldots$ are disjoint sets such that $\overline{\overline{T}} < \overline{\overline{\bigcup_{i=1}^{\infty} E_i}}$, then $\overline{\overline{T}} < \overline{\overline{\bigcup_{i=1}^{n} E_i}}$ for some n. A second was the necessary and sufficient condition that $\mathfrak{n} - \mathfrak{m}$ is always defined when $\mathfrak{m} < \mathfrak{n}$:

(4.3.8) If $\mathfrak{m} < \mathfrak{n}$, then there is a unique $\mathfrak{p}$ such that $\mathfrak{n} = \mathfrak{m} + \mathfrak{p}$.

The four remaining equivalents were cancellation laws, two for addition and two for multiplication:

(4.3.9) $\mathfrak{m} + \mathfrak{p} = \mathfrak{m} + \mathfrak{q}$ implies that $\mathfrak{p} = \mathfrak{q}$ or that $\mathfrak{p} \leq \mathfrak{m}$ and $\mathfrak{q} \leq \mathfrak{m}$.

(4.3.10) If $\mathfrak{m} + \mathfrak{m} < \mathfrak{m} + \mathfrak{n}$, then $\mathfrak{m} < \mathfrak{n}$.

(4.3.11) If $\mathfrak{p}^{\mathfrak{m}} < \mathfrak{q}^{\mathfrak{m}}$, then $\mathfrak{p} < \mathfrak{q}$.

(4.3.12) If $\mathfrak{m}^{\mathfrak{p}} < \mathfrak{m}^{\mathfrak{q}}$ and $\mathfrak{m} \neq 0$, then $\mathfrak{p} < \mathfrak{q}$.

On the other hand, Tarski noted that the Axiom was not needed to establish either the converse of (4.3.10) or the related proposition that $\mathfrak{m} + \mathfrak{m} > \mathfrak{m} + \mathfrak{n}$ implies $\mathfrak{m} > \mathfrak{n}$.

These six equivalences, which Tarski considered as completing his article of 1924, were given without proof in an important paper summarizing many set-theoretic results that he and Lindenbaum had recently obtained [1926, 311–312]. An article entitled "L'arithmétique des nombres cardinaux," containing detailed proofs for these theorems, was supposed to appear in the next volume of *Fundamenta Mathematicae*, but for unknown reasons

[2] Sierpiński 1921, 393–394; *cf.* the discussion of Hartogs' work in 3.4.

[3] Personal communication from Tarski (10 March 1982).

it was never published. These equivalences, like most of the more than one hundred theorems given in the joint paper of 1926, remained undemonstrated in print until two decades later. In fact Sierpiński, who was acquainted with the joint paper, independently discovered a proof that (4.3.8) implies the Axiom, and eventually published this result [1947].

The joint paper also contained a proposition which Lindenbaum claimed to be equivalent to the Axiom and which bore a close relation to the Trichotomy of Cardinals. While $\bar{\bar{A}} \leq \bar{\bar{B}}$ if and only if there is a one–one mapping of A into B, Tarski defined $\bar{\bar{A}} \leq * \bar{\bar{B}}$ to mean that A is empty or that there is a mapping of B onto A. In the presence of the Axiom the notions of $\leq$ and $\leq *$ were equivalent. Relative to the rest of Zermelo's system, Lindenbaum asserted, the Axiom was equivalent to the Trichotomy of Cardinals with $\leq$ replaced by $\leq *$:

(4.3.13) $\mathfrak{m} \leq * \mathfrak{n}$ or $\mathfrak{n} \leq * \mathfrak{m}$.[4]

However, the proof of this equivalence too was not published until much later, and then by Sierpiński rather than Lindenbaum, who had been killed in 1941.[5]

In the context of their relation $\leq *$, Lindenbaum and Tarski [1926] discussed two propositions whose deductive strength relative to the Axiom remained uncertain:

(4.3.14) If $\mathfrak{m} \leq * \mathfrak{n}$, then not $\mathfrak{n} < \mathfrak{m}$.

(4.3.15) If $\aleph_0 \leq * \mathfrak{n}$, then $\aleph_0 \leq \mathfrak{n}$.

They pointed out that (4.3.15) is equivalent to the proposition (1.3.3) that every Dedekind-finite set is finite, and hence to a restricted form of the Trichotomy of Cardinals:

(4.3.16) $\mathfrak{m} \leq \aleph_0$ or $\aleph_0 \leq \mathfrak{m}$.

Furthermore, they added, (4.3.15) was equivalent to restricted versions of (4.3.6) and of (4.3.9) respectively:

(4.3.17) If $\mathfrak{m} + \aleph_0 < \mathfrak{n} + \aleph_0$, then $\mathfrak{m} < \mathfrak{n}$.

(4.3.18) If $\aleph_0 + \mathfrak{p} = \aleph_0 + \mathfrak{q}$, then $\mathfrak{p} = \mathfrak{q}$ or $\mathfrak{p}, \mathfrak{q} \leq \aleph_0$.

As for (4.3.14), it implied three consequences of the Axiom: the existence of a non-measurable subset of $\mathbb{R}$, the existence of an uncountable subset of $\mathbb{R}$

[4] Lindenbaum and Tarski 1926, 301, 312.

[5] Sierpiński 1948.

lacking a perfect subset, and Sierpiński's proposition (4.1.5) that $\mathbb{R}$ cannot be partitioned into two sets of smaller power than $\mathbb{R}$. Two years later, Tarski demonstrated that (4.3.14) also yields $\aleph_1 \leq 2^{\aleph_0}$.[6] Lindenbaum and Tarski mentioned only in passing a proposition which is apparently stronger than (4.3.14), since it is obtained from (4.3.14) by the Trichotomy of Cardinals, and of which (4.3.15) is a special case:

(4.3.19) If $\mathfrak{m} \leq * \mathfrak{n}$, then $\mathfrak{m} \leq \mathfrak{n}$.

In fact, (4.3.19) can easily be shown equivalent to the Partition Principle.

Finally, in the same article, Lindenbaum and Tarski explored the relationship between the Axiom of Choice and three forms of the Generalized Continuum Hypothesis:

(4.3.20) For every infinite cardinal $\mathfrak{m}$ there is no $\mathfrak{n}$ with $\mathfrak{m} < \mathfrak{n} < 2^{\mathfrak{m}}$.

(4.3.21) For every ordinal α there is no $\mathfrak{n}$ with $\aleph_\alpha < \mathfrak{n} < 2^{\aleph_\alpha}$.

(4.3.22) For every ordinal α, $2^{\aleph_\alpha} = \aleph_{\alpha+1}$. [1926, 313–314]

Without the Axiom, they asserted, one could demonstrate that (4.3.21) and (4.3.22) are equivalent and that (4.3.20) is equivalent to the conjunction of (4.3.22) and the Axiom. Consequently, the form (4.3.20) of the Generalized Continuum Hypothesis implies the Axiom, a surprising result whose proof was only published much later by Sierpiński [1947a].

Some of the lemmas that Lindenbaum and Tarski used to obtain these results shed light on the connection between the Axiom and (4.3.20). If a particular cardinal $\mathfrak{m}$ satisfies (4.3.20), then it will be said that CH($\mathfrak{m}$) holds:

(4.3.23) If CH($\mathfrak{m}$) and CH($\mathfrak{m} + \mathfrak{n}$) hold, then $\mathfrak{m} \leq \mathfrak{n}$ or $\mathfrak{n} \leq \mathfrak{m}$.

(4.3.24) If CH($\mathfrak{m}$), CH($2^{\mathfrak{m}}$), and CH($2^{2^{\mathfrak{m}}}$) all hold, then $\mathfrak{m}$, $2^{\mathfrak{m}}$, and $2^{2^{\mathfrak{m}}}$ are all alephs.

(4.3.25) If CH($\mathfrak{m}^2$) and CH($2^{\mathfrak{m}^2}$) hold, where $\mathfrak{m} \neq 0$, then $\mathfrak{m}$ and $2^{\mathfrak{m}}$ are alephs. [1926, 315]

Thus, in particular, if there is no cardinal $\mathfrak{n}$ such that $\aleph_0 < \mathfrak{n} < 2^{\aleph_0}$ or such that $2^{\aleph_0} < \mathfrak{n} < 2^{2^{\aleph_0}}$, then $\mathbb{R}$ can be well-ordered. Eventually, Specker [1954] published a proof of (4.3.24) and (4.3.25).

While the joint article of Lindenbaum and Tarski had little influence until Sierpiński explored it after the Second World War, Tarski's paper of 1924

---

[6] Tarski in Sierpiński 1928b, 227. Proofs for the other stated consequences of (4.3.15) were independently obtained by Sierpiński, who first published them in his [1947b].

intrigued the Rumanian mathematician Gabriel Sudan. After designating two cardinals as consecutive if no cardinal is strictly between them, Sudan expressed his proposition as follows:

(4.3.26) If $\mathfrak{m}$, $\mathfrak{n}$, and $\mathfrak{p}$ are infinite, and if $\mathfrak{n}$ and $\mathfrak{p}$ are equal or consecutive, then $\mathfrak{m}\mathfrak{n}$ and $\mathfrak{m}\mathfrak{p}$ are equal or consecutive. [1939]

Sudan's proof that (4.3.26) is equivalent to the Axiom relied heavily on the techniques found in Tarski's paper.

When in 1938 Tarski again discussed propositions that imply the Axiom, it was in the context of inaccessible cardinals. He termed a cardinal weakly inaccessible if it equalled $\aleph_\alpha$ for some regular limit ordinal α.[7] Early in the century, Hausdorff had first raised the question of whether or not such a cardinal existed [1908, 444], and a few years later had concluded that it would be so large as to have little relevance to ordinary set theory [1914, 131]. Yet the opposite turned out to be the case, for the existence of large cardinals could even affect $\mathbb{R}$.[8] In 1930 Sierpiński and Tarski introduced what they called strongly inaccessible cardinals, those uncountable cardinals which are not the product of fewer cardinals of lesser power.[9] Defined in this way, the notion presupposed the Axiom, but in 1938 Tarski gave several alternative formulations which did not, such as the following: The cardinal of set M is strongly inaccessible if M is not the union of fewer than $\overline{\overline{M}}$ sets of power less than $\overline{\overline{M}}$ and if M is not equipollent to the power set of any set. By means of the Axiom, Tarski showed every strongly inaccessible cardinal to be weakly inaccessible, but his proof of the converse relied, as it had to do, on the Generalized Continuum Hypothesis [1938a].

From these investigations Tarski discovered three further equivalents to the Axiom. He established the equivalence between the last two and the Axiom by using Zermelo's remaining postulates, but the first required Fraenkel's Axiom of Replacement as well (see 4.9):

(4.3.27) For every set A there is a set B such that if $X \subseteq B$ and not $\overline{\overline{A}} \leq \overline{\overline{X}}$, then $X \in B$.[10]

(4.3.28) If $\mathfrak{m}$ is any infinite cardinal such that $\mathfrak{n} \leq \mathfrak{m}$ and $\overline{\overline{M}} = \mathfrak{m}$, then the set $\{X \subseteq M\colon \text{not } \mathfrak{n} < \overline{\overline{X}}\}$ is equipollent to $\mathfrak{m}^{\mathfrak{n}}$.

(4.3.29) For every set A there exists a set M which is equipollent to the set $\{X \subseteq M\colon \text{not } \overline{\overline{A}} \leq \overline{\overline{X}}\}$. [1939, 180–181]

---

[7] An ordinal is *regular* if it is not cofinal with any smaller ordinal.

[8] See 4.9, 4.11, and the Epilogue.

[9] Sierpiński and Tarski 1930, 292.

[10] Tarski 1938b, 201.

The unusual form of these equivalents sprang from their relationship to strongly inaccessible cardinals. For in 1938 Tarski had introduced a new set-theoretic axiom, which he called the Axiom of Inaccessible Sets. In effect, this axiom (4.3.30) asserted that for every set A there exists a larger set B whose cardinal is strongly inaccessible. However, he based the actual form of this axiom on a different characterization that he gave for such cardinals:

(4.3.30) For every set A there exists a set B such that

(a) $A \in B$;
(b) if $X \in B$ and $Y \subseteq X$, then $Y \in B$;
(c) if $X \in B$, then the power set of X is a member of B;
(d) if $X \subseteq B$ and $\overline{\overline{X}} < \overline{\overline{B}}$, then $X \in B$.

In **ZF**, Tarski deduced the Axiom of Choice from his Axiom of Inaccessible Sets. By modifying Zermelo's second proof of the Well-Ordering Theorem, Tarski first showed that every set B which satisfies (d) can be well-ordered. Since the given set A satisfied (a) and (b), it followed that $\overline{\overline{A}} < \overline{\overline{B}}$ and hence that an arbitrary set A can be well-ordered as well.[11] The equivalence of (4.3.27)–(4.3.29) to the Axiom of Choice was derived by related techniques.[12] Indeed, (4.3.27) was a modification of (d), while (4.3.28) and (4.3.29) emerged from an alternative version of Tarski's axiom.

Thus through Tarski's research, and to a lesser extent through that of Lindenbaum, the role played by the Axiom of Choice in cardinal arithmetic came to be much better understood. By 1939 Tarski had shown sixteen propositions on cardinals to be equivalent to the Axiom and thereby had significantly advanced Sierpiński's program of research. A decade earlier he had stressed that, leaving aside propositions about infinite cardinal products and cardinal exponentiation, "at present there are only a few theorems in cardinal arithmetic for which one of two possibilities cannot be shown to hold: either the independence of the theorem from the Axiom of Choice or its complete equivalence to the same" [1929, 52]. Even more fundamental was the result that the Axiom could be deduced from two general set-theoretic propositions which at first appeared to have little to do with it: the Generalized Continuum Hypothesis (4.3.20) and the existence of strongly inaccessible cardinals greater than any given power. Later developments in set theory were to multiply greatly the importance of such large cardinals, and the structure of the hierarchy of large cardinals (inaccessible, measurable, *etc.*) would depend to a large extent on whether one accepted the Axiom.

[11] Tarski 1938a, 84–87.

[12] Tarski 1938b, 198–201, and 1939, 180–181.

4.4 Zorn's Lemma and Related Principles

The history of maximal principles, such as Zorn's Lemma, is strewn with multiple independent discoveries of fundamentally similar propositions.[1] From time to time such multiple discoveries have occurred in mathematics and the other sciences, but it is rather surprising that so many mathematicians failed to notice the same type of proposition in publications by other mathematicians. Unlike many other multiple independent discoveries, those involving maximal principles were in no sense simultaneous, since they extended over three decades. The whole affair suggests that mathematics possessed an inadequate information retrieval system at the time, despite the existence of abstracting journals.

The first explicit formulation of a maximal principle occurred in an article that Hausdorff published in 1909. As noted in 3.4, neither his first formulation (very close to what was later called Zorn's Lemma) nor his second in the *Grundzüge* of 1914 (eventually termed Hausdorff's Maximal Principle) found an audience until other mathematicians had independently rediscovered such principles. Significantly, Hausdorff did not propose either of his formulations as an axiom, and he had no interest in replacing the Well-Ordering Theorem by a different technique, as some later mathematicians preferred to do.

When the second edition of Hausdorff's *Grundzüge* appeared in 1927, he considered such formulations once again. There he introduced for the first time the term "maximal set" (*Maximalmenge*) to describe what is here called an $\subseteq$-maximal element, *i.e.*, a member B of a family of sets such that B is not properly included in any member of the family. On that occasion he stated a maximal principle, and its corresponding minimal principle, by means of transfinite ordinals—an approach which, like his terminology, was reminiscent of his article of 1909. In particular, he defined a family A of sets to be extendable above by limits if, for each of its subfamilies well-ordered by $\subseteq$, there exists an upper bound in A with respect to $\subseteq$; if $\subseteq$ was replaced by $\supseteq$, then he spoke of A as extendable below by limits. Thus his maximal and minimal principles took the following form:

(4.4.1) If A is a non-empty family of sets and is extendable above by limits, then A contains a *Maximalmenge* [$\subseteq$-maximal element] B.

(4.4.2) If A is a non-empty family of sets and is extendable below by limits, then A contains a *Minimalmenge* [$\subseteq$-minimal element] B. [1927, 174]

---

[1] See Campbell 1978 for a reconstruction, differing somewhat from the present account, of the early history of maximal principles. For a brief discussion of these principles, see Rubin and Rubin 1963, 10–11, as well as Fraenkel and Bar-Hillel 1958, 68–69.

In fact, (4.4.1) was essentially the proposition (3.4.1) that he had proved in 1909, but now he showed how the conclusion of (4.4.1) could be strengthened so that B includes every member of a given subset of A well-ordered by $\subseteq$. This addition was close to his formulation of 1914 and indicates how in 1927 his approach amalgamated those of his earlier works.

While Hausdorff applied his two theorems (4.4.1) and (4.4.2) to derive some known results, he made no attempt to do so systematically. After introducing these theorems late in his book merely as applications of transfinite ordinals, he pointed out that from (4.4.1) could be obtained a Hamel basis for the real numbers and, from (4.4.2), Janiszewski's theorem on an irreducible continuum containing two given points (see 3.4). Indeed (4.4.2), the first minimal principle that Hausdorff formulated, appears to have been inspired by Janiszewski's work. Although the young Warsaw mathematician Kazimierz Kuratowski, who was likewise influenced by Janiszewski, had established such a minimal principle five years earlier, in 1927 Hausdorff remained unaware of Kuratowski's principle. Finally, Hausdorff mentioned that (4.4.1) could be used to obtain a proposition deduced earlier in his book from the Well-Ordering Theorem: In every metric space X and for every positive real r, there is an $\subseteq$-maximal subset Y of X such that the distance between any two points in Y is at least r.[2] In all probability he did not suspect that (4.4.1) or (4.4.2) was equivalent to the Axiom.

Hausdorff's casual handling of maximal principles helps to account for the fact that few, if any, mathematicians applied (4.4.1) or (4.4.2) before similar propositions were independently rediscovered. Surprisingly, Hausdorff believed that these two theorems could be demonstrated only by means of transfinite ordinals [1927, 173], whereas Kuratowski had intentionally used them to avoid transfinite ordinals [1922, 88]. To some degree both Hausdorff and Kuratowski were correct. A natural way to deduce a maximal principle, such as (4.4.1), from the Axiom was to use transfinite induction, though neither the Axiom nor transfinite ordinals then needed to appear in a particular application of a maximal principle. In another sense, however, Hausdorff was mistaken, since Kuratowski had demonstrated one such principle, namely (4.4.4), by using the Axiom but no ordinals, in the style of Zermelo's second proof of the Well-Ordering Theorem. Like Hausdorff, Kuratowski did not stress the importance of his propositions (4.4.3) and (4.4.4) on minimal and maximal elements, which, following Janiszewski, he termed irreducible and saturated sets:

(4.4.3) Let S be the family of those subsets of a set E which have a given property P. Suppose that E is in S and that, for every non-empty subset T of S well-ordered by $\supseteq$, $\bigcap T$ is in S. Then S contains an irreducible [$\subseteq$-minimal] set.

[2] Hausdorff 1927, 173–174, 126, 120.

(4.4.4) Let S be the family of those subsets of a set E which have a given property P and which include a given set A. Suppose that A is in S and that, for every non-empty subset T of S well-ordered by $\subseteq$, $\bigcup T$ is in S. Then S contains a saturated [$\subseteq$-maximal] set. [1922, 88–89]

For Kuratowski the propositions (4.4.3) and (4.4.4), which he deduced explicitly from Zermelo's axiomatization, were merely particular instances of his general method for avoiding transfinite ordinals in the proof of theorems whose statement did not refer to such ordinals. This tendency to exclude transfinite ordinals whenever possible—especially in topological theorems—was encouraged by various mathematicians of the time, such as Lindelöf [1905, 183], Brouwer [1911, 138], and Sierpiński [1920, 1]. Kuratowski's method for doing so required a certain function on the power set of the set in question, a function whose existence he deduced (when necessary) from the Axiom of Choice.[3] Thus he made no attempt to replace either the Axiom or the Well-Ordering Theorem by some other principle, and he applied (4.4.3) only to derive Brouwer's 1911 theorem on the existence of minimal closed sets in $\mathbb{R}^n$ having a given property.[4]

In 1930 another Warsaw mathematician, Edward Szpilrajn, employed Kuratowski's proposition (4.4.4), and referred in passing to Hausdorff's result (4.4.1), in order to establish what was later called the Order Extension Principle:

> If a relation S partially orders a set E, then S can be extended to a relation which orders E.

Szpilrajn, who knew that (4.4.4) relied on the Axiom, could find no way of avoiding it here.[5] There was a good reason for this, since it was later shown that in **ZF** the Order Extension Principle requires the Axiom of Choice.[6]

Though he mentioned Hausdorff and Kuratowski, Szpilrajn did not cite a third mathematician who had also introduced such a maximal principle, Salomon Bochner of Munich. Writing in 1928 about abstract Riemann manifolds, Bochner had stated a "set-theoretic lemma" that was almost identical with Hausdorff's theorem (3.4.1) except for being expressed in terms of an arbitrary partial order instead of inclusion. Curiously, Bochner did not cite Hausdorff's article [1909], where (3.4.1) occurred, but his book [1914], where it did not, and then did so only to give a reference for trans-

---

[3] Kuratowski 1922, 79, 87–88.

[4] Brouwer 1911, 138; see 3.4.

[5] Szpilrajn 1930. After the Second World War he changed his family name to Marczewski.

[6] That the Order Extension Principle is not a theorem of **ZF** follows from the fact that the Ordering Principle implies $\mathbf{C}_{<\aleph_0}$ [Kuratowski in Tarski 1924a, 78] and that there is a model of **ZF** in which $\mathbf{C}_2$ fails [Cohen 1963a,b]; *cf.* Jech 1973, 117.

finite induction.[7] Thus it is probable that, while influenced by Hausdorff, Bochner arrived independently at a maximal principle.

Another mathematician who appears to have formulated a minimal principle independently of Hausdorff, Kuratowski, and Bochner was the topologist R. L. Moore, who taught at the University of Texas. In 1932 Moore published a volume on the foundations of point-set topology, and based his treatment on postulates containing the undefined terms "point" and "region." At the beginning of his book he discussed the Axiom of Choice, stated as Zermelo had done in 1908, which Moore included "among the fundamental propositions of the logic of classes" [1932, 1]. Much later in his book, he established a theorem which was in effect a minimal principle [1932, 84]:

(4.4.5) Suppose G is a [non-empty] family of sets. If, for every subcollection H of G ordered by inclusion, there is some A in G such that $A \subseteq B$ for every B in H, then some member M of G has no proper subset in G [*i.e.*, M is $\subseteq$-minimal].

However, he did not go on to obtain any consequences of (4.4.5), which he had deduced from the Well-Ordering Theorem.

Finally, Zorn's Lemma, the maximal principle which has become the most widely known and about whose origins we are the best informed, was published in 1935:

(4.4.6) If A is a family of sets such that the union of every chain $B \subseteq A$ is in A, then there is a member C of A which is not a proper subset of any member of A [*i.e.*, C is $\subseteq$-maximal].[8]

In contrast to the mathematicians discussed thus far, Max Zorn did not regard (4.4.6) as a theorem but as an *axiom*. He named it the "maximum principle" and earnestly hoped that it would supersede the Well-Ordering Theorem in abstract algebra. By means of his principle, he claimed, the proofs of many results in algebra—such as those of Steinitz concerning algebraic closure and the transcendence degree of a field—became shorter and more algebraic.[9]

Evidently Zorn was not acquainted with previous maximal principles. Many years later he acknowledged having read [Kuratowski 1922] but claimed not to have noticed any maximal principle there.[10] Nevertheless, we may wonder whether Kuratowski influenced Zorn (as Hausdorff might well have influenced Bochner) in a subliminal way.

---

[7] Bochner 1928, 408–409.

[8] A family B of sets is a *chain* if it is ordered by inclusion.

[9] Zorn 1935, 667; *cf.* 4.5.

[10] Zorn in Campbell 1978, 83.

What were the beginnings of Zorn's principle? According to his later reminiscences, he first formulated it at Hamburg in 1933, where Claude Chevalley and Emil Artin then took it up as well. Indeed, when Zorn applied it to obtain representatives from certain equivalence classes on a group, Artin recognized that Zorn's principle yields the Axiom of Choice.[11] By late in 1934, Zorn's principle had found users in the United States who dubbed it Zorn's Lemma. In October, when Zorn lectured on his principle to the American Mathematical Society in New York, Solomon Lefschetz recommended that Zorn publish his result.[12] The paper appeared the following year.

Mathematicians quickly put Zorn's Lemma to a variety of uses. Zorn himself, in contrast to Hausdorff and Kuratowski, employed it to demonstrate many known theorems in algebra and thereby to avoid the Well-Ordering Theorem. At that time algebraists were relying on the Axiom chiefly in the guise of the Well-Ordering Theorem, even though some of them felt uncomfortable doing so. Part of this discomfort stemmed from the use of a non-constructive assumption (see 4.5), but much of it came instead from the belief that the Well-Ordering Theorem and the concomitant use of transfinite induction were transcendental tools inappropriate to algebra. This second kind of discomfort, felt in a similar way by topologists, prompted Zorn to put forward his maximum principle. Soon it proved useful in topology as well as algebra. A good part of its success was due to a few mathematicians who quickly disseminated it.

In France, where the Axiom had been so poorly received three decades earlier, Zorn's friend Chevalley introduced the maximum principle to the Bourbakists,[13] and the first volume published by the polycephalic Bourbaki referred to the following version of it as "Zorn's Theorem" [1939, 36–37]:

(4.4.7) If E is a partially ordered set such that every ordered subset of E has an upper bound in E, then E contains a maximal element.

After deducing (4.4.7) from the Axiom, Bourbaki stated Zorn's principle (4.4.6) as a corollary. Furthermore, Bourbaki gave a related maximal principle as a second corollary, namely (4.4.8). Here Bourbaki defined a family M of sets to have finite character if $X \in M$ is equivalent to the statement that every finite subset of X is a member of M:

(4.4.8) If a [non-empty] family M of subsets of a set A has finite character, then M contains a maximal element. [1939, 37]

[11] In his paper [1935, 669], Zorn mentioned this implication but did not prove it.

[12] Zorn in Campbell 1978, 84.

[13] Zorn in *ibid.*, 84. A group of young French mathematicians wrote collectively under the pseudonym of Nicolas Bourbaki; see Dieudonné 1970.

In 1940, also influenced by Zorn, the Princeton topologist John Tukey deduced from the Axiom four variants of what he termed Zorn's Lemma, and sketched a proof of their equivalence to the Axiom. One was essentially Bourbaki's (4.4.7), and another was very close to (4.4.8), later named the Teichmüller–Tukey Lemma. It is uncertain whether the Bourbakists obtained these two theorems from Tukey, or vice versa. The use of the term "finite character" by both Tukey and Bourbaki, as well as the Bourbakists' contacts at Princeton, render it unlikely that the two formulations were made independently. Tukey's third variant also concerned finite character, while his fourth was as follows:

(4.4.9) Every partially ordered set A has an ordered subset B which contains all its upper bounds (if any) in A.

Following Zorn's lead, Tukey insisted that such a maximal principle should be taken as an axiom to replace the Well-Ordering Theorem. However, Tukey went beyond Zorn by giving a proof that (4.4.7) implies the Trichotomy of Cardinals and hence is equivalent to the Axiom of Choice [1940, 7–8].

As the work of Bourbaki and Tukey illustrates, by 1939 mathematicians were exploring a variety of maximal principles and their consequences, no longer as independent rediscoveries of such principles, but as research branching from a common stem: Zorn's Lemma. Studies of such variants and their equivalence to the Axiom proliferated after the Second World War (see the Epilogue). Nevertheless, there remained one final independent rediscovery, due to the German algebraist Oswald Teichmüller [1939], then a *Privatdozent* at Berlin.

Teichmüller's article resembled Zorn's in several respects. In particular, Teichmüller formulated three maximal principles and indicated how the algebraist could replace the Axiom, as well as the Well-Ordering Theorem, by any of them. Two of these were versions of the Teichmüller–Tukey Lemma (4.4.8), found independently, but each lacked the term "finite character." The third was as follows:

(4.4.10) For every set M and every set S of propositional functions $P_s(x_1, x_2, \ldots, x_{n_s})$, there exists an $\subseteq$-maximal subset A of M such that for every P_s in S and for every $x_1, x_2, \ldots, x_{n_s}$ in A, $P_s(x_1, x_2, \ldots, x_{n_s})$ is true.[14]

Since Teichmüller showed each of his maximal principles to be equivalent to the Axiom, the use of these principles was a matter of convenience or taste

[14] Teichmüller 1939, 570–571. One of these maximal principles Teichmüller credited to Erhard Schmidt, whose conversations with Zermelo in 1904 had assisted him in proving the Well-Ordering Theorem.

rather than of deductive strength. Teichmüller remained indifferent to axiomatic set theory, in contrast to Zorn, but displayed a greater interest in the deductive strength of special cases of his maximal principles. In this vein Teichmüller wished to discover when restricted forms of his principles and of the Axiom would suffice for a given purpose. He went further than Zorn, not only by proving that the Axiom is equivalent to various maximal principles, but also by the range of his applications of these principles in algebra and analysis (see 4.5).

It is worthwhile to compare the different maximal and minimal principles, discussed in this section, from both a logical and a historical perspective. Logically, as seen from Zermelo's system without the Axiom of Choice, each of these principles is equivalent to the Axiom. Moore's minimal principle of 1932 is identical with Zorn's later and more famous maximal principle, except for the replacement of inclusion by its converse and for the use of a bound instead of union. Hausdorff's earliest version of 1909, Kuratowski's of 1922, and Bochner's of 1928 were very similar to each other and required merely that one consider subfamilies well-ordered by inclusion. In contrast, Moore and Zorn considered all the subfamilies ordered by inclusion, rather than just the well-ordered ones. All the same, the apparently weaker versions of Hausdorff and Kuratowski are equivalent to Zorn's. Finally, the seemingly greater generality obtained by formulating a maximal principle in terms of an arbitrary partial order instead of inclusion, as was done by both Bochner in 1928 and Bourbaki in 1939, is somewhat misleading. In fact, every partial order can be represented by inclusion—as Kuratowski had already realized [1921, 162].

Historically, the interrelations among these principles differ greatly from the logical ones. At times, logically equivalent principles have suggested very dissimilar consequences. Like the Trichotomy of Cardinals and the Well-Ordering Theorem, such equivalent principles have possessed quite different degrees of intuitive plausibility, which varied from one period and one community to another. Until Zorn, the maximal principles were regarded as theorems, and not especially important ones, rather than as a type of axiom. They achieved prominence because two mathematicians, Zorn and Teichmüller, independently recognized that such a principle could be used to circumvent the Well-Ordering Theorem in algebra. This circumvention was not motivated principally by a desire to evade a dubious assumption, but by a wish to formulate a principle more directly applicable to abstract algebra. During the 1930s algebraists frequently demanded maximal sets of various kinds, such as a maximal proper ideal in a given ring or a basis for a vector space. Thus at the time a maximal principle was a natural way for an algebraist to reformulate the Axiom.

Why, then, did maximal principles not attain prominence at an earlier date? Since the deductive strength of such principles was no greater, and no less, than that of the Axiom, their use as a substitute for the Well-Ordering Theorem was largely a matter of convenience. Steinitz had applied this theorem liberally in algebra, but by the 1930s some of his successors considered it, together with transfinite ordinals, to be a transcendental device and not properly a part of algebra (*cf.* 3.5). Hence there arose in algebra a perceived need to replace both ordinals and the Well-Ordering Theorem by a more algebraic device. Yet since maximal principles had been formulated previously only as peripheral theorems in set theory, algebraists remained ignorant of them until Zorn illustrated how useful they could be.

4.5 *Widening Applications in Algebra*

Before Zorn published his Lemma in 1935, algebraists used the Axiom mainly in the guise of the Well-Ordering Theorem. Hence they were able to apply transfinite induction, as well as definition by transfinite recursion, to arbitrary sets. The first instances of such a procedure in algebra had occurred in 1905, when Hamel generated a basis for the vector space $\mathbb{R}$ over the rational numbers, and in 1910, when Steinitz published his research on algebraic and transcendental extensions of fields. Somewhat later the Axiom served briefly in graph theory—then on the border between algebra and set theory—and in the theory of integral domains (see 3.5).

In time, however, the main algebraic concept of relevance to the Axiom turned out to be maximality. As this concept found uses in more and more parts of algebra, so too did the Well-Ordering Theorem. After 1918, when modern algebra grew increasingly abstract, Zermelo's Theorem was employed in group theory, ring theory, Boolean algebra, and the theory of lattices, and found new applications in both linear algebra and field theory.[1] Most of the algebraists who used abstract methods accepted the Axiom with nonchalance and without discussion. As will soon be apparent, there was a significant exception to this trend. For in algebra, as in analysis, attempts were under way to reformulate mathematics constructively.

Hamel bases proved useful to both analysts and algebraists, since their disciplines shared a common border that included the theory of vector spaces. In 1924 the Norwegian analyst Ralph Tambs Lyche relied on a Hamel basis to derive a condition equivalent to the general solvability of Abel's functional equation.[2] Five years later Celestyn Burstin, an Austrian, used such a basis to establish that every vector space over $\mathbb{R}$ with the power of the

[1] For a detailed discussion of the Axiom's applications to rings, fields, and vector spaces, see Warner 1965, 770–797.

[2] Tambs Lyche 1924. A solution to Abel's equation was a real function g such that $g(f(x)) = g(x) + k$, where f was a given real function and k was some non-zero constant.

continuum can be ordered so as to satisfy the Archimedean Axiom.[3] While Hamel had used his basis to obtain discontinuous real functions f such that $f(x + y) = f(x) + f(y)$ for all x and y, in 1937 his German compatriot Erich Kamke investigated the existence, which had already been shown by Lebesgue in 1907, of discontinuous complex functions f such that both $f(z_1 + z_2) = f(z_1) + f(z_2)$ and $f(z_1 z_2) = f(z_1)f(z_2)$ for all z_1 and z_2. Rather than using a basis for the vector space $\mathbb{C}$ of complex numbers over the field $\mathbb{Q}$ of rationals, Kamke employed what was in effect a transcendence basis for $\mathbb{C}$ over $\mathbb{Q}$, a notion that had originated with Steinitz.[4] On the other hand, the general concept of a basis for a vector space had emerged from the researches of Hausdorff, who in 1932 deduced the following proposition from the Well-Ordering Theorem:

(4.5.1) Every vector space has a basis.[5]

Drawing on Hausdorff's researches, Teichmüller proved a result more directly applicable to analysis: Every Hilbert space has an orthonormal basis.[6]

In contrast to Hamel bases, which served to link linear algebra with analysis, Steinitz' influence was confined to abstract algebra. There the Well-Ordering Theorem proved exceptionally applicable in two branches which grew out of Steinitz' research on abstract fields: the theory of real fields and the theory of rings. A real field was a field in which -1 was not a sum of squares, and a real closed field was a real field none of whose algebraic extensions was a real field. The German algebraists Emil Artin and Otto Schreier [1927], then at Hamburg, introduced these two concepts in order to generalize the algebra of the field of real numbers.

As elaborated by Artin and Schreier, the theory of real fields depended for its deepest results on the Well-Ordering Theorem. From it they obtained the proposition that every real field has at least one algebraic extension which is also a real closed field. Furthermore, they deduced that every algebraically closed field K has at least one real closed subfield F such that K can be obtained from F by adjoining a root i of $x^2 + 1 = 0$. In these two

---

[3] Burstin 1929, 296–298. A vector space V is *ordered* if there is a relation $<$ ordering the set V such that if $a < b$ and c is any element of V, then $a + c < b + c$. The order satisfies the Archimedean Axiom if $0 < a < b$ implies the existence of an integer n such that $b < na$.

[4] Kamke 1937, 149. A *transcendence basis* for a field F over a subfield K is a maximal subset of F algebraically independent with respect to K.

[5] Hausdorff 1932, 295. Jech and Sochor [1966, 354] showed that the Axiom was essential to (4.5.1) by giving a model of **ZF** in which some vector space over a denumerable field lacked a basis. They used the work of Läuchli [1962, 8], who had established an analogous result for **ZFU**.

[6] Teichmüller 1936, 76–77. A *Hilbert space* over the real (or complex) numbers is a vector space, endowed with a dot product, such that V is also a complete metric space. A basis for a Hilbert space is *orthonormal* if the dot product of any two distinct vectors in the basis is zero and, of a basis vector with itself, is one.

results, the use of the Axiom was essential.[7] Finally, they showed every real field to be an ordered field, a requirement if the real fields were to provide an adequate algebraic generalization of the real numbers.[8]

Ring theory, the second branch to develop from Steinitz' investigations, grew in a more complex fashion, but the Well-Ordering Theorem also proved vital there. The two chief contributors were Emmy Noether, then a *Privatdozent* at Göttingen, and Wolfgang Krull of Freiburg im Breisgau, where Zermelo had taken up residence. In 1923 Krull employed Steinitz' methods to classify rings. Via the Well-Ordering Theorem, Krull extended Steinitz' results by demonstrating that every commutative ring with unit can be extended to an algebraically closed ring with unit.[9]

Krull relied on Noether's earlier researches concerning ideals in rings, and for such ideals the Well-Ordering Theorem was a fundamental tool.[10] A few years later Noether published an article generalizing the theory of ideals, as previously developed for algebraic number fields, and applied Zermelo's Theorem in doing so [1927, 45–46]. In 1929 Krull wished to extend the chief results of ideal theory still further, namely to rings in which the Ascending Chain Condition failed.[11] To attain this goal, he derived two propositions from Zermelo's Theorem:

(4.5.2) Let A be a multiplicatively closed subset of a commutative ring S and let I be an ideal disjoint from A. Then there exists a maximal ideal including I and disjoint from A.

(4.5.3) In a commutative ring, every proper ideal can be extended to a maximal prime ideal.[12]

It was apropos of Boolean rings, which Krull did not consider specifically, that (4.5.3) proved to be particularly significant.[13]

In 1936 the American algebraist Marshall Stone, then at Harvard, contributed his influential findings on the representation of Boolean rings.

---

[7] For each result Läuchli [1962, 6–7] gave a model of **ZFU** in which it was false, while Jech and Sochor [1966, 354] did likewise for **ZF**.

[8] Artin and Schreier 1927, 91. An *ordered field* is a field on which there is an order $<$ such that if $0 < a$ and $0 < b$, then $0 < a + b$ and $0 < ab$.

[9] Krull 1923, 110–111.

[10] A subset I of a commutative ring S is an *ideal* if for every a,b in I and every c in S, both $a + b$ and ac are in I.

[11] The *Ascending Chain Condition* states that every set of ideals which extend a given ideal and which are ordered by inclusion is finite.

[12] Krull 1929, 732, 736–737. A *proper ideal* is an ideal other than the whole ring. A proper ideal I is *prime* if whenever ab is in I, then a is in I or b is in I.

[13] A *Boolean ring* is a ring, with at least two elements, in which $a^2 = a$ for every a. If a Boolean ring has a unit, then it is called a *Boolean algebra*.

Stone, who did not cite Krull, demonstrated a theorem which nevertheless follows from (4.5.3) and which was the keystone of his own work, namely, the Boolean Prime Ideal Theorem:

(4.5.4) In a Boolean ring there exists at least one prime ideal.[14]

From (4.5.4) Stone deduced a proposition equivalent to it and later known as the Stone Representation Theorem: Every Boolean ring is isomorphic to a ring of sets, *i.e.*, a family of sets with symmetric difference as addition and intersection as multiplication. Unlike many German algebraists, he carefully pointed out where the Axiom occurred in his demonstrations. One could hardly hope to avoid the Axiom, he believed, in the proof of his two principal results.[15] In this he was correct (see the Epilogue).

Stone's paper generated a wave of research on Boolean rings and related topics. One instance occurred in ring theory, where the American algebraist Neil McCoy used the Stone Representation Theorem in the course of deducing a lemma connected with Krull's theorem (4.5.2): If b is a non-zero element in a ring S, then there exists an irreducible ideal not containing b.[16] From this lemma McCoy derived a representation theorem for rings: Every ring is isomorphic to a subring of a direct sum of irreducible rings.[17] Meanwhile, from a more set-theoretic standpoint, Tarski was studying the number of prime ideals found in rings of sets [1939a].

As fate would have it, both the Stone Representation Theorem and the Boolean Prime Ideal Theorem had been discovered independently by other mathematicians, a situation that soon embroiled Stone in priority disputes.[18] In fact another Harvard algebraist, Garrett Birkhoff, had found a more general version of the Stone Representation Theorem about the same time as Stone discovered his, and had published it three years earlier: Every distributive lattice is isomorphic to some ring of sets with finite union and finite intersection as meet and join respectively.[19]

Later Birkhoff observed [1940, 80] that his representation theorem for distributive lattices had been inspired by the researches of Tarski [1930]

---

[14] Stone 1936, 78, 100–104.

[15] *Ibid.*, 101, 105.

[16] McCoy 1938, 490. An ideal I is *irreducible* if there do not exist ideals J_1 and J_2, both unequal to I, such that $J_1 \cap J_2 = I$.

[17] *Ibid.*, 489–490. A ring is *irreducible* if it is not the direct sum of two of its proper ideals. The *direct sum* of a set $\{A_b : b \in B\}$ of rings is the ring of functions f with domain B such that $f(b)$ is in A_b for all b in B.

[18] For a discussion of the early history of the Boolean Prime Ideal Theorem, see Stone 1938.

[19] Birkhoff 1933, 461. A *lattice* is a set with two operations, meet $\wedge$ and join $\vee$, which are commutative, associative, idempotent, and such that $a \vee b = a$ if and only if $a \wedge b = b$. If each of meet and join is distributive relative to the other, then the lattice is termed *distributive*. On the history of lattice theory, see Mehrtens 1979.

and his young compatriot Stanisław Ulam [1929], then at Lwów, on abstract measure theory. From the Axiom, Tarski had deduced that:

(4.5.5) On every infinite set there exists a two-valued additive measure such that each singleton has measure zero.

To do so, Tarski had shown that every proper ideal on a ring of sets can be extended to a prime ideal [1930, 42–43]. He did not employ this terminology, however, since he was motivated by set-theoretic rather than algebraic considerations. Extending Tarski's results, Sierpiński proved that the existence of such a measure, even on $\mathbb{N}$, yields the existence of a set which is not Lebesgue-measurable [1938]. Much later, two other Polish mathematicians, Jerzy Łoś and Czesław Ryall-Nardzewski, showed that (4.5.5) is equivalent to the Boolean Prime Ideal Theorem [1955], while James Halpern, one of Tarski's graduate students at Berkeley, established that the Boolean Prime Ideal Theorem is weaker than the Axiom.[20] In this way it gradually became clear that the Boolean Prime Ideal Theorem, and the existence of a non-trivial two-valued additive measure on arbitrary sets, were non-constructive propositions depending on the Axiom but having a smaller degree of non-constructivity. In fact, like the Axiom, the Boolean Prime Ideal Theorem turned out to be equivalent to a wide variety of propositions in algebra, logic, set theory, and topology (see the Epilogue).

The Axiom entered group theory, as it had entered ring theory, via a mathematician influenced by Steinitz. In this case the mathematician was Otto Schreier, who applied the Axiom to group theory in the same year that he collaborated with Artin on the theory of real fields. Previously, Jakob Nielsen had shown that if H is a finitely generated subgroup of a free group, then H is also a free group.[21] Schreier extended Nielsen's theorem by establishing that every subgroup of a free group is free, a proposition that depended necessarily on the Axiom.[22] Here Schreier used the Axiom to deduce that if two free groups have equipollent sets of generators, then the groups are isomorphic [1927].

Like the Nielsen–Schreier Theorem, a second path leading from Steinitz to group theory extended a result previously derived without the Axiom to a more general one obtained with its help. The algebraist Bernhard Neumann, then at Cambridge University, used the Axiom to deduce the lemma that if G is a group, g is an element of G, and H is a subgroup of G

---

[20] Halpern [1964] gave this proof for **ZFU**. The proof that the Boolean Prime Ideal Theorem does not imply the Axiom in **ZF** was due to Halpern and Levy [1971].

[21] A *free group* is a group in which a formal product of elements equals the identity element only if such an equality follows from the group axioms. A subgroup is *finitely generated* if it is generated from a finite subset of the group.

[22] Läuchli [1962, 11–12] gave a model of **ZFU** in which there was a free group whose commutator subgroup was not free. This result was transferred to **ZF** by Jech and Sochor [1966, 354].

not containing g, then there exists a maximal subgroup including H but not containing g. From this lemma Neumann obtained his principal result, previously established for finite groups, that the intersection of the maximal proper subgroups of a group G is identical with the group consisting of the non-generators of G.[23] Two decades later this theorem, as further generalized to every algebraic system, was proved equivalent to the Axiom.[24] Once again the notion of maximality was seen to be thoroughly intertwined with the Axiom.

In contrast to the algebraists mentioned above, who accepted the Axiom of Choice without distress, the Dutch mathematician Bartel van der Waerden reversed himself twice on this matter. In 1930 he published a textbook, *Moderne Algebra*, which quickly became very influential and which was cited by many of those algebraists. There he employed the Well-Ordering Theorem to develop field theory in particular.[25] Yet, in the second edition of 1937, some of his Dutch colleagues who were intuitionists persuaded van der Waerden to abandon the Axiom so as to treat abstract algebra more constructively. Furthermore, he now regarded the Well-Ordering Theorem and transfinite induction as transcendental methods and hence as inappropriate to the finitary operations of algebra [1937, vi]. In this vein he restricted Steinitz' theorems concerning algebraic closure to countable fields (on which a well-ordering was automatically available) and discussed only finite transcendence bases. Similarly, he limited Artin's and Schreier's theorems on real fields to the countable cases.[26] Like many other critics of the Axiom, however, van der Waerden found that he needed it in order to obtain a desired result, in this instance the uniqueness of the algebraic closure of a countable field, and so he used it explicitly [1937, 205]. At the urging of many mathematicians he reinstated the Axiom, transfinite induction, and Steinitz' general theory of fields in his third edition [1950, v].

Prior to this vacillation over the Axiom, van der Waerden's first edition had supplied both Zorn and Teichmüller with fertile applications for their maximal principles. Thus Zorn cited *Moderne Algebra* for the theorem on extending an ideal in a ring to a maximal ideal, that on extending a real field to a real closed field, and Steinitz' results on the existence of an algebraic closure and a transcendence basis for any given field. As Zorn pointed out, all of these propositions followed directly from his maximal principle [1935, 667–668].

Independently, Teichmüller gave the same applications for his own maximal principle, as well as others from linear algebra. In addition, he recognized

---

[23] Neumann 1937, 120–122. An element g in a group G is a *non-generator* of G if whenever $H \cup \{g\}$ generates G, then H generates G.

[24] Schmidt 1953, 36–37. An *algebraic system* consists of a set A together with some set of algebraic operations on A.

[25] Van der Waerden 1930, 194–206, 231–235.

[26] Van der Waerden 1937, 202–206, 210–212, 239–243.

the fundamental similarity between deducing the existence of a basis for a vector space and that of a transcendence basis for a field. He also indicated how his principle yields the existence of an orthonormal basis for any Hilbert space. On the other hand, he employed the Well-Ordering Theorem to obtain the proposition that if A is a family of subspaces of a Hilbert space such that any two members of A have only zero in common, then the dimension of the subspace generated by the union of A is the sum of the dimensions of all the subspaces in A. Teichmüller noted his inability to deduce this proposition from his maximal principle, although he knew that it was theoretically possible to do so, and concluded that his principle was less applicable to point-set topology than to algebra.[27] Here he was mistaken, however, for the Teichmüller–Tukey Lemma proved to be quite valuable in topology as well.

Finally, in contrast to the results discussed above, there occurred two quasi-algebraic developments which involved the Axiom but which did not derive from either Hamel or Steinitz. The first was a proposition in graph theory, later named the Infinity Lemma, that Dénes König had obtained [1927, 121]:

(4.5.6) Suppose that an infinite graph A includes disjoint subgraphs $A_1, A_2, \ldots$ such that each point of A_{n+1} is joined to some point in A_n. Then there exists an infinite branch $a_1, a_2, \ldots$ such that each a_n is in A_n.

König applied his Infinity Lemma to the coloring of maps and to game theory. However, it had originated a year earlier in a set-theoretic version when he attempted to generalize Bernstein's result of 1901 that $2\mathfrak{m} = 2\mathfrak{n}$ implies $\mathfrak{m} = \mathfrak{n}$. On this occasion König had deduced from the Axiom a version of (4.5.6) related to Bernays' later Principle of Dependent Choices (1.1.2): If $A_1, A_2, \ldots$ are finite non-empty sets and if S is a relation such that for every b in A_{n+1} there is some a in A_n with aSb, then there exists a sequence $a_1, a_2, \ldots$ such that for every n, a_n is in A_n, and a_nSa_{n+1}.[28]

The second development occurred on the frontier between algebra, analysis, and set theory: Stefan Banach's researches at Lwów on functional analysis. In 1929 Banach established a fundamental result later known as the Hahn–Banach Theorem: Let p be a real-valued functional on a complete normed vector space V over $\mathbb{R}$ and suppose that for every x and y in V, $p(x + y) \leq p(x) + p(y)$ and that $p(tx) = tp(x)$ for all positive real t; then there is an additive functional f such that for every x in V,

$$-p(-x) \leq f(x) \leq p(x).$$

[27] Teichmüller 1939, 568–573.

[28] D. König 1926, 120–122. In the Principle of Dependent Choices, where the finiteness of A_n was inessential, the hypothesis was converse to that given above: For every a in A_n there is some b in A_{n+1} with aSb.

To obtain this result, Banach relied on the Well-Ordering Theorem.[29] Much later, the Hahn–Banach Theorem was deduced from the Boolean Prime Ideal Theorem and shown to be equivalent to the following proposition:

(4.5.7) On every Boolean algebra there exists an additive real-valued measure m with $m(0) = 0$ and $m(1) = 1$.[30]

The complexity of the historical developments surrounding the Axiom necessitates a thematic treatment, even though every such division is somewhat artificial. Indeed, while certain investigations in algebra developed from previous algebraic research, others were strongly influenced by more distant areas of mathematics. An instance of the latter sort was Birkhoff's representation theorem for distributive lattices, which grew out of Tarski's and Ulam's set-theoretic research on abstract measure theory. This research, in turn, had arisen from real analysis via Lebesgue's theory of integration, itself strongly influenced by set-theoretic methods. Likewise, Stone's algebraic research led him to topologize Boolean algebras [1934]. A complex mathematical tapestry was being woven, whose borders were imprecise and continually shifting.

Yet in this tapestry certain motifs stand out. One of them is the significance of Hamel's research, which eventually led to the proposition that every vector space has a basis, even if the space itself is not finite-dimensional. Within the research that utilized Hamel's concepts, there was a continuing interplay between linear algebra and analysis—as evidenced by the work of Banach, Hausdorff, and Teichmüller. A second motif was Steinitz' preponderant influence in developing abstract algebra and in disseminating the Axiom. From his research there emerged Artin's and Schreier's theory of real fields, as well as much of Krull's and Noether's investigations in ring theory. Here the notion of ideal, particularly of maximal ideal, proved to be central. In Boolean algebra, Stone explored the ramifications of his Representation Theorem and of the Prime Ideal Theorem, found independently in different forms by Birkhoff and Tarski. Lastly, Schreier and Neumann were stimulated by Steinitz in their investigations of free groups and of maximal subgroups respectively.

Despite such fundamental results derived from the Axiom, van der Waerden abandoned his earlier support for it in 1937 and attempted to formulate a constructive, if constricted, version of abstract algebra. As a

[29] Banach 1929, 226–228. A vector space V is *normed* if there is a function s from V to $\mathbb{R}$ such that for all x and y in V and for all a in $\mathbb{R}$: $s(x) \geq 0$; $s(x) = 0$ implies $x = 0$; $s(ax) = |a| s(x)$; $s(x + y) \leq s(x) + s(y)$. A functional p is *additive* if for every a, b in $\mathbb{R}$ and for every x, y in V, $p(ax + by) = ap(x) + bp(y)$. On the prehistory of the Hahn–Banach Theorem, especially as it concerns Eduard Helly and Hans Hahn, see Hochstadt 1980.

[30] Łos and Ryll-Nardzewski 1951, 233–237; Luxemburg 1969, 131.

result, the struggle between the Axiom's supporters and its constructivist critics began to influence the development of algebra. Ironically, the first edition of van der Waerden's text had supplied Zorn and Teichmüller with many applications for their maximal principles. In 1950, when pressure from mathematicians prompted van der Waerden to reinstate in *Moderne Algebra* both the Axiom and Steinitz' results, the direction of future research was clear. Algebraists insisted that the Axiom, whether in the guise of Zorn's Lemma or of the Well-Ordering Theorem, had become indispensable to their discipline.

4.6 Convergence and Compactness in General Topology

While the generalization of geometry beyond the Euclidean spaces $\mathbb{R}^n$ originated with Italian geometers such as Ascoli, Arzelà, and Volterra, it first received a coherent and abstract treatment in Fréchet's thesis of 1906. As Fréchet took steps to delineate functional analysis, and thereby to distinguish it from the calculus of variations, he isolated two important types of generalized spaces: those, which he called L-spaces, where the notion of limit was based on an axiomatization of convergent sequences and, among the L-spaces, those on which a distance function could be defined.[1] Hausdorff gave the latter the name of metric spaces in 1914 when he pondered the role of point-sets within abstract set theory. Preferring to investigate metric spaces rather than L-spaces, Hausdorff offered an alternative generalization for the former: topological spaces, based on the concept of neighborhood [1914, 213]. In effect, he was introducing what are now termed Hausdorff spaces, those topological spaces in which any two distinct points p and q belong to some disjoint open sets A_p and A_q.

As was seen in 2.4, Fréchet relied heavily but implicitly on the Denumerable Axiom to obtain theorems about convergence in L-spaces. Hausdorff, on the other hand, applied the Axiom of Choice consciously and liberally in his topological theorems, but he rarely remarked where he used it. In 1927, borrowing a term from Fréchet, Hausdorff defined a topological space to be separable if it had a countable dense subset, and then obtained the following result by means of the Denumerable Axiom:

(4.6.1) Every subspace of a separable metric space is separable.

---

[1] Fréchet 1906, 5–6, 30. Such a distance function $d: X \times X \to \mathbb{R}$ had to satisfy four postulates for every p, q, r in the space X: (i) $d(p, q) \geq 0$; (ii) $d(p, q) = 0$ if and only if $p = q$; (iii) $d(p, q) = d(q, p)$; (iv) $d(p, q) + d(q, r) \geq d(p, r)$.

More generally, he used the Axiom to demonstrate that every metric space has a dense subset of least power.[2] In fact, (4.6.1) may be false without the Axiom, since a model of **ZF** was later found in which some subspace of $\mathbb{R}$ is not separable.[3]

Among the topologists influenced by Fréchet or Hausdorff, and very few were not, the Axiom was usually applied with nonchalance. In their monographs on general topology Sierpiński and Kuratowski, both of whom had carefully pointed out the Axiom's occurrences in set theory, often made arbitrary choices without special mention. Defining the notion of topological space by means of open sets, Sierpiński implicitly used the Axiom to show that if a topological space X has a countable basis, then every open cover of X has a countable subcover.[4] In other words, he obtained the following result:

(4.6.2) If a topological space X is second countable, then X is Lindelöf.

Moreover, the Axiom appeared implicitly when he deduced two theorems, previously shown in $\mathbb{R}^n$, for more general spaces. These theorems, which required the Axiom even if the space was $\mathbb{R}$ (*cf.* 1.2.2), connected notions defined by neighborhoods to the corresponding notions defined by sequences:

(4.6.3) In a second countable Hausdorff space, p is a limit point of a set A if and only if p is a sequential limit point of A.

(4.6.4) In a second countable Hausdorff space X, a function $f: X \to X$ is continuous at a point p if and only if f is sequentially continuous at p.[5]

Kuratowski, who defined topological spaces via a closure operator, pointed out the Axiom's role in (4.6.2), but he did not remark its use in theorems dealing with analytic sets, nor in the following result which he contributed:

(4.6.5) Every second countable topological space is separable.[6]

---

[2] Hausdorff 1927, 125–127. A subset M of a topological space X is *dense* if every point of X is either a member of M or a limit point of M.

[3] Jech 1973, 142.

[4] Sierpiński 1934, 46. A family A of open sets is a *basis* for a topological space X if every open set in X is the union of some subcollection of A. If X has a countable basis, then X is called *second countable*.

[5] *Ibid.*, 55–57. A function $f: X \to Y$, where X and Y are topological spaces, is *continuous* at a point p of X if, for every open set V containing $f(p)$, there is an open set U containing p such that $f''U$ is a subset of V. The function f is *sequentially continuous* at p if (a) implies (b): (a) Every open set containing p also contains all but finitely many members of the sequence $p_1, p_2, \ldots$; (b) Every open set containing $f(p)$ also contains all but finitely many members of the sequence $f(p_1), f(p_2), \ldots$. Both (4.6.3) and (4.6.4) would remain in force if Sierpiński had presupposed only a *first countable* topological space, *i.e.*, one in which for every point p there is a countable family A of open sets such that every open set containing p is the union of a subfamily of A.

[6] Kuratowski 1933, 102–103, 249.

Yet Kuratowski's research, like that of Hausdorff, reveals how closely the notion of separability was linked to the Denumerable Axiom. Indeed, Kuratowski's proposition (4.6.5) can be false even for a subspace of the real line if the Denumerable Axiom fails.[7]

Shortly after 1920 there arose among Luzin's students at Moscow a new school of topologists.[8] Foremost among them were Paul Alexandroff and Paul Urysohn, whose researches focused on compactness and metrization. In contrast to Luzin, whose sympathies remained constructivist (see 4.11), Alexandroff and Urysohn were strongly influenced by Hausdorff's ideas and employed the Axiom of Choice quite freely.

In a memoir that was accepted by *Fundamenta Mathematicae* in 1923 but was printed elsewhere six years later for unknown reasons, Alexandroff and Urysohn generalized Fréchet's version of compactness (which would now be termed sequential conditional compactness) to one in which denumerability and sequences were no longer central. Their notion, originally called bicompactness, became so fundamental that outside the Soviet Union it later pre-empted the name of compactness: A topological space X is said to be compact if every open cover possesses a finite subcover. Thus, in effect, the Heine–Borel Theorem had been converted into a definition. A property which Alexandroff and Urysohn proved equivalent to it was a generalized form of the Bolzano–Weierstrass Theorem: Every infinite subset of X has a complete accumulation point. To establish this equivalence, they relied once more on the Axiom.[9]

Although Urysohn died in a swimming accident in 1924, several of his posthumous articles, which Alexandroff soon published, greatly stimulated the development of general topology. In one such article of 1925, Urysohn attacked the problem, posed earlier by Fréchet, of determining the largest class of spaces that have non-constant continuous functions. As a partial solution, Urysohn obtained what later became known as Urysohn's Lemma:

(4.6.6) If A and B are disjoint closed subsets of a normal Hausdorff space E, then there exists a continuous real-valued function f on E such that $f(x) = 0$ for x in A and $f(x) = 1$ for x in B.[10]

During the proof Urysohn made a sequence of arbitrary dependent choices, as he had to do.[11] His next paper applied (4.6.6) to solve in part the problem

---

[7] Jech 1973, 142. At times these Polish topologists relied heavily not just on the Axiom but also on the Continuum Hypothesis (1.5.2). See Sierpiński 1928c, 1934a, 1937a, as well as Kuratowski and Sierpiński 1926.

[8] Phillips 1978, 298. See also Arboleda 1979.

[9] Alexandroff and Urysohn 1929, 8–10. In a topological space, p is a *complete accumulation point* of a set A if, for every open set V containing p, $V \cap A$ is equipollent to A.

[10] A topological space is *normal* if, for every pair of disjoint closed sets A and B, there are disjoint open sets U and V including A and B respectively.

[11] Urysohn 1925, 290–292. Läuchli [1962, 15–18] gave a model of **ZFU** in which Urysohn's Lemma failed, while Jech and Sochor [1966, 354] did likewise for **ZF**.

of determining which topological spaces are metrizable, that is, homeomorphic to a metric space:

(4.6.7) A Hausdorff space X is homeomorphic to a separable metric space if and only if X is normal and second countable.[12]

From these researches of Alexandroff and Urysohn there resulted a seminal article by a third member of the Moscow school of topologists: Andrei Tychonoff. In 1930, after investigating the problem of embedding a Hausdorff space in a compact space, Tychonoff published his principal theorem and thereby generalized a result of Urysohn's: Every normal Hausdorff space with a basis of power $\mathfrak{m}$ is homeomorphic to a subspace of a compact space. In this case, the compact space was the product of $\mathfrak{m}$ closed real intervals [0, 1]. Furthermore, Tychonoff solved his original problem by demonstrating that a Hausdorff space X can be embedded in a compact space if and only if X is completely regular, a property which weakened slightly that given by Urysohn's Lemma.[13] Later, any completely regular Hausdorff space became known as a Tychonoff space. To obtain these results, Tychonoff relied on Urysohn's Lemma and thus implicitly on the Axiom. Yet the result for which he is best known, later named Tychonoff's Theorem, did not occur there in its general form:

(4.6.8) The product of any family of compact spaces is compact.

Rather, Tychonoff [1930] proved only that the product of any number of copies of the closed interval [0, 1] is compact. Fortunately, his method of proof via complete accumulation points lent itself to generalization.

One topologist who undertook such a generalization was the Czechoslovakian Eduard Čech, then at Brno. In 1937 Čech was apparently the first to recognize that Tychonoff's Theorem lay hidden in the latter's paper of 1930 and to demonstrate it in full generality. For, in contrast to the Moscow topologists, Čech did not restrict himself to Hausdorff spaces but stated his results for arbitrary topological spaces. Although Čech's method of proof [1937, 830], like Tychonoff's, relied on complete accumulation points, it was supplemented by transfinite induction and an explicit use of the Well-Ordering Theorem. In 1950 Tychonoff's Theorem, which would be more

---

[12] Urysohn 1925a, 310–311. The same year Tychonoff found that one could require merely that the space be *regular* rather than normal, that is, that any closed set and any point not in it could be separated by open sets. Two topological spaces X and Y are *homeomorphic* if there is a bijection $f: X \to Y$ such that f and its inverse are both continuous.

[13] A topological space X is *completely regular* if, for any closed set A and any point p not in A, there is a real-valued continuous function f on X such that, for every x in A, $f(x) = 1$, while $f(p) = 0$.

accurately described as the Čech–Tychonoff Theorem, was shown to be equivalent to the Axiom.[14]

With an incisive understanding of the other generalizable features of Tychonoff's paper, Čech set forth the concept and properties of $\beta(S)$, later known as the Stone–Čech compactification of a Tychonoff space S. What Čech extracted from Tychonoff's article was the idea that, by embedding a Tychonoff space S in a product of closed real intervals $[0, 1]$, one obtained the compact space $\beta(S)$ as the closure of S in this product. On the other hand, Čech proved as well that $\beta(S)$ is determined up to homeomorphism by the requirements that $\beta(S)$ be a compact Hausdorff space, that S be dense in $\beta(S)$, and that every bounded continuous real-valued function on S be extendable to such a function on $\beta(S)$. Recognizing the dependence of his results on the Axiom, Čech thought it unlikely that even a single point in $\beta(\mathbb{N})-\mathbb{N}$ could be defined constructively.[15] In fact, the existence of the Stone–Čech compactification for an arbitrary Tychonoff space was later found to be equivalent to the Boolean Prime Ideal Theorem and hence to depend on the Axiom.[16] As a result of Čech's work and the related investigations of Stone, mathematicians undertook a thorough study of compactification.[17]

In the same year that Čech's article appeared, two generalizations of convergence—directed sets and filters—were developed independently of each other in the United States and France. A few years later, these two generalizations would provide a way to represent compactness as well. Both approaches emerged from the classical description of convergence in terms of sequences. In 1922 the first approach had been formulated jointly by two mathematicians, E. H. Moore of the University of Chicago and H. L. Smith of the University of the Philippines. They intended it for use in what they called general analysis, but unfortunately they expressed it in rather cumbersome language. Only in 1937 did Birkhoff render this approach suitable for topology.

Birkhoff considered the three fundamental topological notions to be convergence, closure, and neighborhood. While in first countable Hausdorff

[14] Kelley [1950] established that Tychonoff's Theorem, even restricted to T_1 spaces, is equivalent to the Axiom. However, if restricted further to Hausdorff spaces, Tychonoff's Theorem is only equivalent to the Boolean Prime Ideal Theorem, as Łos and Ryll-Nardzewski [1955, 51] showed. (A topological space X is T_1 if, for every distinct p and q in X, there are open sets A and B containing p and q respectively such that p is not in B nor q in A.) Ward [1962] showed that the following Weak Tychonoff Theorem is equivalent to the Axiom: The product of any number of copies of a given compact space is compact.

[15] Čech 1937, 823–824. Somewhat similar but less perspicuous research on $\beta(S)$ was carried out by Stone [1937, 383, 475–476]. A point in $\beta(\mathbb{N})-(\mathbb{N})$ can be regarded as a non-principal ultrafilter on $\mathbb{N}$, and such an ultrafilter cannot be proved to exist in **ZF**, since this would yield a non-measurable set; *cf.* Jech 1973, 7, and footnote 19 below.

[16] Rubin and Scott 1954; see the Epilogue.

[17] See, for example, Wallman 1938 and Alexander 1939.

spaces all three were interdefinable, many important function spaces lacked an appropriate notion of convergence [1937, 39]. Through revising the generalized limit introduced by Moore and Smith, Birkhoff obtained such a notion in which denumerability was no longer essential. First, Birkhoff supposed that $\{x_a\}_{a \in A}$ was some directed set, that is, a one–one function with a transitive relation $\leq$ on a set A such that for every a, b in A there was some c in A with $a \leq c$ and $b \leq c$; then $\{x_a\}_{a \in A}$ converged to x (symbolically, $x_a \to x$) if for every open set V containing x there existed some b in A such that $b \leq c$ implied $x_c \in V$. Clearly, this approach generalized the classical notion of sequential limit point by means of a partial order.

Birkhoff's first theorem formed the basis for his subsequent results: In a Hausdorff space X, B is an open set if and only if no directed set of points in $X—B$ converges to a point in B. Not surprisingly, this theorem involved an implicit use of the Axiom. As a corollary he deduced, for Hausdorff spaces X and Y, the equivalence of the usual topological notion that a function $f: X \to Y$ is continuous at a point x, and the following generalization of sequential continuity: If $x_a \to x$ in X, then $f(x_a) \to f(x)$ in Y [1937, 39–41].

Meanwhile Henri Cartan proposed to the Paris Academy of Sciences a different but related generalization of convergence: filters. He defined a filter on a set E to be a non-empty family F of subsets of E such that

(i) the empty set is not in F;
(ii) the intersection of any two members of F is in F; and
(iii) if $A \subseteq B \subseteq E$ and A is in F, then B is also in F. [1937]

A central notion was that of a maximal filter, or, as Cartan termed it, an ultrafilter. As for the convergence of a filter F, he defined a point p to be a limit point of F if F contains every neighborhood of p. By means of the Well-Ordering Theorem and transfinite induction, Cartan formulated and proved a fundamental result later known as the Ultrafilter Theorem:

(4.6.9) Every filter can be extended to an ultrafilter.[18]

From this proposition he deduced that a Hausdorff space E is compact if and only if every ultrafilter on E has a limit point. Thus a characterization of compactness similar to the Bolzano–Weierstrass Theorem was successfully transferred to spaces much more general than $\mathbb{R}^n$ by filters, as well as by complete accumulation points. It is significant that Cartan made no mention of Boolean algebras. Evidently he was unaware that a filter is the dual of an

---

[18] Cartan 1937a. This proposition should be contrasted with what was later called the Weak Ultrafilter Theorem, which was strictly weaker than (4.6.9) but was not provable in **ZF**: There is a non-principal ultrafilter on any infinite set. On the other hand, it can easily be shown in **ZF** that on any set there is a principal ultrafilter. See Jech 1973, 132, Blass 1977a, and footnote 19 below.

ideal in a Boolean algebra and that the Ultrafilter Theorem is equivalent to the Boolean Prime Ideal Theorem.

In 1940 Cartan's characterization of compactness enabled Bourbaki to give a proof of Tychonoff's Theorem, restricted to Hausdorff spaces, by using ultrafilters rather than complete accumulation points:

(4.6.10) The product of any family of compact Hausdorff spaces is a compact Hausdorff space.

Moreover, Bourbaki developed filters as a device suitable for defining a topological space, and, in particular, for characterizing pointwise convergence—a characterization which, on the other hand, did not involve the Axiom. To deduce the Ultrafilter Theorem, Bourbaki employed Zorn's Lemma and remarked that the existence of a non-principal ultrafilter on an infinite set, and even on $\mathbb{N}$, required the Axiom.[19]

Over the next two decades, filters served French topologists as their preferred form of generalized convergence, while Americans confined themselves to directed sets.[20] At Princeton, Tukey [1940] lauded the theory of directed sets as more intuitive than that of filters, since it remained closer to sequential convergence. Furthermore, he rendered the former theory as potent as the latter, though somewhat more complicated. For directed sets he even defined an analogue of the ultrafilter, called an ultraphalanx, such that a space X is compact if and only if every ultraphalanx on X converges. As he remarked, if there is a non-trivial ultraphalanx on $\mathbb{N}$, then there exists a non-measurable subset of $\mathbb{R}$.[21] Eventually, mathematicians recognized that these two methods of convergence (filters and directed sets) were essentially equivalent.[22] However, to establish this equivalence, the Axiom came into play.

As the theory of abstract spaces emerged from the classical analysis of $\mathbb{R}^n$ and culminated in general topology, the Axiom of Choice appeared more and more frequently. At first the Axiom was used principally to investigate sequential convergence, but with the rise of metric spaces it also assumed a role in Hausdorff's work on separability and dense subsets. When the Moscow school undertook research on compactness and metrization, the Axiom proved vital to its central results, particularly Urysohn's Lemma and Tychonoff's Theorem. Since the existence of the Stone–Čech compactification of a space relied on Tychonoff's Theorem, the Axiom became indispensable

---

[19] Bourbaki 1940, 20–37, 63. An ultrafilter F on a set E is *principal* if F consists of all subsets of E containing a certain point p; otherwise F is *non-principal.*

[20] Bartle 1955, 551.

[21] Tukey 1940, 33–36. This remark was essentially the same as Sierpiński's comment [1938] that if there exists a two-valued additive measure on $\mathbb{N}$ such that singletons have measure zero, then there exists a non-measurable set; see p. 231 above.

[22] See Willard 1970, 81–82.

there as well. Lastly, generalized convergence, as embodied in filters and directed sets, relied essentially on the Axiom. In fact, the Ultrafilter Theorem was later shown equivalent to the Boolean Prime Ideal Theorem and to the existence of the Stone–Čech compactification, all of which came to be recognized as non-constructive assumptions weaker than the Axiom.

Topologists not only employed the Axiom with ease but often took it for granted. None exhibited the qualms of conscience which affected van der Waerden in algebra—despite the fact that several of the topologists were students of Luzin, an ardent constructivist. Perhaps topologists sensed that it was inappropriate, when engaged in so abstract an enterprise, to insist on constructive methods. Perhaps the enchanting landscape of abstraction and generalization freed them from any lingering constructivist scruples. Whatever the reason, they did not seriously entertain the question of what general topology might become in the absence of the Axiom. Like so many mathematicians before and afterward, the use of arbitrary choices became second nature to them—if not, indeed, a reflex.

4.7 Negations and Alternatives

As awareness grew that significant theorems in many branches of mathematics used the Axiom, it was natural for the Axiom's critics to seek methods for eliminating it. First of all, they could scrutinize proofs with sufficient care to ensure that the Axiom did not occur there, while leaving in abeyance those theorems whose known demonstrations required the Axiom. This method was less certain than it might appear, however, since the Axiom lay hidden in propositions which initially seemed unrelated to it.

A paramount example from analysis was the Countable Union Theorem, on which the theories of Borel sets and Lebesgue measure depended. As Sierpiński pointed out [1918, 127], if the Denumerable Axiom were discarded, then the Countable Union Theorem might fail and the real line might be a countable union of countable sets. If so, he noted, then Lebesgue measure would not be countably additive and its intricate superstructure of theorems would collapse. Unfortunately, Lebesgue did not fully appreciate the seriousness of this challenge to his theory of measure.

The challenge was better understood by the Italian analyst Leonida Tonelli, who in 1921 proposed to replace Lebesgue's theory of integration by an alternative theory not utilizing the Axiom. Tonelli's integral resembled the one previously suggested by W. H. Young and, in the presence of the Axiom, was identical to Lebesgue's. Because Tonelli rejected the Axiom, however, he made no attempt to demonstrate that his integral was countably additive. Instead, he contented himself with proving that if f is a Tonelli-integrable function and if a sequence $E_1, E_2, \ldots$ of measurable sets converges to a set E, then the sequence of integrals of f on $E_1, E_2, \ldots$ converges to the integral of f on E.[1] In effect Tonelli, who pursued these questions a little further in a later article [1923], was expanding the scope of his own earlier attempt [1913] to circumvent the Axiom in the calculus of variations.

Not all the theorems that a mathematician might wish to preserve could be separated from the Axiom by some new technique, such as Cipolla [1913] had introduced for sequential limit points (see 3.6) or as Tonelli had proposed for the integral. Thus a second method for avoiding the Axiom was to state an alternative postulate, one that was more nearly constructive. If such a postulate constituted a special case of the Axiom, it would only restrict rather than negate Zermelo's assumption. Such a restriction would not permit the derivation of new propositions unobtainable from the Axiom, but hopefully it would permit the most important theorems of real analysis to be deduced. In fact, real analysis was much more suitable for such a quasi-constructive approach than either abstract algebra or general topology, since $\mathbb{R}$ had a countable dense subset, the rationals, on which there existed a constructive well-ordering. On the other hand, a partial negation of the Axiom might imply a different structure for the real line or for other fundamental mathematical objects. Such a partial negation arose not in analysis but in the Cantorian theory of ordinal numbers—a subject to which we shall return shortly.

Using a logical apparatus adopted from Peano, Beppo Levi was the first to suggest a quasi-constructive restriction of the Axiom suitable for analysis. In mathematics, Levi argued, we must always operate within a particular "deductive domain," such as that of the natural numbers or of the real numbers, where certain sets were understood by "a unique and irreducible act of thought" [1918, 306]. From these sets, which he named primary sets but did not specify further, the other sets of the domain were to be defined. An element could be arbitrarily chosen from each set in a finite family of primary sets, but, he insisted, for an infinite family such selections could be carried out only by means of a rule. Nevertheless, it was possible to extend the given deductive domain to a new domain by adjoining additional primary sets. In this way a set of choices from an infinite family might occur in the extension of a deductive domain even though such a set did not exist in the original domain [1918, 305–308]. Here Levi diverged from Peano by permitting infinitely many arbitrary choices, although only in such an extension and only under restricted conditions.

On the other hand, Levi emphasized, the unlimited Axiom of Choice must be rejected as meaningless since it contradicted the essential nature of mathematical analysis. This was not immediately evident, he admitted, since no logical contradiction would result from accepting the Axiom. Yet, in effect, Zermelo's 1904 proof of the Well-Ordering Theorem assumed

[1] Tonelli 1921, vi, 143–154.

what was to be shown. Furthermore, like any other hypothesis, the Axiom could not be applied in the deductive domain of the real numbers unless it followed there from the postulates.[2] Apparently Levi was arguing here from the categoricity of the postulates for $\mathbb{R}$.

Levi, while not granting the unrestricted use of arbitrary choices, postulated their existence under certain conditions by means of his Principle of Approximation: Assuming that D is a deductive domain which includes that of $\mathbb{R}$, let $A_1, A_2, \ldots$ be a sequence of subsets of $\mathbb{R}$. Suppose for a given set M that there is a function $f: E \to M$, where E is a given set of choice functions for the sequence $A_1, A_2, \ldots$. Further, let $d: M \times M \to \mathbb{R}$ be a function such that $d(x, y) = 0$ if and only if $x = y$. Lastly, assume that for each positive real r there is some integer n such that for every u and v in E agreeing on their first n elements, $d(f(u), f(v)) < r$. Then every member of the range of E is in some natural extension of D.[3] This was the unwieldy assumption which Levi offered as an alternative to the Axiom of Choice.

As an application of his principle, Levi augmented the results in Peano's paper of 1890 on differential equations, where arbitrary choices had been intentionally avoided, by noting that in general every system of ordinary differential equations admits an infinite number of solutions with given boundary conditions. More to the point was Levi's proof that his principle implies Lindelöf's Covering Theorem (1.7.7): Every family of open intervals covering a subset A of $\mathbb{R}$ has a countable subfamily covering A. On the other hand, Levi observed that his principle was not equivalent to the Axiom and hence did not justify all the latter's uses, particularly those involving cardinality. Indeed he correctly believed that his principle, which followed from the Denumerable Axiom, did not yield the existence of Hamel's discontinuous real functions f such that $f(x + y) = f(x) + f(y)$.[4]

In 1923, responding to an article that Hilbert had published the previous year on the foundations of logic, Levi stressed the importance of the Principle of Approximation. Although Levi granted only a restricted validity to the Axiom, he disagreed with those constructivists who founded mathematics solely on the natural numbers—a path which, he insisted, would lead to the destruction of analysis.[5] Rather, he put forward his principle as a compromise intended to reconcile diverse mathematical factions. As a further application of his principle, Levi [1923] showed that without the Axiom it yielded a function minimizing a given sequence of functions in the calculus of variations.

For a decade Levi's principle remained unused by any other mathematician. This was not surprising since Levi's arguments in favor of the Principle

---

[2] Levi 1918, 309–311.

[3] *Ibid.*, 312–313. Levi did not state what he meant by a natural extension of a deductive domain.

[4] *Ibid.*, 313–323.

[5] *Ibid.*, 305; Levi 1923, 166–167.

of Approximation were far from compelling. Even the statement of his principle remained vague, for it was unclear what he understood by a deductive domain. However, in 1931 Tullio Viola, then at Bologna, published an article applying this principle to analysis. Accepting Levi's argument that the principle could reconcile constructive and non-constructive approaches to mathematics, Viola went beyond Levi in illustrating its usefulness. In particular, Viola deduced from this principle the equivalence of limit point and sequential limit point in $\mathbb{R}^n$. It also followed that a subset A of $\mathbb{R}^n$ is infinite if and only if A is Dedekind-infinite, and that every uncountable A contains at least one of its condensation points. Lastly, he noted, the principle implied that if A is bounded, then there exists a countable set which contains all the isolated points of A and whose derived set is identical with the derived set of A.[6]

These applications led Viola to estimate the deductive strength of the Principle of Approximation. If transformations which altered distances were permitted, as Viola was not altogether willing to do, then the Principle of Approximation yielded the Denumerable Axiom for subsets of $\mathbb{R}^n$ in any natural extension of $\mathbb{R}^n$ [1931, 292]. In fact, though he did not say so, these two assumptions were equivalent. It seems that Viola's preference for the former assumption was based on his attraction to the idea of approximation, rather than on the greater clarity or self-evidence of Levi's principle.

Viola's research stimulated Levi to publish an article [1934] on his principle in *Fundamenta Mathematicae*, to which Viola added a detailed study of its applications. Levi's primary concern was to determine the permissible extensions of a deductive domain. Since an argument which was existential in one domain could be made constructive in an extension, he insisted that we must ascertain the consistency of the extension or, at least, exhibit certain of the new primary sets within the old domain. In this respect, he continued, Zermelo's argument of 1904 was flawed because a particular well-ordering would have rendered the Axiom unnecessary. Indeed, to ensure that every set can be well-ordered, the deductive domain had to be extended infinitely many times—a procedure that Levi disallowed. Consideration of the real numbers suggested to him that among the legitimate extensions of a deductive domain were those for which the new elements could be approximated in the old domain. In this way he justified his Principle of Approximation.

While Levi [1934] viewed his principle metamathematically and philosophically, Viola concentrated on its potential uses in measure theory. In 1921 Tonelli had already been concerned that the countable additivity of Lebesgue measure depended on the Axiom. A decade later Giuseppe Scorza-Dragoni, then at Rome, endeavored to free measure theory in part from the Axiom [1930]. Soon afterward, Viola demonstrated that the

[6] Viola 1931, 287–293; *cf.* (1.8.1).

Principle of Approximation yielded the countable additivity of Lebesgue measure [1932]. In this vein, Viola [1934] wished to determine those theorems in measure theory that could use Levi's principle as a substitute for the Axiom. Of special interest were propositions about set functions—those real-valued functions whose arguments were subsets of $\mathbb{R}$. In particular, the Principle of Approximation implied Lebesgue's theorem that every countably additive set function is bounded and hence of bounded variation. It also yielded proposition (4.1.22) that every measurable set is the union of countably many closed sets together with a set of measure zero. Finally, Vitali's Covering Theorem followed as well.[7]

After 1934, interest in the Principle of Approximation declined. Viola [1934] had claimed that Levi's principle did not suffice to prove Luzin's Theorem: Every measurable function is continuous on an interval when one neglects a set of arbitrarily small positive measure.[8] Yet soon afterward Scorza-Dragoni showed that in fact it did suffice [1936]. Subsequently a few Italian mathematicians mentioned Levi's principle, but none investigated it further.[9] Those Italian mathematicians who mistrusted the Axiom preferred to determine where it could be avoided altogether, as for families of closed sets in many subspaces of $\mathbb{R}^{\mathbb{N}}$.[10] Outside Italy, the Principle of Approximation generated no interest at all.

A second attempt to formulate alternatives to the Axiom was undertaken independently of Levi by the American logician Alonzo Church [1927]. Above all, Church wished to construct a logic in which the Axiom was false. In this vein he regarded both the Axiom and his three mutually exclusive alternatives A, B, and C as postulates of logic.[11] By couching his alternatives in terms of Cantor's second number-class Z_2, Church showed the influence of his mentor and colleague at Princeton, Oswald Veblen. Cantor had defined a fundamental sequence for a limit ordinal α to be a strictly increasing sequence of ordinals whose limit, or least upper bound, was α [1897, 219]. In 1908 Veblen considered the problem of determining a function f which assigned a fundamental sequence to every limit ordinal in Z_2. This problem concerned Veblen because Hardy's construction [1903] of a subset of $\mathbb{R}$ with power aleph-one depended on such a function (see 1.6). Though Veblen did not solve this problem, he constructed such an f whose domain was restricted to all limit ordinals less than Cantor's first epsilon-number.[12]

---

[7] Viola 1934, 78–92. Vitali's theorem was as follows: Suppose that a bounded measurable subset A of $\mathbb{R}$ is covered by an infinite family of open intervals, each of arbitrarily small length; then, given a positive real number r, there exists a disjoint countable subfamily covering A almost everywhere and whose total length is less than $m(A) + r$.

[8] *Ibid.*, 78.

[9] Cassina 1936, 60; Faedo 1940, 263.

[10] Cassina 1936, 65–81; Faedo 1940, 263–276.

[11] Church 1927, 178–179, 191. However, he did not clearly delineate the sense in which both his postulates and the Axiom belonged to logic, rather than to a mathematical theory of ordinals; *cf.* 4.8.

[12] An *epsilon-number* is an ordinal α such that $\alpha = \omega^\alpha$.

For unknown reasons, Veblen [1908] did not mention the fact that the Axiom would yield the existence of the desired function on Z_2, nor, on the other hand, did he object to the Axiom.

In 1927 Church studied the existence of such functions axiomatically. While Cantor had defined the second number-class Z_2 to be the set of all denumerable ordinals, Church designated W_2 as the least class containing ω and closed under the operations of successor and of taking the limit of a fundamental sequence.[13] In the presence of the Denumerable Axiom, or even of the Countable Union Theorem, W_2 and Z_2 were identical. Otherwise, Cantor's second number-class might be a proper subset of W_2.

Postulate A, Church's first alternative to the Axiom, asserted that:

(4.7.1) There exists a function which assigns a fundamental sequence to every limit ordinal in W_2.

This postulate followed from $\mathbf{C}^{\aleph_1}$, the Axiom of Choice for families of power $\aleph_1$, but Church found no reason to believe that $\mathbf{C}^{\aleph_1}$ and (4.7.1) were equivalent. Nevertheless, for W_2 the consequences of A agreed with those given by the Axiom: Every ordinal in W_2 was denumerable, and hence W_2 equaled Z_2; the family of countable subsets of $\aleph_1$ had the power of the continuum; and finally, $\aleph_1 \leq 2^{\aleph_0}$.[14]

Church believed that postulate B, which implied that $\mathbf{C}^{\aleph_1}$ was false, was consistent with the Denumerable Axiom but not equivalent to it. Postulate B stated that:

(4.7.2) There does not exist a function which assigns a fundamental sequence to every limit ordinal in W_2, but for each α in W_2 there exists a function assigning a fundamental sequence to every limit ordinal less than α.

Although postulates A and B each implied that W_2 and Z_2 had the same members, B gave the additional result that the continuum could not be well-ordered. Likewise, B yielded a number of consequences about fundamental sequences in W_2.[15]

Postulate C, the most radical negation of the Axiom among Church's alternatives, was the following proposition:

(4.7.3) For some β in W_2 there exists no function assigning a fundamental sequence to every limit ordinal less than β.

---

[13] If von Neumann's definition of ordinal is used, Z_2 is simply ω_1. However, this definition had not been widely adopted when Church wrote his article. Strictly speaking, Church redefined the second number-class to be W_2, but his results will be expressed here in Cantor's terms.

[14] Church 1927, 187–191.

[15] *Ibid.*, 187, 191–196.

From postulate C it followed that Z_2 was a proper subset of W_2. Indeed, the first uncountable ordinal ω_1 then belonged to W_2, since ω_1 was the least such β given by the postulate. Besides being the limit of a denumerable sequence of denumerable ordinals, ω_1 was also an epsilon-number, and there occurred $\aleph_1$ epsilon-numbers before it. Moreover, the Denumerable Axiom failed for some family of subsets of $\mathbb{R}$, a result which implied both that the continuum could not be well-ordered and that Lindelöf's Covering Theorem (1.7.7) was false. Although postulate C yielded the result that W_2 contained at least one initial ordinal besides ω, it did not specify how many initial ordinals occurred there.[16] Letting Ω denote the ordinal of W_2, Church established that W_2 contained either a greatest initial ordinal or else $\aleph_\Omega$ such initial ordinals.[17] In the latter case, it could happen that W_2 consisted of all infinite ordinals, a possibility which Church did not envision.[18]

Despite his innovative research, Church's alternatives attracted little attention for three decades. In contrast to Levi, he proposed these alternative postulates from what he described as an experimental viewpoint, rather than from constructivist opposition to the Axiom. Yet the lack of response indicated the difficulties awaiting anyone who negated the Axiom in classical, non-intuitionistic mathematics. The first of these was consistency. Church believed that his three postulates were consistent, as they were proved to be almost four decades later, but he could do no more than state his belief.[19] Like other mathematicians of the time, he lacked a technique to carry out such a consistency proof. Indeed, he did not even refer to Hilbert's *Beweistheorie* or to Fraenkel's proof that the Axiom is independent (see 4.8 and 4.10).

On the other hand, the Axiom itself had not yet been shown to be consistent, or even relatively consistent. Thus what Church needed was not so much a formal demonstration of consistency, as a firm belief in both the consistency and the usefulness of his alternatives. The proposition that the continuum could not be well-ordered, a consequence of either postulate B or C, might be consistent but was most unlikely to prove useful. Furthermore, the conclusion that ω_1 was the limit of a sequence of denumerable ordinals, as implied by postulate C, contradicted the theorems as well as the spirit of Cantor's theory of ordinals.

But, more importantly, were such consequences of postulate C true? Cantor had presupposed that there existed a unique set theory, just as there existed a unique theory of real numbers. By contrast, Church inquired

[16] An *initial ordinal* is an ordinal α such that if γ is smaller than α, then the cardinality of γ is less than that of α.

[17] Church 1927, 196–208.

[18] This possibility was first noticed by Ernst Specker [1957, 200] in his study of Church's alternatives to the Axiom.

[19] The consistency of each postulate relative to **ZF** was shown by Petr Hajek [1966]; see also Gitik 1980.

whether there might be two or more distinct second number-classes in different universes of discourse, just as there existed different consistent geometries, hyperbolic and Euclidean.[20] Unfortunately, he failed to confront the question of what would serve as a foundation for mathematics, and for logic, if set theory were many and not one.

On the other hand, Levi's Principle of Approximation was at least as certain to be consistent as the Axiom, and could be useful in analysis. Viola had shown that Levi's principle sufficed to develop some, but not all, of measure theory. Yet after 1936 no one pursued the study of this principle either. Perhaps such indifference was due to the principle's awkward formulation or to its metamathematical character. More probably, however, mathematicians saw no need for Levi's principle if they accepted the Axiom. And if they did not, the Principle of Approximation remained objectionable as well.

In sum, the opponents of the Axiom did not investigate alternative postulates, except for Levi in real analysis and for Russell in regard to mediate cardinals (see 3.6). We can only conclude that Sierpiński's writings, in particular, were not taken seriously enough by constructively inclined analysts such as Borel and Lebesgue. No alternative to the Axiom of Choice was entertained by a sizeable number of mathematicians until 1962 when two Poles, Jan Mycielski and Hugo Steinhaus, proposed the Axiom of Determinateness.[21] In this sense, and in this sense alone, the Axiom of Choice remained unchallenged.

4.8 The Axiom's Contribution to Logic

The historical relationship between the Axiom of Choice and mathematical logic forms part of a larger historical nexus between set theory and logic.[1] During the years between the two World Wars, three aspects of the relationship between the Axiom and logic became prominent. The first of these, the subject of the present section, involves the uses to which the Axiom was

[20] Church 1927, 18. Three decades later Fraenkel elaborated this analogy between geometry and set theory. For Fraenkel, absolute geometry corresponded in some sense to **ZF**, Euclidean geometry to **ZFC**, and non-Euclidean geometry to **ZF** plus some principle negating the Axiom; see Fraenkel and Bar-Hillel 1958, 80. The analogy was developed further by Cohen and Hersh [1967].

[21] See the Epilogue. The Axiom of Determinateness (**AD**) implies very weak forms of the Axiom of Choice but also yields the negation of stronger forms. In particular, it follows from **AD** that every subset of $\mathbb{R}$ is Lebesgue-measurable.

[1] For a detailed discussion of the historical interplay between set theory and mathematical logic in the period 1870–1935, see Moore 1980.

put in an emerging mathematical logic. The second, treated in 4.9, deals with the place of the Axiom in the various axiomatic systems for set theory, and also sets the stage for the third aspect, which is essentially metamathematical. This aspect, discussed in 4.10, concerns the investigations of the Axiom's consistency and independence relative to the different systems for set theory.

The lack of a well-established boundary between set theory and logic made it all but inevitable that the Axiom's critics would question its logical status. On the one hand, some of them doubted that the Axiom could be deduced in a reasonable system for logic. Thus Peano rejected the Axiom in 1906 on the grounds that a finite, but not an infinite, number of arbitrary choices could be made within his *Formulaire*. To make infinitely many arbitrary choices, he claimed, would violate the requirement that a proof have finite length [1906, 147]. Lebesgue later made a similar claim [1918, 238]. On the other hand, some critics wondered whether the Axiom itself was actually a new postulate of logic. Writing two years after Peano, Wilson believed that logic as then formulated was probably incomplete, and so he entertained the possibility of adjoining the Axiom to it. Yet, from the categoricity of Huntington's postulates for the real numbers, Wilson concluded that Zermelo had no right to introduce an additional postulate asserting some property, such as well-orderability, to hold for $\mathbb{R}$.[2]

These views illustrate the essential fluidity of mathematical logic early in the twentieth century. Distinctions that logicians now regard as fundamental were often blurred or missing altogether. Today logic is considered to possess levels—the propositional calculus with its Boolean connectives (and, or, not) but with no quantifiers, then the first-order predicate calculus with quantifiers ranging only over individuals, next the second-order predicate calculus with quantifiers ranging over predicates as well as quantifiers ranging over individuals, and so on. These levels now appear quite distinct, though cumulative, and their expressive power differs greatly. When Peano and Wilson wrote, however, first-order logic did not possess the clarity and primacy which it subsequently acquired, and excellent logicians such as Frege, Russell, and Schröder moved freely from quantifying over individuals to quantifying over predicates with no sense of crossing a watershed.[3]

A second distinction not observed consistently at the time but now regarded as vital was that between syntax (including the notions of formula, proof, and theorem) and semantics (including the notions of model, logical consequence, and truth). In fact, there existed two traditions of mathematical

[2] Wilson 1908, 437; *cf.* 3.6.

[3] Quantifiers ranging over individuals were carefully distinguished from quantifiers ranging over predicates by Frege, who treated both kinds of quantifiers within a single logic [1893]. When formulating the theory of types, Russell introduced what he called "first-order propositions" as his second logical type [1908, 238]. However, Russell's first-order propositions, whose quantifiers ranged only over individuals, did not constitute first-order logic since even the sentence $\exists x \exists y \exists z (x \in y \;\&\; y \in z)$ could not be a first-order proposition in his sense.

logic, one emphasizing semantic considerations and the other syntactic notions. The first tradition, originating with George Boole, was elaborated by Peirce and Schröder in their researches on the algebra of logic. Indeed, the very word algebra reveals the predominance of the semantic approach. The second tradition developed from the work of Frege and Russell, but attained a greater clarity in the contributions of Hilbert and his school. As we shall see, these two traditions eventually fused in the research of Gödel and Tarski. Not until Gödel [1930] did mathematical logicians carefully and uniformly observe the distinction between a syntactic notion such as consistency (no contradiction can be derived from the given axiom system) and the corresponding semantic notion of satisfiability (the given axiom system has a model).

Schröder found an able successor in Leopold Löwenheim, then at Berlin, who was the first to treat first-order logic as a self-contained branch of logic. Löwenheim studied sentences (*Relativausdrücke*) in which the quantifiers ranged over relations, as well as sentences (*Zählausdrücke*) in which the quantifiers were restricted to individuals. In addition, he permitted his sentences to contain denumerable strings of quantifiers and of relations. Despite this wealth of expressive power, he formulated a theorem which revealed a severe limitation in the strength of first-order logic: If a first-order sentence A is true in every finite domain but not in every domain, then A is false in some denumerable domain. This result, as modified by Skolem to become the Löwenheim–Skolem Theorem, eventually transformed the character of mathematical logic. Löwenheim's proof of his result utilized quantifiers which ranged over every individual in the given domain and which were, in a sense, semantic rather than syntactic. In this vein he relied on a distributive law which stated that a sentence $\forall x \exists y A(x, y)$ is true in a domain if and only if $\exists y_\lambda \forall x A(x, y_\lambda(x))$ is true there, where λ ranges over all individuals of the domain.[4] When Skolem gave a new demonstration of Löwenheim's result five years later, he was particularly careful to recast this portion of the proof.[5]

Skolem, like Löwenheim, accepted Schröder's system as the basis for his logic. While retaining countably many quantifiers in a given sentence, Skolem circumvented Löwenheim's second-order semantic quantifiers through the introduction of new first-order relations—in effect, Skolem functions. Yet Skolem continued to obscure the boundary between syntax and semantics, especially between consistency and satisfiability, by stating Löwenheim's result in the following form: A first-order sentence is either contradictory or else satisfiable in a countable domain. Had Skolem established what he stated, he would have provided first-order logic with the

[4] Löwenheim 1915, 450–456. To Löwenheim, the quantifier $\exists y_\lambda$ was a possibly uncountable sequence of individual existential quantifiers, one for each individual in the given domain.

[5] Skolem 1920, 1.

Completeness Theorem. What he actually proved is the following proposition:

(4.8.1) If a first-order sentence is satisfiable in a set M, it is satisfiable in a countable subset S of M.

Skolem then generalized this proposition to countably many infinitely long sentences, each having countably many individual quantifiers and possibly uncountably many predicates. The Axiom served in an essential way to generate S from the Skolem functions.[6] The proposition (4.8.1), as generalized to any countable family of first-order sentences, became known as the Löwenheim–Skolem Theorem.

In the course of extending Löwenheim's result, Skolem did not record any of its consequences for axiomatic set theory. By 1922, however, Skolem's perspective on set theory had darkened. Choosing to emphasize the set-theoretic paradoxes, he criticized Zermelo's system and proposed several modifications (see 4.9). Furthermore, he proved without using the Axiom a weaker version of the Löwenheim–Skolem Theorem: If a countable family of first-order sentences is satisfiable, then it is satisfiable in the domain $\mathbb{N}$ of natural numbers. With this weaker theorem in hand, he concluded that Zermelo's system could be satisfied in $\mathbb{N}$ despite the fact that it implied the existence of uncountable sets. Later this observation became known as the Skolem paradox. For Skolem, it illustrated the relativism inherent in set-theoretic notions, since whether a set A was countable or uncountable depended on whether A was viewed from inside or outside the given model of set theory.[7] Abandoning sentences with infinite strings of quantifiers and repudiating set theory as a foundation for mathematics, Skolem warmly embraced a Kroneckerian finitism.[8] His belief that first-order logic is the proper setting for mathematics proved to be profoundly influential.

In contrast to Skolem, who developed semantics but lacked a clear sense of syntactic questions, Hilbert and his school all but reduced logic to syntax during the 1920s. Hilbert's interest in foundations had originated with his *Grundlagen der Geometrie* [1899]. During his Paris address of 1900, he posed the problem of demonstrating the consistency of the axioms for $\mathbb{R}$. At the same time he emphasized the utility of the axiomatic method, his belief that every well-formulated mathematical problem can be solved, and his conviction that the consistency of a set S of axioms implies the existence of a

[6] *Ibid.*, 4–13.

[7] Skolem 1923, 217–225. It should be stressed that Skolem was the first to propose that set theory, which seemed to require quantifiers over sets of individuals, over sets of sets of individuals, and so forth, could be mirrored within first-order logic. As we shall see in 4.9, Zermelo believed that to do so created immense distortions.

[8] See, for example, Skolem 1923a, 1–3, 38.

model for S [1900, 264–266]. These were themes to which he returned two decades later.

When he responded in 1922 to the intuitionistic criticisms by which Brouwer and Weyl were then buffeting classical analysis, Hilbert sought to reformulate each foundational question so clearly that its answer would be unique. He especially wished to deduce the Principle of the Excluded Middle, via an unspecified function τ, and to render some form of the Axiom of Choice as obvious as the statement that $2 + 2 = 4$.[9] These two tasks soon merged into one.

The following year Hilbert clarified his goals as well as his methods. It was vital, he insisted, to dispel the fog of doubt obscuring the certainty of mathematical reasoning. To this end he began to collaborate with Paul Bernays on a theory of mathematical proof (*Beweistheorie*) that would serve as a consistent foundation—for analysis as well as for set theory—by treating mathematics and its underlying logic as strings of uninterpreted symbols. In this syntactic fashion, proofs would become such strings having a certain form and obtained by certain mechanical rules from given initial strings, the axioms. Meaning entered only at the metamathematical level, where the permissible methods were finitistic and combinatorial.[10] Hilbert made no attempt, however, to spell out his metamathematics in detail.

As the cornerstone for his *Beweistheorie*, he turned to an unexpected source. Dismissing the critics of Zermelo's Axiom, Hilbert wrote grandly:

> The essential idea on which the Axiom of Choice is based constitutes a general logical principle which, even for the first elements of mathematical inference, is necessary and indispensable. When we secure these first elements, we obtain at the same time the foundation for the Axiom of Choice. Both are done by means of my proof theory. [1923, 152]

What Hilbert had in mind was a new postulate for logic. Since he considered this postulate to go beyond a finitistic logic, he named it the Transfinite Axiom:

$$A(\tau_x(A(x))) \to A(y).$$

In effect it asserted that if a proposition $A(c)$ is provable, where c is the value given by the function τ operating on the formula $A(x)$, then $A(y)$ is provable for every value of y. Thus τ selected a value c for which $A(c)$ would be unprovable if this could occur at all. By means of the Transfinite Axiom, Hilbert introduced the universal and existential quantifiers: $\forall y A(y)$ was defined to be $A(\tau_x(A(x)))$, and $\exists y A(y)$ to be $A(\tau_x(\sim A(x)))$. In 1904 he had treated his quantifiers, which then ranged specifically over the positive integers, as infinite conjunctions $A(1)\ \&\ A(2)\ \&\ \cdots$ and infinite disjunctions

---

[9] Hilbert 1922, 157, 176–177.

[10] Hilbert 1923, 151–155. His article was originally titled "The Axiom of Choice in Mathematical Logic"; see **DMVA 31** (1922), 101.

$A(1) \vee A(2) \vee \cdots$. Yet in 1923 he intentionally avoided such infinite conjunctions and disjunctions by introducing the Transfinite Axiom.[11]

Hilbert's enduring concern was to obtain a syntactic consistency proof for his formal system of logic (which, as he formulated it, included both equality and the natural numbers) and above all for the Transfinite Axiom together with its corollary, the Principle of the Excluded Middle. To execute such a proof, he regarded the function τ as acting on a hierarchy of levels, first on functions f from $\mathbb{N}$ to $\mathbb{N}$, next on functions from $\mathbb{N}^{\mathbb{N}}$ to $\mathbb{N}^{\mathbb{N}}$ (*i.e.*, real functions), and so forth. For the first level he arrived at the desired proof by defining $\tau(f)$ to be the least m such that $f(m) \neq 0$, and otherwise to be 0. While a consistency proof for the second level remained beyond his grasp, he demonstrated that the Transfinite Axiom yielded the Axiom of Choice restricted to families of subsets of $\mathbb{R}$.[12]

Hilbert's bold proposal to found logic on the Transfinite Axiom produced a mixed reaction. Attempting to establish the consistency of number theory, his colleague Wilhelm Ackermann employed this axiom and asserted without proof that it implies the Axiom of Choice [1924, 10, 36]. In Italy, on the other hand, Cipolla insisted that Hilbert's assumption lacked any discernible meaning since the function τ was vague. Indeed, Cipolla argued that the Transfinite Axiom is equivalent to a very general form of the Axiom of Choice [1924, 24–29]:

(4.8.2) For every class M of non-empty sets there is a function which assigns to each A in M an element of A.

If M was taken to be the class of all sets, then (4.8.2) became what was later called the Axiom of Global Choice.[13] Cipolla's class form of the Axiom implies Zermelo's but is not equivalent to it—a fact not recognized at the time.[14] Actually, Cipolla's proof showed Hilbert's axiom to be equivalent, not to (4.8.2), but to a still more general form of the Axiom of Choice:

(4.8.3) There exists a function σ such that $\sigma(C) \in C$ for every non-empty class C.

---

[11] Hilbert 1904, 178, and 1923, 152–157. On the early history of infinitary logic (*i.e.*, logic which permits infinitely long formulas or rules of inference with infinitely many premises), see Moore 1980, 97–101, 124–130.

[12] Hilbert 1923, 157–165.

[13] This term was coined by Azriel Levy; see Levy in Fraenkel *et alii* 1973, 72. Earlier, Levy had used the term Axiom of Universal Choice [1961].

[14] Easton [1964] showed that (4.8.2) does not follow from the Axiom of Choice in Gödel–Bernays set theory. Felgner [1971a] established that (4.8.2) does not generate any new theorems about sets, beyond those given by **ZFC**, but only new theorems about proper classes; see also Gaifman 1975. Thus a conservative extension of **ZFC** is obtained by adjoining (4.8.2) to it. (More generally, a formal system T' is an *extension* of a formal system T if the primitive symbols of T' include all the primitive symbols of T and if every theorem if T is a theorem of T'; then T' is a *conservative extension* of T if every theorem of T' that can be expressed in the primitive symbols of T is also a theorem of T.)

From this equivalence Cipolla concluded that Hilbert's axiom was just as illegitimate as Zermelo's.

When Hilbert returned to his Transfinite Axiom in 1926, he reformulated it in such a way that it more nearly paralleled the Axiom of Choice:

$$A(y) \rightarrow A(\varepsilon_x(A(x))).$$

This ε-axiom, as he later called it, stated in effect that if $A(c)$ held for some value c, then $\varepsilon_x(A(x))$ was one such c. Hilbert employed the ε-axiom, which he regarded as a variant of the Axiom of Choice, in his abortive attempt to demonstrate the Continuum Hypothesis by means of the definability of number-theoretic functions. At the same time he treated propositions about infinite sets as a type of ideal element, analogous to the points at infinity which were adjoined to the Euclidean plane in order to obtain projective geometry [1926]. Soon afterward, he cited the ε-axiom as the source of all such ideal propositions, and continued to rely on it to parry Brouwer's thrusts against the Principle of the Excluded Middle.[15] In 1929 Skolem gave some support to Hilbert's position by demonstrating that if a first-order theory is satisfiable, it remains satisfiable after the addition of both the Principle of the Excluded Middle and all the first-order instances of the second-order ε-axiom.[16]

Hilbert's attitude toward foundational questions diverged from Skolem's not only in his syntactic emphasis but also in his sympathy for Zermelo's axiomatization. In 1930 Hilbert argued that the chief task of foundations was to discover a proof for the substantive, non-logical (*inhaltliche*) assumptions underlying Zermelo's system and, as a beginning, to demonstrate the consistency of the axioms for the real numbers. If mathematics were ultimately founded on *inhaltliche* assumptions such as Zermelo's, Hilbert insisted, then it would lose its absolute certainty; instead, each mathematical expression must be treated as a concrete object. An important first step in Hilbert's program had been achieved when Ackermann settled the consistency of the ε-axiom as restricted to number-theoretic predicates. For higher levels, however, the problem remained open. Hilbert noted that a proof for the consistency of the ε-axiom, limited to predicates acting on number-theoretic functions, would imply the consistency of the Principle of the Excluded Middle for sets of real numbers as well as that of a weak form of the Axiom of Choice [1930, 1–3].

In contrast to Luzin and Sierpiński, who measured the Axiom's strength by the cardinality of the sets on which it acted, Hilbert preferred a measure based on definability. Thus the weak form, mentioned above, was the Axiom restricted to families of sets of real numbers, while the strong form satisfied the additional condition that if two subsets A and B of $\mathbb{R}$ are equal, then the

---

[15] Hilbert 1928, 67–68, 81.

[16] Skolem 1929, 28–49.

choice function selects the same element from both A and B, no matter how A and B are defined.[17] Later mathematicians used definability in a different fashion to gauge the strength of the Axiom as applied to a given family M of sets: the complexity of the quantifiers needed to define M.[18]

In 1928 Hilbert and Ackermann published a slim volume, *Grundzüge der theoretischen Logik*. Based on courses that Hilbert had given at Göttingen between 1917 and 1922, it discussed both first-order and higher-order logic. This treatise, much concerned with syntax but little with semantics, soon prompted mathematicians to try their ingenuity on two fundamental problems. The first of these, which grew out of Hilbert's conviction that every well-formulated mathematical problem is solvable, was called the *Entscheidungsproblem*, or Decision Problem. It asked for an algorithmic procedure determining whether any given first-order (or higher-order) sentence is satisfiable.[19] As a partial solution, the young English logician Frank Ramsey contributed an algorithm deciding the satisfiability of each first-order sentence, all of whose existential quantifiers preceded its universal ones at the beginning of the sentence. To obtain this result, he established a combinatorial lemma about finite sets. However, he first demonstrated the following infinite version of it, later known as Ramsey's Theorem, by relying explicitly on the Axiom [1929]:

(4.8.4) Suppose that the family of all m-member subsets of an infinite set A is partitioned into sets $S_1, S_2, \ldots, S_n$; then there is an infinite subset B of A such that all the m-member subsets of B belong to the same S_i.

Ramsey's Theorem later attained considerable importance in infinite combinatorics. Although its proof does not require the Axiom when A is $\mathbb{N}$, Ramsey's Theorem depends on the Axiom whenever A is an arbitrary infinite set.[20]

The second problem posed by Hilbert and Ackermann was to establish the completeness of first-order logic: If a first-order sentence is true in every domain, then it is deducible from the axioms of first-order logic.[21] The mathematician who succeeded in proving this Completeness Theorem was the one who subsequently showed it to be extremely likely that the *Entscheidungsproblem* lacked a positive solution: Kurt Gödel. In his doctoral dissertation of 1930, written under Hans Hahn at Vienna, Gödel exhibited

[17] Hilbert 1930, 3–6.

[18] See Shoenfield 1961, 134–135.

[19] Hilbert and Ackermann 1928, 72–81.

[20] Kleinberg [1969] established that Ramsey's Theorem cannot be proved in **ZF** but that it follows from proposition (1.3.3) that every Dedekind-finite set is finite. See also Blass 1977.

[21] Hilbert and Ackermann 1928, 65–68.

a more profound understanding of the distinction between syntax and semantics, as well as their interconnections, than had his predecessors. Thus, in particular, Gödel recognized clearly and observed carefully the distinction between the consistency of a formal system and its satisfiability, an essential first step in proving the Completeness Theorem [1930].

It was such judicious distinctions which soon led Gödel to establish, by means of his Incompleteness Theorem [1931], that semantic and syntactic notions were *not* equivalent in general. If a first-order theory was axiomatized and could express the arithmetic of the natural numbers, then there was a sentence true in the theory but which could not be deduced from the axioms of the theory. Furthermore, in his Second Incompleteness Theorem, he showed that no such theory could prove its own consistency—unless the theory was actually inconsistent [1931, 196].

In the course of deriving the Completeness Theorem for first-order logic, Gödel weakened its hypothesis to "true in every countable domain" and employed König's Infinity Lemma—in effect, a restricted form of the Axiom (see 4.5). Since Gödel established the Completeness Theorem only for languages with countably many symbols, he did not require such an assumption. Later he noted that in his proof he had not wished to avoid the Axiom.[22] As a corollary, he deduced what came to be known as the Compactness Theorem for first-order languages with countably many symbols: A denumerable set S of first-order sentences is satisfiable if and only if every finite subset of S is satisfiable. Finally, he made precise a long-standing belief of Hilbert's by demonstrating that every countable consistent set of first-order sentences has a model.[23]

As we have seen, foundational research in the 1920s confined logic principally to first-order languages with countably many symbols, and hence to languages with denumerable sets of sentences. In a similar vein Skolem underlined the importance of countable models. The uncountable re-entered mathematical logic in two guises, one semantic and one syntactic. Beginning in 1926, Tarski conducted a seminar at the University of Warsaw on model-theoretic methods such as the elimination of quantifiers.[24] To Skolem's paper of 1934, which exhibited a non-standard model of arithmetic, Tarski appended an editorial note stating that six years earlier he had extended the Löwenheim–Skolem Theorem in the following fashion: If a consistent set S of first-order sentences cannot be satisfied in a finite model, then S has both a denumerable and an uncountable model; hence S cannot be categorical [1934, 161]. At the time Tarski did not publish a demonstration because he believed himself to lack a precise expression for his results and especially for the notion of satisfiability. Hence it is problematical whether he then possessed a proof of the proposition that if a countable set of sentences

---

[22] Van Heijenoort 1967, 510–511; Wang 1970, 24.

[23] Gödel 1930, 350–359.

[24] Vaught 1974, 159.

has an infinite model, then it has a model of every infinite cardinality.[25] This proposition, later termed the Löwenheim–Skolem–Tarski Theorem, was eventually shown to be equivalent to the Axiom.[26]

Although his primary interest, like that of Tarski, lay in semantics, the Russian logician Anatolii Malcev introduced languages with uncountably many symbols for both propositional and first-order logic [1936]. After referring to Skolem, Malcev proved an extension of the Löwenheim–Skolem Theorem that was different from Tarski's. Malcev's extension was later called the Downward Löwenheim–Skolem Theorem:

(4.8.5) Every model of an infinite set S of first-order sentences has a submodel whose power is at most that of S.[27]

Two decades later (4.8.5) was likewise shown to be equivalent to the Axiom.[28]

Malcev generalized the Compactness Theorem, which Gödel had stated for languages with countably many symbols, to arbitrarily large sets of sentences in propositional logic. The proof which Malcev gave relied on the Well-Ordering Theorem. He also demonstrated that for every set S of first-order sentences there exists a set T of sentences of propositional logic such that S is satisfiable in first-order logic if and only if T is satisfiable in propositional logic. While the Compactness Theorem for first-order languages followed directly from these two results, he did not state it in his paper.[29]

In fact, a certain carelessness in Malcev's proofs earned him the opprobrium of several later logicians.[30] When he formulated the Compactness Theorem for first-order logic [1941] and applied it to algebra, he referred to his earlier paper for a demonstration, even though the theorem was not mentioned there. More damaging was Malcev's conflation of syntactic and semantic notions, particularly consistency and satisfiability—an error which he probably inherited from Skolem but which a careful reading of Gödel's article [1930] should have corrected.[31]

Outside the Soviet Union the Compactness Theorem came to occupy a central position only through the research of Leon Henkin more than a decade later. Working with Tarski at Berkeley, Henkin [1954] established that the Completeness Theorem for arbitrarily large first-order languages is equivalent to the same theorem restricted to propositional logic, to the Compactness Theorem for such first-order languages, and to the Boolean

[25] Vaught [1974, 160–161] argues that Tarski probably had such a proof.

[26] Vaught 1956, 262–263; see also Rubin and Rubin 1963, 69–71.

[27] Malcev 1936, 329–331.

[28] Vaught 1956, 262–263.

[29] Malcev 1936, 323–329. His second result was, in effect, a semantic generalization of Herbrand's Theorem [1930].

[30] See the review of Malcev 1941 by Henkin and Mostowski [1959].

[31] Malcev 1936, 323–329.

Prime Ideal Theorem. Consequently, all these theorems of logic involved the Axiom in an essential way. Similarly, the Löwenheim–Skolem–Tarski Theorem did not gain full recognition until the Berkeley school of model theorists emerged under Tarski's direction in the 1950s.

Skolem, however, had already stressed certain consequences of the Löwenheim–Skolem Theorem during the 1920s and 1930s.[32] Central among these was the relativity of all mathematical concepts, especially those involving cardinality. Even though Skolem understood that this relativity affected the finite as well as the uncountable, he continued to aim his objections at the latter.[33] To Paul Bernays, on the other hand, the fact that neither the finite nor the uncountable could be characterized by first-order languages suggested instead the inadequacy of this level of logic.[34] Later it was recognized that one could characterize first-order logic, as opposed to other levels of logic, by the fact that it satisfied both the Compactness Theorem and the Löwenheim–Skolem Theorem.[35]

Although Gödel's Second Incompleteness Theorem dashed Hilbert's hopes of establishing the consistency of arithmetic and set theory finitistically, Bernays continued to develop Hilbertian proof theory. In 1939 Bernays employed the ε-axiom to obtain for each first-order theory a conservative extension without quantifiers, thus showing the relative consistency of the ε-axiom.[36] During the 1950s the Bourbakists followed Hilbert by basing their logic on the ε-axiom.[37] Most contemporary logicians, however, do not introduce a form of the Axiom so directly into first-order logic but utilize it in their metamathematics.

In either case, the Axiom has become thoroughly woven into the fabric of mathematical logic. The Completeness Theorem, the Compactness Theorem, and the Löwenheim–Skolem–Tarski Theorem—all essential to the development of first-order logic—depend on the Axiom. It was the emergence of uncountable languages and the importance of uncountable models that made this dependence unavoidable. Skolem's scruples about the uncountable, so reminiscent of Borel's and Poincaré's, were overwhelmed by the incoming tide of higher infinities. Once again the Cantorian legacy rendered the Axiom useful, indeed indispensable, and hence acceptable to a large majority of mathematicians and logicians.

---

[32] See, for example, Skolem 1929, 49.

[33] Skolem 1941, 37–38.

[34] Bernays in Skolem 1941, 50. Zermelo [1932] also argued that first-order logic could not do justice to the needs of mathematical reasoning; see Moore 1980, 125–128, and 4.9.

[35] See Barwise 1974, 259.

[36] Hilbert and Bernays 1939, 9–33, 130–149. As A. Levy has pointed out, it follows from Bernays' result that the ε-axiom is a conservative extension of **ZF**, and hence does not imply the Axiom of Choice in the language of **ZF**. On the other hand, if one permits terms of the form $\varepsilon_x(A(x))$ to appear in the Axiom of Replacement, then the ε-axiom implies the Axiom of Choice. See Levy in Fraenkel *et alii* 1973, 72–73.

[37] Bourbaki 1954, 14–16, 35–36.

4.9 Shifting Axiomatizations for Set Theory

In 1904, when Zermelo proposed the Axiom of Choice, he was still working largely within a Cantorian tradition where logic was not seen to be relevant to the set-theorist. Indeed, as a Platonic realist, Cantor did not in any sense view his set theory from an axiomatic perspective. The first transformation in the nature of set theory occurred around 1906 when Zermelo, now operating primarily within a Hilbertian axiomatic tradition and motivated by a desire to secure his Axiom of Choice against its numerous critics, embedded the Axiom in a postulate system for set theory as a whole. Thus axiomatic set theory was separated from Cantor's original creation, which von Neumann later dubbed naive set theory. However, various mathematicians criticized Zermelo's system, and some of them, including Fraenkel [1921] and Skolem [1923], suggested both internal modifications and additional axioms. During the 1920s and 1930s the logical underpinnings of Zermelo's system were also re-examined and transformed. This second transformation, carried out principally by Skolem, Gödel, and Bernays, made first-order logic the basis for set theory. Although it gradually became the orthodox view to regard axiomatic set theory as embedded in first-order logic, at the time Zermelo vehemently opposed such an approach. Nevertheless, the metamathematical results on the consistency and independence of the Axiom, which form the subject of 4.10, all relied on first-order logic.

The first transformation, from Cantor's naive set theory to Zermelo's axiomatic set theory, originally received the approval of only three German mathematicians—Hessenberg [1909], Jacobsthal [1909], and Hartogs [1915]—and of no one outside Germany. In fact, Zermelo's system elicited private objections from Jourdain and Russell, as well as public criticisms from Poincaré [1909a], Weyl [1910], and Schoenflies [1911]. These criticisms, diverse though they were, primarily concerned the consistency of Zermelo's system and the ambiguity in his notion of *definit* property—a notion essential to the Axiom of Separation (see 3.3).

The second transformation in the nature of set theory was initiated by one of these critics, Hermann Weyl. Unlike his fellow critics, he believed that Zermelo's system was essentially correct and required only a satisfactory definition of *definit* property. Weyl, unable to state a definition that he found adequate, nevertheless described a property as *definit* if it was built up from the relations $\in$ and $=$ by a finite number of unspecified definition-principles [1910, 112–113]. Over the next few years Weyl succeeded in formulating a definition of *definit* property which he considered satisfactory. In 1917 he defined a *definit* property to be a proposition that was either of the form $x \in y$ or $x = y$, or was obtained from such propositions by finitely many uses of negation, conjunction, disjunction, existential quantification, and substitution of a constant for a variable. However, as he sought to avoid the notion of natural number in this definition, it became forbiddingly convoluted. He concluded that to seek such a definition not using the natural numbers was a scholastic pseudoproblem and that set theory must be founded on the natural numbers rather than conversely. Hence he rejected the efforts of Dedekind and Zermelo to build the natural numbers within set theory.[1] After arriving at such a constructivistic view of mathematics, Weyl soon embraced Brouwer's intuitionism.

Because Weyl's revised notion of *definit* property had little direct influence on set theorists and logicians, the second transformation of set theory did not take effect until several years later, chiefly through the efforts of Skolem. In August 1922, Skolem delivered a paper on Zermelo's system to the Fifth Congress of Scandinavian Mathematicians. Published a year later, the paper modified Zermelo's notion of *definit* property in a way analogous to the revision previously suggested by Weyl.

The essential difference between the two revisions was that Skolem explicitly restricted quantification to individuals, while Weyl did not. Thus Skolem treated Zermelo's Axiom of Separation as an axiom schema within first-order logic. As Skolem pointed out, the Löwenheim–Skolem Theorem implied that Zermelo's system had a denumerable model even though the system yielded the existence of uncountable sets.[2] This result, later called the Skolem paradox, was thoroughly repugnant to Zermelo, who believed that the Löwenheim–Skolem Theorem revealed the limitations of first-order logic rather than the inadequacy of axiomatic set theory [1932, 85].

One question that Skolem raised about Zermelo's system, the possible existence of infinite descending $\in$-sequences

$$\cdots \in A_3 \in A_2 \in A_1,$$

had already been broached by Dimitry Mirimanoff of Geneva in 1917. Even earlier Schoenflies had disallowed sets A such that $A \in A$ by postulating that for all objects A and B, $A \in B$ and $B \in A$ cannot both hold. Yet Schoenflies made no attempt to exclude sets A, B, C such that $A \in B \in C \in A$, much less to rule out infinite descending $\in$-sequences in general [1911, 245]. Mirimanoff, on the other hand, permitted infinite descending $\in$-sequences but endowed them with a special status. He designated any set A_1 that occurred at the beginning of some infinite descending $\in$-sequence as an extraordinary set, while all other sets were said to be ordinary sets.[3]

Mirimanoff's principal aim was to resolve what he considered the fundamental problem of set theory: to find necessary and sufficient conditions for a set of objects to exist. For the particular case of ordinary sets he believed himself to have a solution, but not for the general case of arbitrary sets. In order to carry out this partial solution, he introduced the notion of rank. He

---

[1] Weyl 1917, 4–6, 36–37.

[2] Skolem 1923, 218–220.

[3] Mirimanoff 1917, 42. Contrary to the claim of Hao Wang [1970, 35], Mirimanoff did not propose an axiom which prohibited descending $\in$-sequences.

defined the rank of each urelement and of the empty set to be zero, while the rank of any other ordinary set A was to be the least ordinal α greater than the ranks of all the elements of A. Consequently, every ordinary set had a rank α, and the collection O_α of all ordinary sets of rank α was a set. At this point he introduced his partial solution to the fundamental problem by asserting that a collection B of ordinary sets is a set if and only if there exists an ordinal greater than the rank of every set in B.[4] However, he did not attempt to determine how large an ordinal existed.

Two aspects of Mirimanoff's work deserve emphasis. First, although he did not introduce an axiom which prohibited infinite descending $\in$-sequences, he did illuminate the structure of those sets that do not form part of such sequences. Hidden within his system but unnoticed at the time were the rudiments of Zermelo's later cumulative type hierarchy.[5] The second aspect involves the Axiom of Replacement, which Mirimanoff formulated in 1917 as follows: If a collection B of ordinary sets is equipollent to a set, then B is also an ordinary set.[6] Surprisingly, Mirimanoff hardly mentioned Zermelo's system, and seems to have obtained the other postulates that he used—the Axiom of Union and the Power Set Axiom—from Julius König's book [1914] on the foundations of set theory. Despite his contributions, Mirimanoff did not directly affect the further development of Zermelo's system any more than did Weyl.[7]

By contrast, the researches of Fraenkel and Skolem around 1922 directly influenced the structure of Zermelo's system in several ways, not the least of which was the addition of the Axiom of Replacement to the postulates for set theory. Tradition has often ascribed this axiom to Fraenkel alone and occasionally to both Fraenkel and Skolem,[8] perhaps because Zermelo adopted a version of it from Fraenkel (see below). Thus the traditional ascription is accurate as to the path by which this axiom entered Zermelo's system, but not as to its first appearance—a fact that illustrates how tangled the phenomenon of multiple independent discoveries can be. While it remains uncertain what motivated Mirimanoff to propose the Axiom of Replacement, both Fraenkel and Skolem arrived at it independently around 1922 after discovering that Zermelo's system did not yield the set

$$\{\mathbb{N}, \mathscr{P}(\mathbb{N}), \mathscr{P}(\mathscr{P}(\mathbb{N})), \ldots\},$$

---

[4] Mirimanoff 1917, 42–45, 48–52. He assumed explicitly that the collection of all urelements is a set.

[5] See Zermelo 1930, discussed below.

[6] Mirimanoff 1917, 49. In a letter of 28 August 1899 to Dedekind, Cantor had stated as a fact (but not as an axiom) that every multitude equipollent to a set is a set; see 1.6.

[7] Later, Mirimanoff [1917] did affect the views of von Neumann on descending $\in$-sequences and thereby indirectly influenced Zermelo's system, as we shall see below.

[8] See, for example, van Heijenoort 1967, 113, 291. On the other hand, Kuratowski [1924] credited this axiom to Mirimanoff and Fraenkel.

where $\mathscr{P}(\mathbb{N})$ is the power set of $\mathbb{N}$. Skolem stated this axiom in the following way: Suppose that $A(x, y)$ is a *definit* property, that is, a first-order formula built up from $\in$ and $=$, and suppose that for every x there is at most one y such that $A(x, y)$ holds; then for each set M there exists a set M_A such that $y \in M_A$ if and only if there is some x in M for which $A(x, y)$ is true.[9] Fraenkel, on the other hand, formulated the axiom somewhat vaguely at first: If M is a set and if M' is obtained by replacing each member of M with some object of the domain, then M' is also a set.[10]

Fraenkel and Skolem differed greatly in their foundational attitudes, a fact which influenced their respective versions of the Axiom of Replacement. Whereas Skolem insisted that Zermelo's system must be viewed from the standpoint of first-order logic, Fraenkel mistrusted mathematical logic in general. In fact, Fraenkel believed that Zermelo's notion of *definit* property was inadequate precisely because it relied on what Fraenkel called "general logic," which he regarded as a very insecure base.[11] In order to avoid general logic, Fraenkel proposed that the notion of *definit* property be replaced by his notion of *definitorisch* function. A function f was said to be *definitorisch* if its value $f(x)$ could be obtained from a set x, the empty set, and the set $\mathbb{N}$ of natural numbers by the finite iteration of the operations of union, power set, and unordered pair.[12] In this vein, Fraenkel revised his Axiom of Replacement to state that if f is a *definitorisch* function and if M is a set, then $f''M$ is also a set.

Unfortunately, expressed in this fashion, Fraenkel's axiom no longer yielded that

$$\{\mathbb{N}, \mathscr{P}(\mathbb{N}), \mathscr{P}(\mathscr{P}(\mathbb{N})), \ldots\}$$

is a set. Indeed, as the young Hungarian mathematician John von Neumann demonstrated in 1928, Fraenkel's revised Axiom of Replacement could be deduced in Zermelo's system. Hence von Neumann insisted that, in order to develop the theory of ordinals and cardinals properly, one had to strengthen the Axiom of Replacement by introducing a notion of function which was more general than Fraenkel's. Von Neumann proceeded to do so.[13]

Von Neumann's set-theoretic research had originally been inspired by Zermelo's system and by Fraenkel's contributions to it. Writing to Zermelo

---

[9] Skolem 1923, 225–226.

[10] Fraenkel 1922, 230–231. On 6 May 1921 he had written a letter to Zermelo, and mentioned in it that he had not yet received a reply to his April letter containing an "example about the independence of the Axiom of Choice." Fraenkel brought a further point to Zermelo's attention: "Let Z_0 be an infinite set (*e.g.*, the one that you designated [1908a]) and let the power set of Z_0 be Z_1, the power set of Z_1 be Z_2, and so on. How does it follow from your theory that $\{Z_0, Z_1, Z_2, \ldots\}$ is a set, so that, *e.g.*, its union exists? If your theory does not yield such a proof, then clearly the existence of, say, $\aleph_\omega$, would not be provable" (letter quoted from Zermelo's *Nachlass*).

[11] Fraenkel 1922c, 101; see also his 1925, 250.

[12] Fraenkel 1925, 254; *cf.* his 1922a, 253–254.

[13] Von Neumann 1928a, 375–377.

on 15 August 1923, von Neumann enclosed a draft of an article on his own axiomatization for set theory, and listed what he believed to be its characteristic features: (i) Sets were obtained from the primitive notions of function and argument; (ii) Fraenkel's Axiom of Replacement was adopted in a strengthened form; and (iii) collections that were "too large" were permitted in the system but were not allowed to be members of any set.[14] Soon afterward, in the course of submitting this article for publication, von Neumann also came into contact with Fraenkel. Three decades later Fraenkel recalled what had occurred:

> Around 1922–23, being then professor at Marburg University, I received from Professor Erhard Schmidt, Berlin (on behalf of the *Redaktion* of the *Mathematische Zeitschrift*) a long manuscript of an author unknown to me, Johann von Neumann, with the title "Die Axiomatisierung der Mengenlehre," this being his eventual doctor[al] dissertation which appeared in the *Zeitschrift* only in 1928 (vol. 27). I was asked to express my view since it seemed incomprehensible. I don't maintain that I understood everything, but enough to see that this was an outstanding work and to recognize *ex ungue leonem*. While answering in this sense, I invited the young scholar to visit me (in Marburg) and discussed things with him, strongly advising him to prepare the ground for the understanding of so technical an essay by a more informal essay which should stress the new access to the problem and its fundamental consequences. He wrote such an essay under the title, "Eine Axiomatisierung der Mengenlehre," and I published it in 1925 in the *Journal für Mathematik* (vol. 154) of which I was then Associate Editor.[15]

When von Neumann published "Eine Axiomatisierung der Mengenlehre" [1925] as an exposition of his system, he stressed above all the third characteristic feature that he had mentioned to Zermelo. Embodied as an axiom, it supplied a precise condition under which a collection was too large to be a set, or was, in Gödel's [1940] terminology, a proper class:

(4.9.1) A class S is a proper class if and only if S is equipollent to the class of all sets.

Thus any class which was cardinally smaller than the class of all sets was a set. Von Neumann pointed out that (4.9.1) implies both the Axiom of Separation and the Axiom of Replacement.[16]

When writing to Zermelo, von Neumann had also observed that (4.9.1) yields the Well-Ordering Theorem. Indeed, by the argument underlying Burali-Forti's paradox, the class S of all ordinal numbers is not a set, and hence from (4.9.1) it follows that S must be equipollent to the class of all sets.

---

[14] Letter from von Neumann to Zermelo in Meschkowski 1967, 271–273.

[15] Letter from Fraenkel to Stanislaw Ulam in Ulam 1958, 10n.

[16] Von Neumann 1925, 223. Later, A. Levy [1968] established that the Axiom of Union likewise follows from (4.9.1) in von Neumann's system.

In this way von Neumann obtained the following strengthened version of the Well-Ordering Theorem:

(4.9.2) The class of all sets can be well-ordered.[17]

In 1928, when he published these observations on (4.9.1) and Burali-Forti's paradox, he remarked that (4.9.1) thus supplied a global choice function F such that $F(A) \in A$ for every non-empty set A.[18] Here von Neumann, as Zermelo had done before him, turned a paradox into a mathematical theorem. Von Neumann could do so precisely because (4.9.1) provided an adequate criterion for proper classes (Cantor's "absolutely infinite multitudes"), a criterion that had been missing from Cantor's and Jourdain's attempted proofs of the Well-Ordering Theorem (see 1.6). Thereby von Neumann established that Cantor was mistaken in his belief that absolutely infinite multitudes, such as that of all ordinals, cannot be consistently treated as unities.

The Axiom of Choice, as a consequence of (4.9.1), enabled von Neumann to define the notions of ordinal number and cardinal number in his system without having to introduce any new primitive concepts. In 1923 he had already proposed a definition of ordinal number as a set: An ordinal number is the set of all ordinal numbers that precede it, where zero is taken to be the empty set. To show that this definition could be made rigorous, he relied on Fraenkel's Axiom of Replacement.[19] Five years later he developed the ordinals as sets within his axiom system and, via the Well-Ordering Theorem, he defined the cardinal of a set A to be the least ordinal equipollent to A.[20] In this fashion, the Axiom of Choice provided an extremely useful simplification of Cantorian set theory.

One feature of Zermelo's system that concerned Fraenkel, Skolem, and von Neumann was its lack of categoricity.[21] Yet their responses differed greatly. In his very first article [1921], Fraenkel sought to render set theory

---

[17] Letter from von Neumann to Zermelo in Meschkowski 1967, 272.

[18] Von Neumann 1928, 726–727. Circa 1958, writing to Ulam about von Neumann's contribution to set theory, Gödel commented on axiom (4.9.1): "The great interest which this axiom has lies in the fact that it is a maximum principle, somewhat similar to Hilbert's axiom of completeness in geometry. For, roughly speaking, it says that any set which does not, in a certain well defined way, imply an inconsistency exists. Its being a maximum principle also explains the fact that this axiom implies the axiom of choice. I believe that the basic problems of abstract set theory, such as Cantor's continuum problem, will be solved satisfactorily only with the help of stronger axioms of *this* kind, which in a sense are opposite or complementary to the constructivistic interpretation of mathematics" [Gödel in Ulam 1958, 13n]. Concerning Hilbert's Completeness Axiom, see p. 266 below.

[19] Von Neumann 1923, 199.

[20] Von Neumann 1928, 727–731.

[21] An axiom system is *categorical* if any two of its models are isomorphic.

categorical by introducing his Axiom of Restriction, inverse to the Completeness Axiom that Hilbert had proposed for geometry in 1899. Whereas Hilbert had postulated the existence of a maximal model satisfying his other axioms (an act possible because these included the Archimedean Axiom), Fraenkel's Axiom of Restriction asserted that the only sets to exist were those whose existence was implied by Zermelo's axioms and by the Axiom of Replacement. In particular, there were no urelements. Fraenkel insisted that the intersection of all the domains satisfying the rest of his axioms constituted a minimal domain in which the Axiom of Restriction held.[22]

Skolem viewed the non-categoricity of Zermelo's system from two perspectives. On the one hand, the Löwenheim–Skolem Theorem revealed that no system of axiomatic set theory, formulated within first-order logic, could be categorical. On the other hand, even if Zermelo's system were strengthened by adjoining the Axiom of Replacement, there would remain other sources of non-categoricity specific to his system. Thus, Skolem continued, if $\mathfrak{B}$ was a domain that satisfied Zermelo's axioms and that contained an infinite descending $\in$-sequence, then the subdomain $\mathfrak{B}'$ of $\mathfrak{B}$ consisting of all those members of $\mathfrak{B}$ which did not begin any infinite descending $\in$-sequence also satisfied the axioms. Moreover, the subdomain $\mathfrak{B}''$ of $\mathfrak{B}'$, consisting of all the members of $\mathfrak{B}'$ whose finite descending $\in$-sequences ended with the empty set rather than with an urelement, likewise satisfied the axioms. Conversely, the domain $\mathfrak{B}$ could be extended to a larger domain by adjoining a new urelement. Whether it was possible to adjoin some subset of $\mathbb{N}$ not contained in a particular denumerable domain was a difficult problem which Skolem left open. Finally, he suspected that Zermelo's system did not settle every question about cardinality and, in particular, did not decide Cantor's Continuum Problem.[23] Not until 1963 were the questions of adding a subset of $\mathbb{N}$ to a denumerable model, and of the independence of the Continuum Hypothesis, settled definitively in first-order logic by Paul Cohen (see the Epilogue).

Aware of the divergent views of Fraenkel and Skolem, von Neumann also examined the possible categoricity of set theory. In order to render it as likely as possible that his own system was categorical, he went beyond Mirimanoff by introducing an axiom which prohibited infinite descending $\in$-sequences. Moreover, von Neumann recognized that his system would surely lack categoricity unless he excluded those large cardinals $\aleph_\alpha$ (later termed weakly inaccessible) which were regular and had a limit ordinal α as index.[24]

---

[22] Fraenkel 1922, 234, and 1922b, 163. He did not appear to realize that, as stated, his Axiom of Restriction conflated the object language with the metalanguage. For a recent discussion of how such an axiom could be formulated in set theory, see A. Levy in Fraenkel *et alii* 1973, 113–119.

[23] Skolem 1923, 229.

[24] Von Neumann 1925, 238–240. Kuratowski [1924], to whom von Neumann did not refer, had already pointed out that Zermelo's system, even if strengthened by the Axiom of Replacement, does not suffice to obtain a weakly inaccessible aleph.

Even after these additions, von Neumann believed that categoricity remained beyond reach. Firstly, he dismissed Fraenkel's Axiom of Restriction as untenable because it relied on the concept of subdomain and hence on inconsistent "naive" set theory. This objection could be overcome, he added, by presupposing a model $\mathfrak{M}$ of Fraenkel's system and then considering its subdomains to be those subclasses of $\mathfrak{M}$ satisfying the axioms. Nevertheless, the system would remain non-categorical because the use of a model $\mathfrak{M}'$, extending $\mathfrak{M}$, might alter which sets were contained in the intersection and might then exclude, in particular, new extraordinary sets whose infinite descending $\in$-sequences were "outside" $\mathfrak{M}$. Secondly, von Neumann analyzed the effect of the Löwenheim–Skolem Theorem on categoricity. He observed that Euclidean geometry was categorical but that set theory, as presently formulated, was not. Yet since geometry was based on set theory, he believed that in all probability there could exist no categorical axiomatization of set theory or, indeed, of any theory with an infinite model. On the other hand, he insisted, a similar critique applied to any axiomatization of the notion of finite set, since a set infinite in one model could be finite in an extension. Consequently, Brouwer's intuitionism was vulnerable as well. Even Hilbert's proof theory was powerless in these matters, von Neumann concluded, since they concerned categoricity rather than consistency.[25]

Excellent though it was, von Neumann's analysis lacked a clear understanding of how first-order and second-order logic diverge in their effects on categoricity. Today it is well known that the Dedekind–Peano postulates for the natural numbers, like Hilbert's axioms for Euclidean geometry and for the real numbers, are categorical in second-order logic. On the other hand, by 1922 it was clear to Skolem that any first-order theory asserting the existence of uncountable sets could not be categorical.[26] As von Neumann's discussion illustrates, however, logicians other than Skolem had not yet fully digested the differences between first-order and second-order logic. Once the effect of first-order logic on categoricity had been grasped, one could inquire whether first-order logic truly met the needs of mathematics.

It was Zermelo, perhaps influenced by Hilbert, who soon argued that first-order logic did not suffice for mathematics and especially for set theory.[27] Little by little, this became a prominent theme in Zermelo's writing during the period 1929–1935. The first stirrings of his thinking on this subject occurred in lectures that he gave at the University of Warsaw, during May and June of 1929, on the foundations of mathematics. "The true

---

[25] Von Neumann 1925, 230–232, 238–240.

[26] The situation for first-order theories with denumerable models was not clarified until later, when Skolem [1933 and 1934] showed that the first-order theory of the natural numbers was not categorical. In particular, he gave a denumerable non-standard model of arithmetic.

[27] Hilbert and Ackermann [1928, 86] had written: "As soon as the object of investigation becomes the foundation of . . . mathematical theories, as soon as we want to determine in what relation the theory stands to logic and to what extent it can be obtained from purely logical operations and concepts, then second-order logic is essential." In particular, they defined the set-theoretic concept of well-ordering by means of second-order, rather than first-order, logic.

mathematics," he emphasized in his fourth lecture, "is infinitistic in its essence and is founded on the assumption of infinite domains; it can be designated precisely as the 'logic of the infinite'."[28] The question remained as to what constituted such a logic.

That same year, stimulated by the research of Fraenkel and von Neumann, Zermelo published his first contribution to set theory and logic in nearly two decades. Appearing in the Warsaw journal *Fundamenta Mathematicae*, Zermelo's article responded to the objections directed against his original notion of *definit* property. His reply rested squarely on second-order logic. While he did not refer explicitly to Skolem or Weyl, his article seems to have been aimed in part at their critiques. Thus he noted that some authors had treated his notion of *definit* property as superfluous because they reduced it to general logic. Yet, Zermelo insisted, "At the time [1908] there did not exist a generally recognized 'mathematical logic,' to which I could appeal, any more than today when every foundational researcher has his own logistical system" [1929, 340].

On the other hand, he was not completely satisfied with either Fraenkel's or von Neumann's revisions of the notion of *definit* property. He rejected Fraenkel's attempt to bypass this notion, by admitting only propositional functions of a certain form, because Fraenkel characterized these *definitorisch* functions by a construction. To proceed in this way, Zermelo believed, was to violate the essence of the axiomatic method. As an alternative, he proposed to axiomatize the notion of *definit* property—a step that he had contemplated in 1908 without executing it in detail. Now he had kind words for von Neumann's efforts to do so by axiomatizing the general notion of function. All the same, he regarded von Neumann's approach as unnecessarily convoluted in that it replaced simple set-theoretic concepts by complicated functions [1929, 339–340].

Zermelo axiomatized the notion of *definit* property, within what was essentially second-order logic, for an arbitrary axiom system rather than for set theory *per se*. A given axiom system had its own primitive relations, such as $\in$ and $=$ in the case of set theory. In any axiom system the primitive relations were taken to be *definit* for all values of the variables. If the propositions P and Q were *definit*, then so were $\sim P$, $P \& Q$, and $P \vee Q$. If $P(x)$ was *definit* for every value of the free individual variable x, then $\forall x P(x)$ and $\exists x P(x)$ were *definit*. Furthermore, if $P(f)$ was *definit* for each propositional function f having only individual variables, then $\forall f P(f)$ and $\exists f P(f)$ were also *definit*. Then the class of *definit* propositions was the intersection of the classes of propositions closed under the logical operations stated above. Zermelo was especially pleased with this axiomatization of *definit* property because it did not use the notion of natural number—a comment possibly aimed at Skolem's finitism.[29]

---

[28] Zermelo in Moore 1980, 136.

[29] Zermelo 1929, 341–344.

Skolem soon responded by publishing a critique of this new characterization of *definit* property. After observing that his article of 1923 already contained a similar characterization, he underlined a vital point of difference. While Zermelo permitted quantification over propositional functions, he did not. Thus he refused to allow the clause whereby $\forall f P(f)$ and $\exists f P(f)$ were taken to be *definit* properties. Furthermore, he argued that this clause was obscure and might even engender Russell's paradox. Did Zermelo, he wondered, intend to characterize such functions of propositional functions by additional axioms? Most important of all, he insisted, was the fact that if functions of propositional functions were considered to remain within first-order logic, then no new sets would be generated beyond those given by his 1923 version of *definit* property. Lastly, he observed that the Löwenheim–Skolem Theorem ruled out any characterization of the notion of *definit* property by means of finitely many axioms.[30] In this way Skolem strongly attacked Zermelo's attempt to base set theory explicitly on a second-order logic.

In 1930, acknowledging Skolem's critique but maintaining his own position, Zermelo proposed Zermelo–Fraenkel set theory (**ZF**) in a form closely related to that used today. He included seven postulates in his new system, which he called **ZF**′, and permitted urelements to occur in it. Three of these postulates—the Axiom of Extensionality, the Power Set Axiom, and the Axiom of Union—were drawn directly from his system of 1908. Two more of them—the Axiom of Pairing and the Axiom of Separation—were modified versions of postulates that had formed part of the earlier system. Now Fraenkel's Axiom of Pairing (if a and b are objects, then $\{a, b\}$ is a set) replaced Zermelo's more inclusive Axiom of Elementary Sets, while the Axiom of Separation was altered to agree with his article of 1929: If M is a set and $P(x)$ is a propositional function (first-order or second-order), then $\{x \in M: P(x)\}$ is a set. Fraenkel's Axiom of Replacement was expressed in similar form: If $y = P(x)$ is a propositional function (first-order or second-order) such that each $P(x)$ is a set or an urelement, then $\{P(x): x \in M\}$ is a set. The seventh and final axiom, probably adopted from von Neumann but perhaps stated independently, was what Zermelo called the Axiom of Foundation: There is no infinite descending $\in$-sequence. For this axiom, Zermelo also offered a second formulation, which he claimed to be equivalent to the first: Every non-empty set A contains an $\in$-minimal element, *i.e.*, an element a such that no member of a belongs to A. Observing that the Axiom of Foundation was satisfied in all known applications of set theory to other branches of mathematics, he stressed its usefulness in understanding the structure of set theory itself.[31]

---

[30] Skolem 1930, 337–341.

[31] Zermelo 1930, 29–31. Mendelson [1958] established that to prove the equivalence of these two forms of the Axiom of Foundation requires the Axiom of Choice.

On the other hand, there were two axioms which had occurred in his system of 1908 but which did not form part of **ZF′**—the Axiom of Infinity and the Axiom of Choice. He omitted the first of these on the grounds that it did not belong to general set theory. As for the Axiom of Choice, he did not include it explicitly in his system because it differed in character from the other axioms and because, so he believed, it could not serve to delimit models of set theory. Nevertheless, he assumed the Axiom as a general logical principle within his metamathematics, and concluded that every set could be well-ordered.[32]

Zermelo's principal result was that any normal domain $\mathfrak{D}$ of **ZF′** (or, as we would say now, any standard transitive model of **ZF′** in second-order logic) can be characterized up to isomorphism by two cardinals, chosen independently of each other.[33] The first was the cardinal of the basis for $\mathfrak{D}$ (the set of all urelements in $\mathfrak{D}$), while the second was the characteristic of $\mathfrak{D}$ (the least ordinal greater than all the ordinals in $\mathfrak{D}$). In particular, he established that the characteristic of a normal domain must be either aleph-zero or else a strongly inaccessible cardinal.[34]

To obtain these theorems, Zermelo introduced what is now called the cumulative type hierarchy for set theory. Suppose Q_0 is the set of urelements in a particular model $\mathfrak{D}$ of **ZF′** with characteristic κ. Zermelo proved that $\mathfrak{D}$ could be decomposed into κ disjoint levels Q_α, where, for $\alpha > 0$, Q_α consisted of all those sets A which did not occur at earlier levels but all of whose members did. The cumulative levels P_α were defined by letting

$$P_1 = Q_0, \qquad P_{\alpha+1} = P_\alpha \cup Q_\alpha, \qquad \text{and} \qquad P_\beta = \bigcup_{\alpha<\beta} P_\alpha$$

for every limit ordinal β less than κ. By means of his cumulative type hierarchy, Zermelo established that any two normal domains with the same characteristic and equipollent bases were isomorphic and that the isomorphism was completely determined by its values on their bases. If two normal domains had equipollent bases but different characteristics, then one was isomorphic to a subdomain of the other.[35]

Here Zermelo offered a vision of the models of set theory, formulated within second-order logic, that differed fundamentally from Skolem's. At the same time Zermelo presented a rationale for his choice of axioms that had been lacking in his system of 1908 and that Mirimanoff had dimly seen in 1917. In effect, Zermelo's new system **ZF′** determined initial segments P_α of the cumulative type hierarchy that were closed under the operations of

---

[32] Zermelo 1930, 31. This was only possible because his object language and metalanguage were in higher-order logic. If both were in first-order logic, then the Axiom of Choice could be true in the metalanguage but false in the object language (see 4.10 and the Epilogue).

[33] A model of set theory is *standard* if the membership relation in the model is the actual relation $\in$. A model $\mathfrak{M}$ is *transitive* if $x \in y$ and $y \in \mathfrak{M}$ imply $x \in \mathfrak{M}$.

[34] Zermelo 1930, 29–34.

[35] *Ibid.*, 36–47.

union and power set. Since any strongly inaccessible ordinal α yielded a model of **ZF**′ in which the Axiom of Infinity was true, **ZF**′ was far from being categorical. Though he believed that one could enforce categoricity as Fraenkel and von Neumann had wished to do, he argued that its absence was a virtue of **ZF**′. Indeed, the presence of ever larger models of **ZF**′, each extending the previous ones, preserved the unlimited validity of set theory, which was essentially an open system.[36] In this spirit he assumed, within the metalanguage, the large cardinal axiom that the class of strongly inaccessible cardinals is equipollent to the class of all ordinals.[37]

Not long afterward, Zermelo came to fear that Skolem's views on the logic appropriate for mathematics would prevail. On 15 September 1931 Zermelo spoke at the annual meeting of the *Deutsche Mathematiker-Vereinigung*, which took place at Bad Elster. His lecture fused polemic with a radically new perspective on the relationship between mathematics and logic. Vehemently rejecting the Löwenheim–Skolem Theorem and all attempts to obtain a denumerable model for set theory, he proposed a new mathematical logic based on his cumulative type hierarchy for **ZF**′. This logic contained no quantifiers, but allowed conjunctions and disjunctions of length α for every ordinal α. Proofs were also allowed to be of length α [1932].

At Bad Elster, Gödel had spoken on his Incompleteness Theorem, a result which Zermelo found distasteful. Indeed, Zermelo believed it to be vital to mathematics that every true sentence be provable. On 21 September, dissatisfied with Gödel's theorem but hoping to dispel any misconceptions about it, Zermelo wrote to Gödel. "I would like to remark," Gödel replied on 12 October, "that I do not see the essential point of my result to be that one can somehow go beyond the limits of a formal system ... but that for any formal system in mathematics there exist propositions which are *expressible* within this system but which *cannot be decided* by the axioms of this system."[38] Gödel granted, on the other hand, that relatively undecidable propositions were always decidable in higher systems, which nevertheless included other undecidable propositions. On 29 October Zermelo responded in turn, arguing that mathematics needed a much richer logic in which, by new methods of proof, every proposition would be decidable.

Here a potentially fruitful dialogue came to an end. While Gödel used Zermelo's cumulative type hierarchy in his later researches on models of set theory (see 4.10), he did not pursue Zermelo's infinitistic conception of logic. Such a conception ran counter to the finitism which increasingly dominated mathematical logic by 1930. Influenced by Hilbert and Skolem, Gödel pursued his logical investigations within the finitistic tradition.

---

[36] This contrasted sharply with the views of other set theorists—including Cantor, Schoenflies [1911], Mirimanoff [1917], and Finsler [1926]—who regarded set theory as a single closed system.

[37] Zermelo 1930, 44–47. This occurred eight years before Tarski proposed his equivalent Axiom of Inaccessible Sets in the object language; see 4.3.

[38] Gödel in Moore 1980, 127. See also Grattan-Guinness 1979, 298–299.

From one perspective, first-order logic could never be truly adequate, since it could not characterize the real numbers, or even the natural numbers, up to isomorphism. Nevertheless, after 1930 mathematical logic became increasingly identified with first-order logic. Set theory was formulated more and more frequently within first-order logic, rather than within the second-order logic preferred by Zermelo. In this spirit Bernays published his own axiomatization for set theory [1937]. Modifying von Neumann's system so as to render it structurally more similar to Zermelo's, Bernays intentionally used first-order logic to express the notion of *definit* property.[39] Soon afterward Gödel simplified Bernays' system somewhat, but likewise insisted that it be embedded in first-order logic.[40] By this point the logical basis for future set-theoretic research had been decided, and henceforth models of set theory would rarely resemble Zermelo's conception. The second transformation of set theory was complete.

4.10 Consistency and Independence of the Axiom

Before significant progress could be made in determining whether or not the Axiom of Choice led to a contradiction and whether or not the Axiom followed from the rest of Zermelo's system, two preconditions had to be met. The system had to be stated in a more precise form, and its underlying logic had to be specified. Because of the mistrust that the system encountered initially, it is not surprising that fourteen years passed before Abraham Fraenkel published an argument for the independence of the Axiom. His argument depended on infinitely many urelements.[1] At the same time, he addressed the problem of rendering the system unambiguous by clarifying Zermelo's notion of *definit* property (see 4.9). In the years that followed, Fraenkel turned to investigating the relative deductive strength of weaker forms of the Axiom as well as of related propositions such as the Ordering Principle. However, he tended to conflate mathematical and metamathe-

[39] Bernays 1937, 65. He noted that he had first presented his system in lectures at Göttingen during 1929–1930.

[40] Gödel 1940. Neither Bernays' system nor Gödel's revision of it permitted urelements. Later, Wang [1949] established that Gödel's system was an extension of **ZF**. About the same time, Novak [1950] proved that if **ZF** is consistent, then so is Gödel's system. Mostowski [1950] improved Novak's result by showing that Gödel's system is a conservative extension of **ZF**.

[1] Fraenkel 1922 and 1922a. In the present context, the *independence* of a proposition P from a system S means that P cannot be deduced in S, or, equivalently, that S and the negation of P are consistent. During April 1921, Fraenkel had sent Zermelo a letter which sketched a version of his argument for the independence of the Axiom; see footnote 10 of 4.9.

matical notions, a failing that reflected the embryonic state of mathematical logic at the time (see 4.8).

The emergence of first-order logic as the accepted underlying logic for mathematics, and for set theory in particular, made it possible to formulate Fraenkel's method and results more precisely. This task was undertaken by two Polish mathematicians, Adolf Lindenbaum and Andrzej Mostowski [1938]. Their work revealed the influence of Gödel, who had recently discovered what he termed the constructible sets. Indeed, Gödel used these sets to establish the relative consistency of both the Axiom of Choice and the Generalized Continuum Hypothesis. Here, on a more limited scale, he continued the work of Hilbert, who had attempted to establish the consistency of set theory by finitistic means.

Fraenkel operated within the Hilbertian axiomatic tradition when he insisted [1921] on the importance of ascertaining whether the axioms in Zermelo's system were independent of each other. On 10 July 1921 Fraenkel completed an article, published the following year in *Mathematische Annalen* [1922], which sketched a proof for the independence of three of these axioms. First of all, he proposed to show that the Axiom of Extensionality was independent by using a domain of ordered sets. While he did not give the details of this argument, he outlined the two independence proofs that he considered the most important—those for the Axiom of Separation and the Axiom of Choice.

In the case of the Axiom of Separation, Fraenkel proposed to generate a domain of sets by starting with the empty set 0 and with the denumerable set $Z = \{0, \{0\}, \{\{0\}\}, \ldots\}$, which he, like Zermelo, used to represent the set of natural numbers. The desired domain was then obtained by iterating the operations of power set, union, and unordered pair any finite number of times on these two sets. As he proceeded to show, Z was a subset of every infinite set in this domain. Since the domain did not contain the set $Z - \{0\}$, it did not satisfy the Axiom of Separation. On the other hand, it satisfied the remaining axioms of Zermelo's system, including the Axiom of Choice.[2]

In a similar fashion Fraenkel sketched his argument for the independence of the Axiom of Choice. He emphasized that a detailed proof would depend on his more precise notion of *definit* property, which he did not state. As in the previous argument, he began with the sets 0 and Z. From them he generated a domain by means of the Axioms of Pairing, Union, Power Set, and also Separation. Then he assumed that in the domain there existed a disjoint denumerable family A of sets $M_1, M_2, \ldots$, each having more than one member, which satisfied the following condition: Every *definit* property true of at least one element in some M_m is true of all the elements of some M_n. A set A satisfying this condition certainly lacked a choice function. Although it was possible, so he believed, to obtain such an A whose elements contained only sets, the approach that he regarded as the simplest was to let the elements of A

[2] Fraenkel 1922, 234–236.

be sets of urelements.[3] Nevertheless, it remained uncertain precisely how Fraenkel intended to generate the desired domain.

Soon Fraenkel published a second article [1922a] giving in more detail a modified version of his argument that the Denumerable Axiom is independent in Zermelo's system. Meanwhile, Fraenkel had been in correspondence with Zermelo on this subject. Indeed, he credited Zermelo with the idea that the members of A needed to consist of only two urelements, as well as with the central role of symmetry in the proof.[4]

Fraenkel began his argument by positing the least domain $\mathfrak{B}$ generated by means of the Axioms of Pairing, Union, Power Set, and Separation from 0, Z, $A = \{\{a_1, \bar{a}_1\}, \{a_2, \bar{a}_2\}, \ldots\}$, and the denumerably many distinct urelements $a_1, \bar{a}_1, a_2, \bar{a}_2, \ldots$. To obtain the least domain $\mathfrak{B}$, he used his metamathematical Axiom of Restriction, which von Neumann was later to criticize (see 4.9). Fraenkel's chief theorem, from which the independence result followed at once, was this:

> For every set M in the domain $\mathfrak{B}$ and for every n, there exists some subset A_M of A, containing all but finitely many members of A, such that if a_n is in A_M, then a permutation of a_n and $\bar{a}_n$ maps M onto itself.

In other words, Fraenkel's theorem stated that every M in $\mathfrak{B}$ was "symmetric" with respect to almost all members of A. Clearly the theorem held when M was either 0, Z, or A. From this point, the demonstration proceeded inductively by applying the operations of union, unordered pair, and power set to a given set M for which the theorem held, and showing that the theorem remained in force for the resulting set. The most difficult step was to establish that the Axiom remained true when the Axiom of Separation was applied to such an M.[5]

Despite the potential fecundity of Fraenkel's permutation method, at the time no one else attempted to exploit it. Returning to it in 1928 at the International Congress of Mathematicians in Bologna, Fraenkel observed that the Ordering Principle is independent of Zermelo's system (excluding the Axiom). Since Fraenkel's earlier argument established the independence of $\mathbf{C}_2^{\aleph_0}$, the Denumerable Axiom restricted to families of unordered pairs, he had only to note Kuratowski's theorem that the Ordering Principle implies $\mathbf{C}_{<\aleph_0}$, the Axiom of Choice for families of finite sets, and hence yields $\mathbf{C}_2^{\aleph_0}$ in particular. It remained an open problem whether $\mathbf{C}_{<\aleph_0}$ was weaker than the Ordering Principle and whether this principle was in

[3] *Ibid.*, 236–237. The independence of Zermelo's postulates other than the Axioms of Separation and Choice was shown at Marburg by Heinrich Vieler [1926] in a doctoral thesis directed by Fraenkel. N. J. Lennes claimed, in the abstract of a lecture [1922], to have a proof for the independence of all Zermelo's postulates other than the Axiom of Choice. However, Lennes published no details then or later.

[4] Fraenkel 1922a, 253–254.

[5] *Ibid.*, 255–257.

turn weaker than the Axiom. Conjecturing that both statements were true, he quite correctly suspected that their proofs would be difficult.[6]

By 1931 Fraenkel believed himself to have demonstrated a closely related theorem:

(4.10.1) The Denumerable Axiom $\mathbf{C}^{\aleph_0}$ is independent of $\mathbf{C}_{<\aleph_0}$.

In other words, $\mathbf{C}_{<\aleph_0}$ does not imply the Denumerable Axiom in Zermelo's system without the Axiom of Choice.[7] Fraenkel's abstract supplied no details, however, and the first published indications did not appear until four years later. On that occasion Fraenkel, who wished to banish urelements from set theory altogether, relied on an even more sophisticated use of them than he had done in 1922. His new model $\mathfrak{F}$ was generated from a special family of sets of urelements by the Axioms of Pairing, Union, Power Set, Separation, and $\mathbf{C}_{<\aleph_0}$—all used any finite number of times. Again he employed his dubious Axiom of Restriction to generate only those sets obtained from the previous axioms, an act which consciously disregarded von Neumann's criticisms.[8]

Whereas Fraenkel's 1922 demonstration had been based on a set of pairs of urelements, here he used a set $Q = \{Q_1, Q_2, \ldots\}$ such that each Q_n contained infinitely many urelements. He proposed to show that there existed no choice function for Q within the model and consequently that the Denumerable Axiom did not follow from $\mathbf{C}_{<\aleph_0}$ in Zermelo's system without the Axiom of Choice. This conclusion resulted from what he claimed as his principal theorem: Every object M in the domain $\mathfrak{F}$ is symmetric with respect to almost all the Q_n. That is, if each p_n is a permutation of Q_n, then M is unchanged by performing all but finitely many of the p_n. Here the chief difficulty was to establish the lemma that if the principal theorem held when M was a disjoint denumerable family of non-empty finite sets, then it continued to hold for any choice set S on M [1935, 51].

In one essential respect the detailed proof of (4.10.1) that Fraenkel later published [1937] deviated from his earlier version of 1935. Restricting himself in 1937 to permutations which interchanged only finitely many elements of Q_n, he replaced symmetry by "half-symmetry" in his principal theorem. Indeed, he conceded that the theorem was false as previously stated, but insisted that it could be rectified if invariance were required only with respect to all even permutations of Q_n.[9] Nevertheless, his argument for (4.10.1) still harbored a serious error, which Lindenbaum and Mostowski pointed out a year later [1938, 30].

---

[6] Fraenkel 1928; see also his 1928a.

[7] Fraenkel 1931; see also his 1932.

[8] Fraenkel 1935, 41, 48–51.

[9] Fraenkel 1937, 7–9. An *even permutation* is a permutation obtained by a finite, even number of transpositions of two elements.

By modifying Fraenkel's method, these two Warsaw mathematicians obtained a fully adequate proof of (4.10.1), as well as other independence results. Lindenbaum, who strongly influenced the young Mostowski's ideas on the Axiom, had suggested that he examine Fraenkel's 1922 proof in order to render it more precise. Mostowski's first efforts in this direction can be found in his doctoral thesis, directed by Tarski, which he completed in 1933.[10] Then, since positions in mathematics were rare in Poland, Mostowski went to Zurich in order to become an actuary. Quickly bored with such trivial mathematics, he attended a seminar by Bernays. Meanwhile, in the fall of 1935, Lindenbaum gave a colloquium at the University of Warsaw on Mostowski's correct demonstration of (4.10.1). Mostowski, now back in Poland, presented a version of this proof to the Warsaw section of the Polish Mathematical Society in January 1936. Two years later he and Lindenbaum published a critique of Fraenkel's proofs as well as a summary of their joint research.[11]

Above all, Lindenbaum and Mostowski objected to Fraenkel's conflation of mathematics and metamathematics. They found it impossible to ascertain whether his concept of function, used in place of Zermelo's notion of *definit* property, was to be taken in the object language or in the metalanguage. As a result, Fraenkel's further constructions remained imprecise. It was vital to distinguish carefully, they stressed, between a set-theoretic axiom system U_0 which was the mathematical object of investigation, and another set theory U which served as the metamathematical means for carrying out such an investigation. With this in mind, they also differentiated between two notions of propositional function—the usual intuitive notion and another notion sufficiently narrow to be definable in U_0 by a finite number of symbols. In the first case, Fraenkel's constructions could be executed with the aid of the satisfaction relation, as recently introduced by Tarski, but the proof was complicated and subtle. Although Fraenkel's demonstration, if reformulated, could be carried out in the second case as well, it was flawed by the serious error mentioned above.[12]

Lindenbaum and Mostowski stated nine propositions that they had shown to be independent of **ZF**. The first two, $\mathbf{C}_2^{\aleph_0}$ and the Ordering Principle, had been proved to be independent by both authors jointly, while the other seven were due to Mostowski alone. In addition to the result (4.10.1), these included the Partition Principle, the theorem (1.7.10) that $\mathfrak{m} + \mathfrak{m} = \mathfrak{m}$ for every infinite $\mathfrak{m}$, and the distinctness of Tarski's five levels of finiteness.[13]

---

[10] Mostowski in Crossley 1975, 12, 44. Mostowski's thesis was officially directed by Kuratowski, since Tarski was not yet a professor. The thesis, on definitions of finiteness in Tarski's system of logic, was published as [Mostowski 1938]. A related article appeared as his 1938a.

[11] Lindenbaum and Mostowski 1938, 31n.

[12] *Ibid.*, 29–30.

[13] *Ibid.*, 27–32. Concerning these five levels, see 4.2. It should be noted that Mostowski's proofs applied to **ZF**, while Fraenkel's were limited to Zermelo's original system.

In particular, it could not be deduced in **ZF** that every Dedekind-finite set is finite. Lindenbaum's and Mostowski's independence proofs all depended, as had Fraenkel's, on the use of denumerably many urelements.

While Mostowski did not publish these independence proofs at the time, he illustrated his method by one result that did appear, on which Fraenkel had labored unsuccessfully:

(4.10.2) The Axiom of Choice is independent of the Ordering Principle. [1939, 203]

Mostowski's proof of (4.10.2) was considerably influenced by his travels. During the summer of 1937, he attended lectures on axiomatic set theory that Gödel offered at the University of Vienna. There Gödel discussed in detail his new result, obtained from his constructible sets, that the Axiom of Choice is relatively consistent (see below). Consequently, Mostowski chose to formulate his published proof of (4.10.2) by using some of Gödel's methods rather than, as he had done earlier, Tarski's formal system of logic.[14]

In his joint article with Lindenbaum, Mostowski had distinguished between an axiom system U_0, the object of investigation, and a system U, the metamathematical tool used in the investigation. When Mostowski published his proof of (4.10.2) in 1939, he let U_0 be Bernays' system altered so as to permit urelements, and he took U to be von Neumann's system. After assuming that von Neumann's system is consistent, Mostowski provided an interpretation for U_0 within it. In particular, those objects which were interpreted as urelements in U_0 were, in fact, sets within U. Thereby he established the relative consistency of set theory with urelements. Although he would have preferred to avoid urelements altogether in his independence proofs, he was unable to do so. Significantly, he recognized that his proof of (4.10.2) depended in an essential way on the fact that the Axiom of Choice held in U. He was unperturbed by this situation, however, since Gödel had shown that the consistency of U without the Axiom implies the consistency of U with the Axiom.[15]

Though Fraenkel had mentioned certain group-theoretic aspects of his permutation method [1937, 3, 9], it was Mostowski who clarified these aspects and who made group theory central to such independence proofs. In order to do so, Mostowski required a series of definitions. First, he let K be $\{\mathbb{N}-\{n\}: n > 0\}$, the set in U which was to be interpreted as the set of urelements in U_0. Then he defined H_0 to be the group of all permutations of K, and took H to be some fixed subgroup of H_0. If M was a set of subsets of K such that $\bigcup M = K$, if M was closed under finite union, and if each member of M was closed under every f in H, then M was termed an H-ring. The aim of his final series of definitions was to define the notions of M,H-distinguished

[14] Mostowski 1939, 204.

[15] *Ibid.*, 204, 212, 251.

element and of M,H-distinguished domain. If A was a subset of K, then $H(A)$ was the group consisting of all permutations f in H which left each member of A fixed. $R_H(A)$ was the class of all sets x such that $|f, x| = x$ for every f in $H(A)$, where $|f, x|$ was defined as follows: $|f, 0| = 0$, $|f, x| = x$ if $x \in K$, and

$$|f, x| = \{|f, y|: y \in x\}$$

for any non-empty set x. Then x was an M,H-distinguished element if there existed some $B \in M$ with $x \in R_H(B)$ and if for every ordinal α,

$$\left(\bigcup^{\alpha} x\right) \cap M \subseteq \bigcup_{B \in M} R_H(B),$$

where

$$\bigcup^{0} x = x \quad \text{and} \quad \bigcup^{\alpha} x = x \cup \bigcup_{\beta < \alpha} \left(\bigcup^{\beta} x\right)$$

for $\alpha > 0$. A non-empty class S of M,H-distinguished elements was an M,H-distinguished domain if for some A in M and for every f in $H(A)$, the propositions $x \in S$ and $|f, x| \in S$ were equivalent. After these preliminaries, Mostowski stated and proved his principal metatheorem: If the notions of set-or-urelement, class, and the empty set are interpreted respectively as M,H-distinguished element, M,H-distinguished domain, and $\mathbb{N}-\{0\}$, then the collection of all M,H-distinguished elements in U is a model for all the axioms of U_0 [1939, 205–221].

By the proper choice of a subgroup H and an H-ring M, Mostowski observed, many independence results were possible. In order to establish (4.10.2) in particular, he introduced a function g from the rationals $\mathbb{Q} = \{r_n: n \geq 1\}$ onto K such that $g(r_n) = \mathbb{N}-\{n\}$. For any $x, y \in K$, he defined $x \prec y$ to hold if and only if $g^{-1}(x) < g^{-1}(y)$, where $<$ was the usual order-relation on the rationals. Then he took H to be the largest subgroup of H_0 which preserved $\prec$, and M to be the set of all finite subsets of K. Consequently, the proof of (4.10.2) reduced to verifying that the Ordering Principle held in the model, and that the Axiom of Choice failed there because the model contained no well-ordering for K.[16]

The problem of establishing the independence of a proposition P from an axiomatic system S for set theory was equivalent to showing the consistency of the negation of P relative to S. Both problems assumed as a hypothesis that the system S is consistent. At first, however, mathematicians hoped to prove or refute the consistency of Zermelo's system without any such hypothesis. Indeed, Zermelo had originally posed this very problem [1908a, 262]. Yet even Hilbert, who wished to show the consistency of set theory within mathematical logic, had not succeeded in establishing as much as the consistency of the axioms for the real numbers (see 4.8).

---

[16] *Ibid.*, 236–250.

There was a substantive reason for Hilbert's limited success, since his metamathematics included at most the arithmetic of the natural numbers. During 1930–1931 Gödel discovered that the axioms for the natural numbers do not suffice to prove their own consistency within first-order logic and, *a fortiori*, that the consistency of Zermelo–Fraenkel set theory cannot be deduced in **ZF** itself.[17] Consequently, a proof for the consistency of **ZF** could come *only* from a still stronger system, whose consistency would be even more subject to doubt. In 1930 Zermelo, attacking the problem of consistency independently of Gödel, found that the cumulative type hierarchy up to the first strongly inaccessible ordinal was a model for **ZF** (see 4.9). Thus if **IN** was the proposition that there is a strongly inaccessible ordinal, then the consistency of **ZF** was a theorem of **ZF** + **IN**. To many mathematicians, however, it was uncertain that such an ordinal exists.

In the wake of Gödel's discovery, the only hope was to establish the *relative* consistency of a proposition vis-à-vis the remaining axioms for set theory. The principal technique was to build a model for both set theory and the proposition inside a given model $\mathfrak{M}$ for set theory, *i.e.*, to construct an inner model of $\mathfrak{M}$. Although Skolem had briefly discussed such inner models in 1923 (see 4.9), their first substantial use came from von Neumann in 1929. Von Neumann showed that if his axioms for set theory have a model, then there is an inner model Π satisfying both his axioms and the assumption of no infinite descending $\in$-sequences. Thus the consistency of his axioms yielded their consistency when supplemented by the Axiom of Foundation. Von Neumann obtained his inner model Π as the image of a function ψ defined by transfinite recursion on the class of all ordinals, but did not use the cumulative type hierarchy often attributed to him.[18]

Wilhelm Ackermann, a follower of Hilbert's approach to logic, settled a more limited but related question in 1937. Ackermann established the consistency of a restricted version of Zermelo's system (excluding the Axiom of Infinity but including the Axiom of Choice) by reducing its consistency to that of elementary number theory, already established by Gerhard Gentzen in 1936.[19] Ackermann carried out his proof within Hilbert's system for first-order logic by representing each set as a natural number in binary notation. The unique representation of each number, as a sum of distinct powers of 2, allowed Zermelo's axioms (except that of Infinity) to be transformed into theorems of elementary number theory [1937].

Yet the essential question was whether Zermelo's system as a whole was consistent, since only by including the Axiom of Infinity did this system attain its true significance. In particular, did the Axiom of Choice, which had been

---

[17] Gödel 1931, 173, 196.

[18] Von Neumann 1929, 236–238.

[19] Gentzen [1936] had used transfinite induction, up to the first epsilon-number in Cantor's second number-class, in order to deduce the consistency of the axioms for the natural numbers.

the subject of so much debate, render the system more likely to be contradictory?

Gödel found that the answer was no. During the fall of 1935, when he visited the Institute for Advanced Study at Princeton, he informed von Neumann that he had recently established the relative consistency of the Axiom by introducing his "constructible" sets. Furthermore, Gödel conjectured that the Continuum Hypothesis would hold for these sets.[20] In his first research announcement, dated 9 November 1938, he defined a set to be constructible if it appeared in Russell's ramified theory of types when extended to include transfinite orders.[21] By means of this definition he proved that if von Neumann's system without the Axiom of Choice is consistent, then the system remains consistent when the Axiom is adjoined. To obtain this result, carried out in first-order logic, he began with a model of von Neumann's system and then showed that the class of constructible sets is an inner model of this system satisfying the Axiom of Choice as well. By this time he had also discovered that three other propositions hold in his inner model of constructible sets and thereby had established the consistency of these propositions relative to von Neumann's system. One was the Generalized Continuum Hypothesis and two were from descriptive set theory: The existence of a subset of $\mathbb{R}$ that has the power of the continuum, lacks a perfect subset, and is the complement of an analytic set; the existence of a non-measurable set E of real numbers such that both E and $\mathbb{R}-E$ are projections of the complements of some analytic sets in the plane.

Though Gödel provided few details in his article of 1938, he made a number of important observations. In particular, he remarked that in the inner model of all constructible sets, the Continuum Hypothesis holds since every constructible subset of $\mathbb{N}$ has an order less than ω_1. His inner model also satisfied the proposition A that every set is constructible. "The proposition A," Gödel concluded,

> added as a new axiom, seems to give a natural completion to the axioms of set theory, in so far as it determines the vague notion of an arbitrary infinite set in a definite way. In this connection it is important that the consistency-proof for A does not break down if stronger axioms of infinity (*e.g.*, the existence of inaccessible numbers) are adjoined to T [von Neumann's system]. Hence the consistency of A seems to be absolute in some sense, although it is not possible in the present state of affairs to give a precise meaning to this phrase. [1938, 557]

Later, proposition A became known as the Axiom of Constructibility.

During the summer of 1937, before he published his results on constructible sets, Gödel gave a course of lectures in axiomatic set theory at the University

[20] Wang 1981, 656–657; 1978, 184.

[21] In 1938 Gödel described the *order* of a set as its order in the transfinite ramified hierarchy of types; *cf.* 1944, 146–147. However, in terms of the constructible hierarchy M_α (discussed below), he called a set *A constructible of order* α if $A \in M_{\alpha+1}-M_\alpha$ [1939, 221].

of Vienna.[22] One of his five or six auditors was Mostowski, who reminisced about this course in 1974:

> I attended a lecture of Gödel, and I think this was the first publication, if one can call a lecture a publication, of the theory of his result on the consistency of the Axiom of Choice. So he had a one-semester course in Vienna on axiomatic set theory, in which he gave axioms for what is now called Gödel–Bernays set theory, and then developed a model. He constructed a model in which the Axiom of Choice was valid. At that time, I am sure that he did not have the consistency proof for the Continuum Hypothesis, because he restricted his lecture exclusively to the Axiom of Choice. The construction went more or less like this: He had these levels of the constructive hierarchy, defined, more or less, as he defined them later in his paper on consistency of the Continuum Hypothesis [1938], but he did not formulate the Axiom of Constructibility. Only he proved that all these levels have well-orderings, so that a well-ordering on a given level can be lifted to the well-ordering of the next level, and also on limit ordinals you can get this well-ordering. So in this way he obtained a model in which each set was well-ordered within the model. That was his construction. He never mentioned, at that time, that he had a proof for [the consistency of] the Continuum Hypothesis. So I must say after I read his publication [1938] . . . , I was very upset that he carried this work so much further, because at that time he had only this very weak result.[23]

It seems that Gödel introduced the Axiom of Constructibility, and obtained the relative consistency of the Generalized Continuum Hypothesis, during the summer of 1938.[24]

In 1939 Gödel published a more detailed summary of his proof that the Axiom of Choice and the Generalized Continuum Hypothesis are relatively consistent. Whereas his previous article [1938] had noted in passing that such relative consistency results could be executed for *Principia Mathematica* and **ZF**, now he sketched the main steps of the proof for **ZF**. First he defined M' to be the set of all those subsets of M definable by a first-order propositional function with $\in$ as its only relation-symbol. In a manner analogous to Zermelo's development of the cumulative type hierarchy, he specified the hierarchy of constructible sets:

$$M_0 = \{0\}, \quad M_{\alpha+1} = (M_\alpha)', \quad \text{and} \quad M_\beta = \bigcup_{\alpha<\beta} M_\alpha \quad \text{if } \beta \text{ is a limit ordinal.}$$

In particular, he established that M_Ω was a model of **ZF** if Ω was the least strongly inaccessible ordinal. Since proposition A implied both the Axiom of Choice and the Generalized Continuum Hypothesis, he was able to show that both held in M_Ω.[25]

---

[22] Mostowski 1939, 204; *cf.* Wang 1981, 656n.

[23] Mostowski in Crossley 1975, 41–42.

[24] Wang 1981, 656.

[25] Gödel 1939, 220–224. One could dispense with an inaccessible ordinal in the metatheory by first arithmetizing the syntax and then showing in number theory that the consistency of **ZF** implies the consistency of both **ZF** and the Axiom of Constructibility.

During the winter of 1938–1939 Gödel lectured on constructible sets at the Institute for Advanced Study, and these lectures soon appeared as a slim but significant booklet [1940]. In it he first published his modified version of Bernays' axiomatization for set theory. The final axiom within Gödel's system was later called the Axiom of Global Choice, and asserted the existence of a choice function for the class of all sets.[26] In place of his previous definition of constructible set, he stated a new but equivalent definition which made use of transfinite recursion as well as of eight elementary operations on sets. Particularly important was his notion of absoluteness. In effect, he termed a set-theoretic concept absolute if, when relativized to the constructible sets, it remained identical with the unrelativized notion.[27] More intuitively, an absolute notion (such as subset but not power set) looked the same both inside and outside the model of constructible sets. Gödel did not use an inaccessible cardinal in these lectures, as he had not in [1938], since in both cases he worked within a set theory having the notion of proper class, such as von Neumann's or his own. In **ZF** there was no such notion and so he relied on a strongly inaccessible ordinal to obtain a model that was a set.

With Gödel's proof that the Axiom of Choice was relatively consistent, a phase in the history of set theory came to an end. Three decades earlier, Zermelo had proposed his axiomatization, in large measure, in order to secure his Axiom of Choice and his demonstration for the controversial Well-Ordering Theorem. Now Gödel had established conclusively that introducing the Axiom into **ZF** did not lead to a contradiction, provided that **ZF** was consistent. Thus, in a sense, Zermelo's efforts to secure the Axiom had been crowned by success. On the other hand, because of Gödel's Second Incompleteness Theorem it was impossible to prove the consistency of **ZF** in any system with less deductive strength or even in **ZF** itself. As a result, the consistency of **ZF** was left to the empirical test of experience. In three decades no one had deduced a contradiction in this system, nor would anyone during the next four decades. Hence, in empirical terms, **ZF** was consistent. Yet this conclusion lay uneasily on the discipline of mathematics, whose practitioners usually considered its truths to be time-independent.

At first, Gödel's results on relative consistency did not appear to reveal any new vistas, and his Axiom of Constructibility was received without enthusiasm. While Gödel [1938] considered this postulate to be a promising addition to the axioms of set theory, only Jean Cavaillès concurred by arguing in print for its acceptance.[28] Meanwhile, Kuratowski and Mostowski studied the effects of the Axiom of Constructibility on projective sets.[29] By

---

[26] Gödel 1940, 6; *cf.* 4.8.

[27] *Ibid.*, 35–38, 42–44. Mathematicians now describe Gödel's concept as absoluteness for L (the universe of constructible sets), since his concept of absoluteness has been generalized to any model of set theory.

[28] Cavaillès 1947, 19–20; this pamphlet was written during the period 1940–1941.

[29] Addison 1959, 338. Mostowski later noted that his manuscripts on this subject were lost during the Second World War; see Mostowski in Crossley 1975, 32–33.

1947, however, Gödel had reversed himself. Now he believed that this axiom, as well as its consequence the Continuum Hypothesis, was false, and that both might be negated by some plausible new postulate—perhaps even by an axiom of infinity stronger than any yet known.[30] Gödel's conjecture was borne out when Dana Scott discovered [1961] that if there exists a measurable cardinal, then the Axiom of Constructibility is false.[31]

As for the Fraenkel–Mostowski method of independence proofs, few significant new results were obtained from it during the 1940s. The sole exception was Mostowski's proof, employing an uncountable set of urelements for the first time, that the Principle of Dependent Choices does not imply the Axiom of Choice [1948]. Ironically, the Fraenkel–Mostowski method yielded its deepest results shortly before it was superseded in 1963.

Its successor was the method of forcing and generic sets, by which Paul Cohen established in first-order logic that the Axiom of Choice could not be deduced in **ZF** [1963a,b]. Since Cohen's method did not require urelements, it could be used to obtain independence results about $\mathbb{R}$ in particular. Indeed, his most celebrated application of it established that the Continuum Hypothesis was not a theorem of **ZFC**. Nevertheless, Cohen's proof of the independence of the Axiom relied heavily on Gödel's constructible sets and on certain features of the Fraenkel–Mostowski method.[32] Today, Gödel's and Cohen's contributions are universally recognized as the highest achievements yet attained in set theory.

---

[30] Gödel 1947, 520–524.

[31] Nevertheless, due in good part to the researches of Jensen, the Axiom of Constructibility regained some of its lustre during the 1970s. Devlin, indeed, argued for its truth [1977, iii–iv]. Concerning measurable cardinals, see 4.11.

[32] A tantalizing question is whether Gödel had previously found a proof for the independence of the Axiom. Speaking in 1966, on the occasion of Cohen's receipt of the Fields Medal for his independence proofs, Alonzo Church expressed one view: "Gödel . . . in 1942 found a proof of the independence of the axiom of constructibility in type theory. According to his own statement (in a private communication), he believed that this could be extended to an independence proof of the axiom of choice; but due to a shifting of his interests toward philosophy, he soon afterwards ceased to work in this area, without having settled its main problems. The partial result mentioned was never worked out in full detail or put into form for publication" [Church 1968, 17]. Hao Wang expressed a second view in an article which Gödel approved for publication in 1976 or 1977: "It was in 1943 when Gödel arrived at a proof of the independence of the axiom of choice in the framework of (finite) type theory. The idea of the proof makes it clear why the proof works. For that reason alone, it would be of interest to reconstruct the proof. It uses intensional considerations. The interpretation of the logical connectives is changed. A special topology has to be chosen. The method looked promising toward getting also the independence of *CH*. But Gödel developed a distaste for the work. . . . He now regrets that he did not continue the work. If he had continued with it, he would probably have gotten the independence of *CH* by 1950, and the development of set theory would have progressed faster" [Wang 1981, 657]. A third view is that of John Addison, who once asked Gödel why he had not published his proof for the independence of the Axiom of Choice. Gödel replied that he feared his proof would lead set-theoretic research in the wrong direction (personal communication from Addison, July 1981).

4.11 Scepticism and Inquiry

Between the two World Wars there was a growing dichotomy in attitudes toward the Axiom of Choice. On the one hand, the Axiom was vigorously applied in many of the most rapidly advancing fields of mathematics. On the other hand, the attitudes expressed toward the Axiom rarely reflected the mathematical advances that had illuminated its role. By and large, these attitudes had crystallized during the first three years after Zermelo proved the Well-Ordering Theorem, and few of those who stated an opinion at that time modified it on the basis of later research. Regrettably, the profound technical advances discussed in this chapter failed to stimulate an equally profound advance in the various philosophies of mathematics.

Some of the technical developments themselves reflected ambiguously on the Axiom, at times militating for it, at times against it. This ambiguity became particularly apparent in measure theory, where the Axiom had been used to establish the countable additivity of Lebesgue measure, on the one hand, and the existence of a set that is not Lebesgue-measurable, on the other. In 1914, by replacing the condition of countable additivity with finite additivity, Hausdorff obtained a weaker version of Lebesgue's Measure Problem. This version asked whether there exists a function m, called a measure, which assigns a non-negative real number to every bounded subset A of $\mathbb{R}^n$ and which satisfies the following conditions:

(a) The n-dimensional unit cube has measure one.
(b) Congruent sets have the same measure.
(c) $m(A \cup B) = m(A) + m(B)$ if A and B are disjoint.

For $n \geq 3$, Hausdorff demonstrated that the answer was no. To do so, he used the Axiom to partition a sphere into sets A, B, C, D such that A, B, C, and $B \cup C$ were all congruent, while D was countable (see 3.7). This result, which came to be known as Hausdorff's paradox, led to the discovery of an even more implausible result: the Banach–Tarski paradox.

The roots of the Banach–Tarski paradox lay in Banach's concern with solving Hausdorff's version of the Measure Problem for the two cases that Hausdorff had left open—the line and the plane. After obtaining such a measure in both cases by means of the Axiom [1923], Banach turned to the problems which Hausdorff's paradox raised. In a joint paper, Banach and Tarski established that in $\mathbb{R}^n$ any two bounded sets with non-empty interiors are equivalent by finite decomposition, provided that $n \geq 3$.[1] Consequently, they observed, any two spheres of different radii, or any two polyhedra, are equivalent by finite decomposition.[2] Later mathematicians would describe

[1] Two sets A and B in a metric space are *equivalent by finite decomposition* if there is some m such that A and B can be partitioned into pieces $A_1, A_2, \ldots, A_m$ and $B_1, B_2, \ldots, B_m$, where A_i is congruent to B_i for $i = 1, 2, \ldots, m$.

[2] Banach and Tarski 1924, 244–245, 260–264.

the following instance of this result as the Banach–Tarski paradox: A sphere of radius r can be decomposed into a finite number of pieces and reassembled into two spheres of radius r. Of course, to phrase the matter in this fashion makes a non-constructive proof appear to be a construction.

Banach and Tarski did not regard their result so psychologically. While they acknowledged that it utilized Hausdorff's seemingly paradoxical decomposition of a sphere, they did not consider their result to detract from the Axiom. In fact, they emphasized, without the Axiom it was not known how to deduce even that two polygons, one of which is properly included in the other, are not equivalent by finite decomposition. They observed that the essential difference between these two results—the one paradoxical and the other intuitive—was that there is an additive congruence-preserving measure on the Euclidean line, and on the Euclidean plane, but no such measure for three or more dimensions.[3]

In 1929 von Neumann shed further light on the Banach–Tarski paradox and, at the same time, proposed a more far-reaching generalization of the Measure Problem. It might appear that the existence of a suitable measure on $\mathbb{R}^2$, contrasted with the non-existence of such a measure on $\mathbb{R}^n$ for $n \geq 3$, revealed a fundamental dichotomy between the nature of the Euclidean plane and that of Euclidean space. As von Neumann hastened to point out, the essential difference lay not in the dimensionality of space but in the underlying group of motions. To clarify the matter, he revised the Measure Problem by letting M be any set, S any subset of M, and G any group of bijections of M onto itself. Then he called m an (M, S, G)-measure if m assigned a non-negative real number to each subset A of M and if m had the following properties:

(i) $m(S) = 1$.
(ii) If f is in G, then $m(A) = m(f''A)$.
(iii) $m(A \cup B) = m(A) + m(B)$ if A and B are disjoint.

Here he altered Hausdorff's problem in two essential respects, first by replacing $\mathbb{R}^n$ with an arbitrary set, and second by considering an arbitrary group instead of Euclidean congruence. After he gave a sufficient condition for an (M, S, G)-measure to exist, he showed that the condition was fulfilled by the group of Euclidean motions in $\mathbb{R}^n$ for $n = 1$ and $n = 2$, but not for $n \geq 3$. On the other hand, he demonstrated that if G had a free subgroup with two generators (as was the case for the group of Euclidean motions in $\mathbb{R}^n$ whenever $n \geq 3$), then an (M, S, G)-measure did not exist. Finally, by varying the

[3] *Ibid.*, 245, 264. Banach and Tarski had independently discovered the Banach–Tarski paradox, and had then decided to write a joint paper on the subject [1924]; however, the more general result on bounded sets with non-empty interiors was due to Tarski alone (personal communication from Tarski, 10 March 1982).

group G on $\mathbb{R}$, he found that an $(\mathbb{R}, S, G)$-measure could fail to exist there too. In all of these results, the Axiom played an essential role.[4]

The Measure Problem was generalized in a different direction by Banach, who regarded the problem as essentially set-theoretic rather than geometric or even group-theoretic. Banach inquired whether there exists a function m that assigns a non-negative real number to each subset A of the closed interval $[0, 1]$ and that satisfies the following conditions:

(I) $m(A) \neq 0$ for some A.
(II) $m(A) = 0$ if A contains exactly one element.
(III) m is countably additive.

By replacing the condition (b) on congruence with the weaker set-theoretic condition (II), whose inclusion was necessary in order to avoid a trivial solution, he was able to reinstate countable additivity. Nevertheless, Banach and Kuratowski then discovered that there exists no such m, provided that the Continuum Hypothesis is assumed [1929].

Soon afterward, Banach recognized that in his problem the interval $[0, 1]$ could be replaced by any set of power $2^{\aleph_0}$ and, furthermore, that his problem really concerned the cardinality of a set on which a measure was desired. Consequently, he proposed what he called the Generalized Measure Problem: Does there exist some set E with a function m that assigns a non-negative real number to each subset A of E and that satisfies conditions (I), (II), and (III)? From the Generalized Continuum Hypothesis he deduced that the power of such an E could not be smaller than the first weakly inaccessible cardinal [1930, 101].

Quickly two further developments ensued. On the one hand, Tarski established that if countable additivity were weakened to finite additivity in the Generalized Measure Problem, then the Axiom ensured that every infinite set E has such a measure m. Moreover, he showed that m could be restricted to take only the values zero and one [1930]. On the other hand, Stanislaw Ulam, then a young graduate student at Lwów, was stimulated by conversations with Kuratowski to improve on Banach's result.[5] Without assuming the Generalized Continuum Hypothesis or even the Continuum Hypothesis, Ulam proved that a set E satisfying the Generalized Measure Problem cannot have a power less than the first strongly inaccessible cardinal, provided that no weakly inaccessible cardinal is less than or equal to $2^{\aleph_0}$ [1930, 150]. Ulam relied on the Axiom in an essential way to obtain this result.

In more recent terminology, the cardinal of a set satisfying the Generalized Measure Problem is called a real-valued measurable cardinal, whereas such a cardinal whose measure takes only the values zero and one is called a

---

[4] Von Neumann 1929a, 78–82, 87, 115–116.

[5] Ulam 1930, 141.

measurable cardinal. Ulam established that any measurable cardinal must be at least as large as the first strongly inaccessible cardinal and that the same is true of any real-valued measurable cardinal if every weakly inaccessible cardinal is greater than $2^{\aleph_0}$. Ironically, it was later found that without the Axiom even $\aleph_1$ could be a measurable cardinal.[6]

Yet the mathematicians who had already formed an opinion about the Axiom were merely confirmed in their opinion by developments such as those that occurred in measure theory. Borel, for example, had viewed Hausdorff's paradox as a definitive reason for rejecting the Axiom in 1914, and three decades later he was still inveighing against the Axiom on similar grounds.[7]

A significant exception was Russell. His attitude toward the Axiom was strongly influenced by Frank Ramsey, who in 1925 proposed to revise Russell's theory of types so as to secure it against certain criticisms. Discarding the complications that protected this theory from semantic antinomies such as Richard's paradox, Ramsey obtained what became known as the simple theory of types. By contrast to Russell, Ramsey viewed the Axiom positively, and interpreted it in a way that did not require a choice function to be definable in the underlying formal system:

> The Multiplicative Axiom, interpreted as it is in *Principia* [*Mathematica*], is not a tautology but logically doubtful. But, as I interpret it, it is an obvious tautology, and this can be claimed as an additional advantage in my theory [of simple types]. It will probably be objected that, if it is a tautology, it ought to be able to be . . . deduced from the simpler primitive propositions which suffice for the deduction of the rest of mathematics. But it does not seem to me in the least unlikely that there should be a tautology, which could be stated in finite terms, whose proof was, nevertheless, infinitely complicated and therefore impossible for us.[8]

Intrigued by Ramsey's viewpoint, Russell modified his position toward the Axiom. In 1927 he wrote: "I have been led by the arguments, first of Dr. H. M. Sheffer, and then of Mr. F. P. Ramsey, to the view that Zermelo's axiom is true; I am therefore less reluctant than I should have been formerly to assume that [space-time] events can be well-ordered."[9] Despite having accepted both the Axiom of Choice and the existence of a well-ordering for the continuum, Russell adopted a neutral position toward these two propositions when he reviewed in [1931] a collection of Ramsey's essays. Six years later, in his preface to the second edition of the *Principles of Mathematics*, Russell interpreted his neutrality by means of possible universes: "It is impossible to prove that there are possible universes in which [the Axiom of

---

[6] This was shown by Jech [1968].

[7] See Borel 1946, 1947, and 1947a.

[8] Ramsey 1925, 382. Hans Hahn, who was closely connected with the Vienna Circle, regarded Ramsey's attempt to show the Axiom to be tautological as a failure (see Hahn *et alii* 1931, 138, and Hahn 1980, 34). All the same, Hahn accepted the Axiom [1921, 25].

[9] Russell 1927, 299–300.

Choice] would be false; but it is also impossible (at least, so I believe) to prove that there are no possible universes in which it would be false" [1937, viii]. As the earlier researches of Fraenkel and the later work of Cohen would show, Russell was correct on the second point but was mistaken on the first.

Unlike Russell, Lebesgue deviated very little from the position which he had originally taken on the Axiom. Writing for *Fundamenta Mathematicae* on mappings between Euclidean spaces, Lebesgue argued that the Countable Union Theorem would not need the Axiom if the term denumerable were taken to mean effectively enumerable. Likewise, he insisted that the union of denumerably many measurable sets is measurable [1921, 260]. Although his first assertion was justified, since an effectively enumerable set has a uniquely determined bijection onto the natural numbers, his second assertion was vague. The most generous interpretation would be that a measurable set ought to be determined by an effectively enumerable set of intervals.

The attitudes of many others besides Lebesgue were little affected by later developments. Ernest Hobson, who had rejected the Axiom in 1905, maintained his opposition to it, while Hadamard argued for the Axiom on grounds that had hardly changed in two decades.[10] Jules Richard also continued to oppose the Axiom for his original reasons [1929].

Luzin, whose first contribution to the debate over the Axiom appeared near the end of the First World War, generally sided with Baire, Borel, and Lebesgue. Nevertheless, Luzin treated the Axiom as a heuristic device for finding theorems, which were then to be proved without it whenever possible.[11] He also used the Continuum Hypothesis in the same fashion [1914]. Like Sierpiński, he considered it important to indicate when a demonstration employed the Axiom and when it did not.[12] On the other hand, he believed that "all the arguments which can be invented to support this axiom [of choice] are psychological in nature . . ." [1927a, 83]. Indeed, as he came to agree more and more with the Empiricists, he began to favor Poincaré's rejection of the actual infinite.[13] Ironically, his philosophical views

[10] Hobson 1921, 169, 238, 248–254; 1926, 380; 1927, 177, 252, 262–269; and Hadamard 1926, 67–69. Hobson's rejection of the Axiom was voiced more gently than before, but he continued to object, for example, to the proposition that the set of all real functions can be ordered.

[11] Luzin's thoughts on the subject were recorded by Sierpiński [1918, 103]; *cf.* 4.1. As an early example of Luzin's use of the Axiom as a heuristic device, see Luzin and Sierpiński 1917a.

[12] See, for example, Luzin 1917. He was mistaken, however, in claiming that certain theorems on analytic sets did not require the Axiom; see 3.6.

[13] Luzin 1930, 322; *cf.* 1947, 198–199. His mistrust of the Axiom led him to propose what he called the Uniformization Problem. A real function f *uniformizes* a subset E of the plane if f is included in E (considered as a relation on $\mathbb{R}$) and if f has the same domain as E. Luzin's Uniformization Problem asked whether some E can be defined such that no "definable" f uniformizes E [1930a,b]. He considered the answer to be yes, since, so he believed, the complement of some analytic set could be defined that could not be uniformized by the complement of *any* analytic set. Nevertheless, his belief was erroneous; see Kondô 1939. On a related question, see Luzin and Novikov 1935.

on the Axiom and the actual infinite deviated from much of his research on projective sets.

The fundamental lack of change in the positions of those for and against the Axiom can best be illustrated by a conference that took place at Zurich during December 1938. The participants in this conference, devoted to the foundations of mathematics, included Bernays, Fréchet, Lebesgue, Sierpiński, and Skolem. Giving consecutive lectures on the Axiom of Choice, Lebesgue and Sierpiński echoed what they had written decades earlier.[14] In his informal and quasi-philosophical lecture, Lebesgue praised Sierpiński as "the man who has best known how to utilize the Axiom of Choice ..." [1941, 111]. All the same, Lebesgue remained a subtle opponent of the Axiom. He noted that when the controversy arose in 1905, the two sides could not make themselves understood because they lacked a common logic. In fact, he insisted, logic itself was at stake.[15]

On the other hand, Lebesgue realized quite well that the issues were not restricted to the underlying logic. Despite the fact that for more than thirty years the Axiom had not led to a contradiction, there remained a feeling of unease. Indeed, he observed, logic could not create confidence in a conclusion unless one had already accepted this conclusion as reasonable. In closing, he argued that "in the studies on the foundations and methods of mathematics, there must be a large place for psychology and even for esthetics" [1941, 122]. He thereby underlined the psychological perspective that he had shared with Baire and Borel since 1905.

In the discussion following Lebesgue's lecture, Bernays suggested that while the Axiom might be somewhat problematical in general, certainly as applied to sets of real numbers it was legitimate. Lebesgue disagreed. Yet he granted that some Polish mathematicians had used the Axiom in a fruitful way by investigating exactly where this hypothesis was needed, and consequently the situation was no longer what it had been.[16]

Sierpiński, in a lecture which primarily summarized research on the Axiom by the Warsaw school, avowed that he was neither for nor against the Axiom. However, he considered it artificial to accept only the Denumerable Axiom, since there was no substantive reason for splitting the Axiom at this cardinality. Bernays suspected that the intensive research on the Axiom by members of the Warsaw school, such as Sierpiński, was an attempt to derive a paradox from the Axiom. On the other hand, Bernays could not believe the Axiom to be merely a hypothesis, rather than true or false. While another participant regarded the Axiom as imprecise, Fréchet insisted that only the precise nature of the Axiom had allowed mathematicians to prove it equivalent to other propositions.[17] In this vein, the chairman of the conference

---

[14] Lebesgue 1941 and Sierpiński 1941.
[15] Lebesgue 1941, 116.
[16] Sierpiński 1941, 134–138.
[17] *Ibid.*, 139–142.

read a brief letter from Gödel, who revealed that he had established the consistency of the Axiom.[18] Nevertheless, Gödel's discovery had little effect on discussions at the conference.

Despite the numerous technical advances, the Axiom remained a rather debatable assumption to many eminent mathematicians and particularly to the participants in the Zurich conference. Yet, as the years passed, the Axiom continued to be used fruitfully in those branches of mathematics undergoing rapid development. Later mathematicians, who had not been involved in the controversy, were increasingly likely to apply the Axiom with no qualms of conscience. Thus in 1977 a functional analyst would write, "In this book the axiom of choice is used without apology or explanation . . ."[19]— a dismissal of debate that can best be understood by realizing how deeply felt had been the qualms of earlier mathematicians.

4.12 *Retrospect and Prospect*

The period between the two World Wars saw the continuation of previous themes in the history of the Axiom of Choice, as well as the emergence of new ones. By and large, the fundamental attitudes toward the Axiom changed little during these years, although their distribution varied. The principal arguments in favor of the Axiom remained those which Zermelo had stated in 1908: The Axiom has been used by many mathematicians, both implicitly and explicitly, a historical fact that can be understood only by granting the Axiom's self-evidence; moreover, the Axiom is necessary to obtain a vast range of important theorems in diverse branches of mathematics. On the other hand, the chief argument against the Axiom continued to be its non-constructive character: The Axiom provided no method for executing the choices which it asserted to exist, or even for defining a choice function uniquely.

While the quasi-philosophical attitudes of mathematicians toward the Axiom changed little, the years 1918–1940 saw major technical advances in the mathematics involving the Axiom. Indeed, as abstract algebra built upon the fundamental discoveries of Steinitz in field theory, the Axiom became an essential tool. Usually it appeared in the guise of the Well-Ordering Theorem, coupled with transfinite induction and with definition by transfinite recursion. The researches of Artin and Schreier on real fields utilized the Axiom in this way to obtain the existence of the real closure of a real field, analogous to Steinitz' algebraic closure of an arbitrary field. In ring theory, which likewise grew from the researches of Steinitz, the Axiom aided Noether and

[18] *Ibid.*, 134.

[19] Brown and Pearcy 1977, 5.

Krull in developing the theory of ideals. They used it especially to show that there exists a maximal prime ideal extending a given proper ideal in a commutative ring. Stone arrived at a similar result in Boolean algebra, where it became known as the Boolean Prime Ideal Theorem, and from it he deduced the Stone Representation Theorem for Boolean algebras.

Within general topology the Axiom proved to be equally essential, though less noticed. Topologists such as Hausdorff and those of the Moscow school tended to employ the Axiom liberally without remarking the fact. Even Kuratowski and Sierpiński, who usually pointed out the appearance of the Axiom in their set-theoretic researches, often used it without special mention in topology. Nevertheless, the Axiom permitted them to obtain results for topological spaces, such as the proposition that separable metric spaces are hereditarily separable, that could be false in the Axiom's absence. Likewise, the Moscow school of topologists utilized the Axiom to deduce Urysohn's Lemma and to characterize separable metric spaces. Above all, the Axiom proved vital in the study of convergence and compactness. In this vein, Tychonoff's Compactness Theorem turned out to be equivalent to the Axiom, while the existence of the Stone–Čech compactification of a space was later shown equivalent to the weaker Boolean Prime Ideal Theorem. Finally, the new methods of filters and directed sets relied heavily on the Axiom for their principal results.

One of the Axiom's most fundamental applications cut across several disciplines: the existence of a set maximal with respect to a given property. As early as 1909, Hausdorff had used the Axiom to derive a maximal principle, and had then applied it in the theory of real functions. Nevertheless, the Well-Ordering Theorem continued to be the standard tool for generating a maximal set in algebra, analysis, topology, and set theory until 1935, when Zorn independently formulated a maximal principle. Unlike Hausdorff and Kuratowski, Zorn proposed his maximal principle (soon called Zorn's Lemma) as an *axiom*. He intended it to supersede the Well-Ordering Theorem in algebra, particularly for results about rings and fields. At the time Zorn claimed but did not prove that the Axiom is equivalent to Zorn's Lemma. On the other hand, Teichmüller independently stated several closely related maximal principles, each of which he showed equivalent to the Axiom. While Zorn had stressed the uses of his principle in abstract algebra, Teichmüller applied his own to both abstract and linear algebra, as well as to the theory of Hilbert spaces.

In mathematical logic the Axiom of Choice played a variety of roles. On the one hand, it became a postulate of logic when Hilbert adopted what he first called the Transfinite Axiom and later the ε-axiom. Indeed, he made the ε-axiom the basis for his *Beweistheorie*, through which he hoped to establish the consistency of analysis and set theory. On the other hand, the Axiom aided Gödel metamathematically when he deduced both the Completeness Theorem and the Compactness Theorem for first-order logic. Since he employed a formal language with countably many symbols, Gödel could have avoided the Axiom if he had wished to do so (he did not). By introducing languages with uncountably many symbols, Malcev rendered the Axiom essential to these two theorems. Though Skolem intentionally avoided the Axiom in his theorem that every satisfiable set of first-order sentences (in a countable language) has a countable model, he could not have done so in proving the stronger result that every model has a countable submodel. In fact, the Löwenheim–Skolem–Tarski Theorem, which generalized Skolem's results, turned out to be equivalent to the Axiom. Thus it eventually became evident that the most fundamental results of mathematical logic need the Axiom.

In contrast to the developments discussed above, where the Axiom was not the focus of investigation, the Warsaw school under Sierpiński analyzed the Axiom's deductive strength. Sierpiński's survey [1918], which initiated this analysis, was an extremely thorough exploration of the interrelations between the Axiom and other propositions. There he clarified the Axiom's role in establishing the equivalence of limit point and sequential limit point, as well as that of continuity and sequential continuity, in $\mathbb{R}^n$. Furthermore, he showed that such an equivalence implies a weak form of the Denumerable Axiom, as does the Countable Union Theorem.

On the other hand, Sierpiński left open a large number of questions concerning the Axiom, particularly vis-à-vis finite sets and cardinal arithmetic. It was Tarski, above all, who pursued these questions. In 1924 he obtained a hierarchy of equivalence classes whose members were definitions of finite set. This hierarchy, which was based explicitly on the Axiom, extended Russell's work on mediate cardinals. Furthermore, Tarski showed a sizeable number of propositions from infinite cardinal arithmetic—such as $\mathfrak{m}^2 = \mathfrak{m}$—to be equivalent to the Axiom. A few years later, he exhibited definitions of finiteness whose equivalence to the usual one was itself equivalent to the Axiom.

There remained the matter of the consistency and independence of the Axiom. In 1922 Fraenkel used urelements to show that the Axiom was independent in Zermelo's system. Later Fraenkel's permutation method of independence proofs was perfected by Lindenbaum and Mostowski, who relied on urelements to prove the independence of the Partition Principle, of $\mathfrak{m} + \mathfrak{m} = \mathfrak{m}$, and of the equivalence of Tarski's five classes of definitions for finiteness. Mostowski also obtained the result, sought earlier by Fraenkel, that the Ordering Principle does not imply the Axiom. Finally, Gödel utilized a new inner model, the constructible sets, to establish the consistency of both the Axiom of Choice and the Generalized Continuum Hypothesis relative to **ZF**. With this result, the principal questions surrounding the Axiom were settled.

Nevertheless, though both Fraenkel and Mostowski would have preferred to avoid urelements in their proofs of the Axiom's independence, at the time no one knew how to do so. That accomplishment had to await more recent investigations, to which we now turn.

Epilogue: After Gödel

> Nothing can better express the meaning of the term "class" than the Axiom of [Separation] and the Axiom of Choice.
>
> Kurt Gödel [1944, 151]

> No doubt [the Axiom of Choice] will always be desirable despite the technical interest of various independence questions involving it and weaker principles. If only it could be deduced from some more primitive principle!
>
> Dana Scott [1974, 214]

During the four decades that have elapsed since Gödel established the relative consistency of the Axiom of Choice and the Generalized Continuum Hypothesis, much mathematical research has focused on the Axiom. These decades are separated into two very different periods by a discovery of the first rank: Paul Cohen's method of forcing. In 1963 he used this method to prove that both the Axiom of Choice and the Continuum Hypothesis are independent from **ZF**, thereby complementing Gödel's results on consistency. During the quarter century prior to 1963, research on the Axiom largely continued the lines of development discussed in Chapter Four, and no conceptual breakthrough occurred. After the discovery of forcing in 1963, set-theoretic investigations concerning the Axiom have taken a distinctly more metamathematical and semantic turn. This brief Epilogue will only attempt to sketch the major themes and developments involving the Axiom since 1940.

5.1 A Period of Stability: 1940–1963

Between the two World Wars, Polish mathematicians at Warsaw and Lwów investigated the deductive strength of numerous propositions relative to the Axiom. Often they wished to learn whether or not some given proposition P is equivalent to the Axiom—that is, as we shall say here, whether P is an equivalent. After the Second World War several of these mathematicians, especially Sierpiński in Warsaw and Tarski at his new location in Berkeley, continued to study such questions. In particular, Sierpiński ventured to demonstrate some of the results in cardinal arithmetic that Lindenbaum and Tarski had stated without proof in their paper of 1926 (see 4.3). Among these results were several propositions which they asserted to be equivalent to the Axiom. During 1947 and 1948 Sierpiński published demonstrations that two of them, (4.3.8) and (4.3.13), were indeed equivalents.[1] In a similar vein he established that the following proposition, a multiplicative version of Leśniewski's (4.1.16), was also an equivalent:

(5.1.1) For infinite $\mathfrak{m}$ and $\mathfrak{n}$, if $\mathfrak{m} < \mathfrak{p}$ and $\mathfrak{n} < \mathfrak{p}$, then $\mathfrak{m}\mathfrak{n} \neq \mathfrak{p}$. [1946]

Furthermore, he demonstrated Lindenbaum's and Tarski's claim that the Generalized Continuum Hypothesis, in the form that for every infinite $\mathfrak{m}$ there is no $\mathfrak{n}$ such that $\mathfrak{m} < \mathfrak{n} < 2^{\mathfrak{m}}$, implies the Axiom of Choice.[2] Lastly, he proved their claim that (4.3.14), a weak form of the Partition Principle, yields both $\aleph_1 \leq 2^{\aleph_0}$ and the existence of a non-measurable subset of $\mathbb{R}$ [1947b].

The deductive strength of propositions relative to the Axiom was of continuing interest to Tarski and his colleagues at Berkeley. In 1948, perhaps stimulated by Sierpiński, he published a demonstration of the result, stated without proof in his joint article with Lindenbaum, that the proposition (4.1.17) in cardinal arithmetic is an equivalent.[3] Soon afterward John Kelley, a Berkeley colleague who cited Tarski's earlier work on the Axiom, found such an equivalent in topology: Tychonoff's Theorem restricted to T_1 spaces [1950]. In response to a question raised by Tarski, his student Robert Vaught showed that a special case of Zorn's Lemma was likewise equivalent to the Axiom:

(5.1.2) Every family of sets has an $\subseteq$-maximal subfamily whose members are disjoint. [1952]

In 1954, returning to an earlier theme in cardinal arithmetic, Tarski considered three ways of stating that a cardinal $\mathfrak{m}$ has an immediate successor $\mathfrak{n}$:

(5.1.3) For every $\mathfrak{m}$ there is some $\mathfrak{n}$ such that (i) $\mathfrak{m} < \mathfrak{n}$, and (ii) there is no $\mathfrak{p}$ such that $\mathfrak{m} < \mathfrak{p} < \mathfrak{n}$.

---

[1] Sierpiński 1947 and 1948.

[2] Sierpiński 1947a. He also refined previous research on the Banach–Tarski paradox; see his 1945 and 1945a. As late as [1965], Sierpiński proved a certain proposition on partitions to be equivalent to the Axiom; *cf.* Sobociński 1964 and 1965.

[3] Tarski 1948, 82–84.

(5.1.4) For every $\mathfrak{m}$ there is some $\mathfrak{n}$ such that (i) $\mathfrak{m} < \mathfrak{n}$, and (ii′) for every $\mathfrak{p}$, if $\mathfrak{m} < \mathfrak{p}$, then $\mathfrak{n} \leq \mathfrak{p}$.

(5.1.5) For every $\mathfrak{m}$ there is some $\mathfrak{n}$ such that (i) $\mathfrak{m} < \mathfrak{n}$, and (ii″) for every $\mathfrak{p}$, if $\mathfrak{p} < \mathfrak{n}$, then $\mathfrak{p} \leq \mathfrak{m}$. [1954, 26, 32]

Tarski established that (5.1.3) can be deduced in **ZF**, whereas (5.1.4) is an equivalent. As for (5.1.5), he remained uncertain of its deductive strength.[4] Not long afterward, Vaught discovered that two central theorems of first-order logic were also equivalent to the Axiom: the Downward Löwenheim–Skolem Theorem (every sentence with a model of power $\mathfrak{m}$ has a model of every infinite power smaller than $\mathfrak{m}$) and the Löwenheim–Skolem–Tarski Theorem (every sentence with a denumerable model has a model of every uncountable power) [1956]. Thus the Axiom turned out to be as indispensable in general topology and mathematical logic as it had previously become in set theory and abstract algebra.

Maximal principles such as Zorn's Lemma remained a particularly important type of equivalent. Though they had been used sporadically by Hausdorff and Kuratowski, such principles came to be widely applied in algebra, analysis, and topology only after Zorn's Lemma sparked the interest of Bourbaki around 1935 (see 4.4). Over the next two decades, maximal principles were widely discussed, especially in the early 1950s.[5] One line of inquiry was to generalize Vaught's maximal principle (5.1.2) by studying propositions of the following form, where S was a given relation: Every family of sets has an $\subseteq$-maximal subfamily F such that every pair of sets A, B in F is in the relation S. The Yugoslavian mathematician Djuro Kurepa found a number of relations S such that the corresponding maximal principle was an equivalent. Among these were the relations of overlap ($A—B \neq 0$ and $B—A \neq 0$ and $A \cap B \neq 0$), non-overlap ($A—B = 0$ or $B—A = 0$ or $A \cap B = 0$), and non-disjointness ($A \cap B \neq 0$). He was unable to establish that incomparability (neither $A \subseteq B$ nor $B \subseteq A$) was such a relation, but only that the Ordering Principle and the maximal principle for incomparability were jointly equivalent to the Axiom [1952]. Later, Ulrich Felgner proved that the maximal principle for incomparability was in fact an equivalent [1969, 228].

In 1960 two American mathematicians, Herman and Jean Rubin, were prompted by Kurepa's research to consider maximal principles of the following form: If a family K of sets has an $\subseteq$-maximal subfamily F such that every pair of sets in F is in the relation S, then K has an $\subseteq$-maximal

[4] Jech [1966a] used forcing to demonstrate that **ZF** does not imply (5.1.5). Later, Truss showed that (5.1.5) implies $\mathfrak{m} = 2\mathfrak{m}$ for every infinite $\mathfrak{m}$ [1973a].

[5] See, for example, Wallace 1944, Fort 1948, Klimovsky 1949, Obreanu 1949, Bourbaki 1950, Kneser 1950, Szele 1950, Witt 1951, Gottschalk 1952, Inagaki 1952, Vaughan 1952, Banaschewski 1953, and Kurepa 1953.

subfamily G such that every pair of sets in G is in the relation T. They established that such a principle was an equivalent when S and T were any distinct relations among disjointness, non-disjointness, overlap, non-overlap, comparability, and incomparability—provided that the Ordering Principle was assumed.[6] In addition, Herman Rubin found two propositions which were equivalent to the Axiom in **ZF** but were weaker in **ZFU**:

(5.1.6) The power set of every well-ordered set can be well-ordered.

(5.1.7) Every set that can be ordered can also be well-ordered.[7]

In 1963 the Rubins published a book summarizing and completing much of the earlier work on equivalents.

A different line of inquiry connected maximal principles and the Axiom to developments in algebra. In 1953 the German mathematician Jürgen Schmidt carefully analyzed the role that the Axiom played in various algebraic theorems and proved several of them to be equivalents, including the following:

(5.1.8) If $\mathfrak{A}$ is an algebra and if b is an element in some basis for $\mathfrak{A}$, then there is an $\subseteq$-maximal subalgebra of $\mathfrak{A}$ not containing b.

(5.1.9) If $\mathfrak{A}$ is an algebra with a finite basis and if $\mathfrak{B}$ is a proper subalgebra of $\mathfrak{A}$, then there is an $\subseteq$-maximal proper subalgebra of $\mathfrak{A}$ which includes $\mathfrak{B}$.[8]

Under the influence of Tarski, Dana Scott was engaged in related algebraic research at Berkeley. There Scott established that the Axiom is equivalent to a proposition on lattices:

(5.1.10) Every lattice with a unit and at least one other element has a maximal ideal. [1954]

Independently of Scott, the Argentinean mathematician Gregorio Klimovsky showed Zorn's Lemma, and hence the Axiom, to be equivalent to (5.1.10)

---

[6] Rubin and Rubin 1960; in certain cases they found that the Ordering Principle could be omitted. H. Rubin [1958] had already extended Kurepa's work, while Chang [1960] generalized that of Vaught.

[7] H. Rubin 1960. Engaged in related research, Kruse proved the following proposition to be an equivalent: If $\mathfrak{p} < \mathfrak{m},\mathfrak{n} < 2^{\mathfrak{p}}$, then either $\mathfrak{m} \leq \mathfrak{n}$ or $\mathfrak{n} \leq \mathfrak{m}$ [1962, 145].

[8] Schmidt 1953, 40–41; *cf.* 29–31, 36. For his later research on equivalents, see Schmidt 1962. An *algebra* $\mathfrak{A}$ is an ordered pair $\mathfrak{A} = (A, \mathscr{F})$, where A is a set and $\mathscr{F}$ is a possibly infinite set of functions $f: A^n \to A$ for one or more values of n. A *subalgebra* of $\mathfrak{A}$ is an algebra $(B, \mathscr{F}')$ such that B is a subset of A and $\mathscr{F}'$ consists of the restriction of each f in $\mathscr{F}$ to B. A *basis* for $\mathfrak{A}$ is a subset C of A such that (i) the only subalgebra including C is $\mathfrak{A}$ itself and (ii) any proper subset of C is included in some proper subalgebra of $\mathfrak{A}$.

weakened to distributive lattices [1958]. Four years later, Klimovsky proved that the Axiom is equivalent to the conjunction of $\mathbf{C}_2$ and the proposition that every group includes an $\subseteq$-maximal Abelian subgroup [1962]. Not long afterward, M. N. Bleicher obtained the equivalence of the Axiom and a number of results on vector spaces, such as the following conjoined with $\mathbf{C}_{<\aleph_0}$ (the Axiom of Choice restricted to families of finite sets):

(5.1.11) If V is a real vector space, then for every subspace S there is a subspace S' such that $S \cap S' = \{0\}$ and $S \cup S'$ generates V.[9]

Influenced by Bleicher, James Halpern proved that a theorem about bases is an equivalent:

(5.1.12) If a subset A of a vector space V generates V, then A includes a basis. [1966]

Scott's result raised the possibility that (5.1.10) would remain equivalent to the Axiom if its hypothesis were strengthened from a lattice to a Boolean algebra—thus giving the Boolean Prime Ideal Theorem. Tarski and his Berkeley colleagues vigorously pursued this question. It was already clear that the Boolean Prime Ideal Theorem could not be deduced in **ZFU**, since Anne Davis had shown that this theorem implies $\mathbf{C}_{<\aleph_0}$, a form of the Axiom known to be independent of **ZFU** (see 4.10).

Indeed, the Boolean Prime Ideal Theorem turned out to have considerable deductive strength. Scott demonstrated that it yields the Order Extension Principle (every partial order can be extended to an order), while Tarski proved that it also implies the proposition of Artin and Schreier that every real field is an ordered field [1954a]. Working closely with Scott and Tarski, Leon Henkin linked the Boolean Prime Ideal Theorem to metamathematics by establishing that it is equivalent to the Compactness Theorem for first-order logic as well as to the Completeness Theorem for either first-order or propositional logic [1954 and 1954a]. Finally, Herman Rubin and Dana Scott jointly showed that each of several topological theorems, including Tychonoff's Theorem restricted to Hausdorff spaces and the Stone–Čech Compactification Theorem, is equivalent to the Boolean Prime Ideal Theorem.[10]

---

[9] Bleicher 1964, 96, 196. Jech improved this result by showing that (5.1.11) is itself an equivalent [1973, 148]. Meanwhile, Bell and Fremlin had obtained an equivalent in the theory of normed vector spaces [1972]. On distributive laws and the Axiom, see Collins [1954] as well as Rubin and Rubin [1963, 118].

[10] Rubin and Scott 1954. The result concerning Tychonoff's Theorem was also discovered by Łos and Ryll-Nardzewski [1955]. Henkin had established in his thesis [1947] that the Compactness Theorem for first-order logic implies the Boolean Prime Ideal Theorem, but it was Tarski who stimulated interest at Berkeley in showing various propositions equivalent to the latter theorem (personal communication from Henkin, 1 April 1982).

Yet it remained uncertain whether the Boolean Prime Ideal Theorem was weaker than the Axiom, as the Polish mathematicians Jerzy Łoś and Czesław Ryll-Nardzewski suspected, or was equivalent to it.[11] In 1961 James Halpern, working under Dana Scott and Azriel Levy, established that the Boolean Prime Ideal Theorem does not yield the Axiom of Choice, and hence is a weaker assumption, in **ZFU**.[12]

Meanwhile, research continued on the deductive strength of various propositions relative to the Boolean Prime Ideal Theorem. In 1961 Jan Mycielski of Wrocław considered the following propositions P_n for each $n > 1$:

(5.1.13(n)) If G is a graph such that every finite subgraph of G can be colored with n colors, then G itself can be colored with n colors.[13]

Here, to color a graph with n colors means to partition the set of vertices into n classes such that no two vertices in the same class are joined by an edge. Mycielski established that, if $n > 1$, then P_{n+1} implies P_n, and P_n itself implies $\mathbf{C}_n$ (the Axiom of Choice restricted to families of n-element sets). Moreover, he proved that P_2 is equivalent to $\mathbf{C}_2$ [1961, 126]. It was not known if some P_n is equivalent to the Boolean Prime Ideal Theorem, only that this theorem yields P_n for every $n > 1$. By means of the Fraenkel–Mostowski method, Levy showed that $\mathbf{C}_n$ does not imply P_n and, furthermore, that P_3 does not follow even from the assumption that for all m, $\mathbf{C}_m$ holds [1963a]. Eventually, Hans Läuchli of Zurich demonstrated that P_n is equivalent to the Boolean Prime Ideal Theorem for each $n > 2$ [1971].

In 1955 two German mathematicians, W. Kinna and K. Wagner, introduced a different kind of proposition as a weakening of the Axiom:

(5.1.14) For every set M there is a function f such that, for each subset A of M with two or more elements, $f(A)$ is a non-empty proper subset of A.

They showed that (5.1.14), later known as the Kinna–Wagner Principle, was equivalent to a proposition about ordinals:

(5.1.15) For every set M there is an ordinal α such that M is equipollent to some subset of the power set of α. [1955, 80]

---

[11] Łoś and Ryll-Nardzewski 1955, 50.

[12] Halpern 1961 and 1964.

[13] Mycielski 1961, 125. Höft and Howard later gave a graph-theoretic equivalent of the Axiom [1973].

From (5.1.14) they deduced the Ordering Principle, and observed in consequence that the Kinna–Wagner Principle cannot be proved in **ZFU**. While they were unable to establish that the Ordering Principle is weaker than the Kinna–Wagner Principle, Mostowski did so by demonstrating that (5.1.14) failed in his 1939 model of the Ordering Principle [1958]. Three years later A. Levy considered the following propositions $Z(n)$, closely related to the Kinna–Wagner Principle:

(5.1.16(n)) For every family M of non-empty sets there is a function f such that, for each A in M, $f(A)$ is a non-empty subset of A having at most n elements.

Levy proved that, for each positive n, $Z(n)$ is equivalent to the Axiom [1962]. On the other hand, as Mostowski had pointed out, it was beyond the scope of known independence techniques to establish that the Kinna–Wagner Principle does not yield the Axiom of Choice [1958, 208].

Nevertheless, the Fraenkel–Mostowski method of independence proofs had progressed considerably since the Second World War. Raouf Doss of Alexandria observed that in Mostowski's model of 1939 the set of urelements was both infinite and Dedekind-finite [1945]. Consequently, the Ordering Principle did not entail proposition (1.1.4) that every infinite set has a denumerable subset and, *a fortiori*, did not imply the Denumerable Axiom. Stimulated by Tarski's introduction of the Principle of Dependent Choices [1948, 96] but apparently unaware that Bernays had formulated the same principle several years earlier [1942, 86], Mostowski used uncountably many urelements to show that this principle does not imply the Axiom. Likewise, he sketched a proof that, for every $\mathfrak{m}$, $\mathbf{C}^{<\mathfrak{m}}$ does not yield $\mathbf{C}_2^{\mathfrak{m}}$.[14]

During 1939 Mostowski had begun to study $\mathbf{C}_n$ and to seek conditions on n and $M = \{m_1, m_2, \ldots, m_k\}$ such that:

(5.1.17) If $\mathbf{C}_m$ holds for every m in M, then $\mathbf{C}_n$ holds.

His results were of two sorts, necessary conditions for (5.1.17) and sufficient conditions for it, since he did not manage to prove some condition to be both necessary and sufficient. His demonstrations that certain conditions on n and M were necessary for (5.1.17) to hold were Fraenkel–Mostowski

[14] Mostowski 1948, 130, but see also the remarks of A. Levy [1965, 224]. The Principle of Dependent Choices was generalized by Levy [1964] to the following Principle of $\aleph_\alpha$-Dependent Choices: Let A be a non-empty set and let S be a binary relation such that, for every $\beta < \omega_\alpha$ and every β-sequence $s = \{a_\gamma : \gamma < \beta\}$ of elements of A, there is some b in A such that sSb; then there is a function $f\colon \omega_\alpha \to A$ such that for every $\beta < \omega_\alpha$, $(f \mid \beta) S f(\beta)$. Here the Principle of $\aleph_0$-Dependent Choices was, in effect, the usual Principle of Dependent Choices.

independence proofs. After obtaining the necessity of (5.1.18) in this way, he conjectured that it was also sufficient:

(5.1.18) For any decomposition of n into a sum of primes

$$n = p_1 + p_2 + \cdots + p_s,$$

there are some $q_1, q_2, \ldots, q_s$ in $\mathbb{N}$ such that

$$p_1 q_1 + p_2 q_2 + \cdots + p_s q_s$$

is in M. [1945, 160]

However, he succeeded in proving (5.1.18) to be equivalent to (5.1.17) only when n was prime or less than 15 or when M was $\{1, 2, \ldots, k\}$ for some k. Analogously, he obtained the result that, for $m < n$, $\mathbf{C}_{\leq m}$ implies $\mathbf{C}_{\leq n}$ if and only if there is no prime between m and n. As particular instances of his research, he showed that $\mathbf{C}_n$ follows from $\mathbf{C}_{kn}$ for any positive k and that $\mathbf{C}_2$ implies $\mathbf{C}_4$ but does not imply $\mathbf{C}_3$ [1945, 138]. The search for conditions equivalent to (5.1.17) was continued by Szmielew [1947] and Sierpiński [1955]. Only in 1970 did Robert Gauntt solve the problem by showing that, while Mostowski's condition (5.1.18) was not sufficient, his sufficient condition (D)—on certain subgroups of permutation groups of $\{1, 2, \ldots, n\}$—was necessary and hence equivalent to (5.1.17).[15]

During the 1950s three mathematicians independently modified the Fraenkel–Mostowski method: Elliott Mendelson, Ernst Specker, and Joseph Shoenfield. The essence of their modification consisted in dispensing with urelements and relying instead on sets which began infinite descending $\in$-sequences. Thus the Axiom of Foundation was violated, but the rest of the Zermelo–Fraenkel or Gödel–Bernays axioms were satisfied. In this fashion Shoenfield established without urelements that the Axiom of Choice cannot be proved from the Ordering Principle in Gödel's system minus the Axiom of Foundation [1955]. Likewise, Mendelson demonstrated without urelements that $\mathbf{C}_2$ was not a theorem of Gödel's system minus Foundation [1956 and 1958]. In a paper written in 1951 but only published six years later, Specker gave a similar argument for the independence of the Denumerable Axiom [1957, 193].

Influenced by Church [1927], Specker also considered a number of possible alternatives to the Axiom of Choice (see 4.7). In addition to Church's three alternatives, which concerned Cantor's second number-class, Specker proposed two others. The first of these was the negation of Sierpiński's proposition (4.1.14):

(5.1.19) $\mathbb{R}$ is a countable union of countable sets.

[15] See Mostowski [1945, 148–149], Gauntt [1970; 1970a], and Truss [1973, 147]. On related questions, see Zuckerman [1969; 1969a].

The second was a generalization of the first:

(5.1.20) The power set of any set with cardinality $\mathfrak{m}$ is the union of countably many sets of cardinality $\mathfrak{m}$.

In this context Specker showed that the consistency of his second alternative implies the existence of an inaccessible cardinal.[16] Thus he suggested a theme—large cardinals—that became important for a later alternative to the Axiom, namely, the Axiom of Determinateness (see 5.2). In fact, the structure of inaccessible, measurable, and other large cardinals depended heavily on whether one accepted the Axiom of Choice or some alternative assumption.

Ironically, the deepest results produced by the Fraenkel–Mostowski method were discovered shortly before forcing superseded it: the independence of the Axiom of Choice from the Boolean Prime Ideal Theorem [Halpern 1961 and 1964]; the independence of $\mathbf{C}_{<\aleph_0}$ from the assumption that for every n, $\mathbf{C}_n$ [Levy 1962]; the independence of the Ordering Principle from $\mathbf{C}_{<\aleph_0}$ [Läuchli 1964]; the independence of $\mathbf{C}_{\aleph_\alpha}$ from $\mathbf{C}_{<\aleph_\alpha}$ if α is a limit ordinal, and of $\mathbf{C}_{\aleph_{\alpha+1}}$ from $\mathbf{C}_{<\aleph_\alpha}$ [Levy 1965]. Perhaps the most striking achievement of this sort was Läuchli's research showing that many results about groups, fields, vector spaces, and topological spaces could not be deduced in set theory without the Axiom [1962]. Among them were the following propositions: The commutator subgroup of a free group is free; every field has an algebraic closure; every real field has a real closure; every vector space has a basis; all the bases of a vector space have the same cardinality; and Urysohn's Lemma.

All the same, neither the Fraenkel–Mostowski method nor the modified version of it due to Specker, Shoenfield, and Mendelson yielded independence proofs about $\mathbb{R}$, or about sets of real functions, or about the Continuum Hypothesis in particular. By remedying this defect, Cohen's method gave rise to an immense body of new research.

5.2 Cohen's Legacy

Paul Cohen was not trained as a set theorist or a logician but as an analyst. Yet he made a conceptual breakthrough that had eluded those most thoroughly schooled in both set theory and mathematical logic. Thereby he invented a technique, called forcing and generic sets, which enabled him to

---

[16] Specker 1957, 174, 199–210; *cf.* Gitik 1980. Also, Specker [1953] showed that the Axiom of Choice is inconsistent with Quine's system "New Foundations" [1937]. For a Fraenkel–Mostowski independence result about the Axiom and Suslin's Hypothesis, see Yesenin-Volpin 1954.

establish the independence of the Axiom of Choice and of the Continuum Hypothesis in **ZF**. This technique proved to have very wide applicability.

The two problems that Cohen attacked differed in the degree to which they had previously been solved. On the one hand, Fraenkel and Mostowski had settled the independence to the Axiom of Choice in set theory with urelements. During the 1950s Mendelson, Shoenfield, and Specker had proved the Axiom's independence relative to a set theory which was free of urelements but which did not satisfy the Axiom of Foundation. Consequently, the independence of the Axiom of Choice in **ZF** itself had not yet been established. Nevertheless, it would have astonished set theorists if someone had shown that the Axiom of Choice was really a theorem of **ZF**. On the other hand, the status of the Continuum Problem was less clear. Skolem had already suspected in 1923 that Zermelo's system did not settle this problem. In 1947 Gödel had conjectured that the Continuum Hypothesis was also independent of his own system for set theory.[1] All the same, no real progress had been made on this question since 1938 when Gödel used the constructible sets to establish that the negation of the Axiom and the negation of the Continuum Hypothesis are both independent of his system.

One intriguing question was whether the independence of the Axiom of Choice and the Generalized Continuum Hypothesis could be shown by using inner models. In particular, if one assumed the consistency of **ZF**, one obtained some model $\mathfrak{A}$ of **ZF** and could hope to find a submodel of $\mathfrak{A}$ that violated some or all of the Axiom of Choice, the Generalized Continuum Hypothesis, and the Axiom of Constructibility. This avenue was not promising, however, since in 1953 J. C. Shepherdson of the University of Bristol proved that no transitive model of Gödel's system could satisfy the negation of the Axiom of Constructibility.[2] *A fortiori*, since the Axiom of Constructibility implied both the Generalized Continuum Hypothesis and the Axiom of Choice, no transitive model of **ZF** could violate any of these three assumptions.

Unaware of Shepherdson's result, Cohen arrived at a similar conclusion in an article establishing the existence of a minimal model for set theory [1963]. Indeed, Cohen found that no first-order formula $R(x)$ of **ZF** can be shown in **ZF** to define a model of both **ZF** and the negation of the Axiom of Constructibility. This result, he observed, indicated that:

> In some sense the problem of independence of the Axiom of Choice is more difficult than that of the consistency, since in the latter case the relation $R(x)$ was taken to be that of constructible sets and the proof that in this collection of sets the Axiom of Choice held required only **Z–F** theory. Of course, [the result] does not preclude the construction of such a model by means of operations going beyond **Z–F** theory, *e.g.* based on an enumeration of statements in **Z–F** theory and the truth or falsity thereof. [1963, 539–540]

[1] Skolem 1923, 229, and Gödel 1947, 522–523.

[2] Shepherdson 1953, 164.

Less than a month after this paper was submitted on 22 March 1963, Cohen discovered the essentials of forcing while using just such an enumeration.

Cohen circulated a preprint, entitled "The Independence of the Axiom of Choice," which contained his solution to the major outstanding problems. By assuming the existence of a countable standard transitive model of **ZF** and the Axiom of Constructibility,[3] he deduced that there exist models of **ZF** such that:

(1) The Axiom of Choice and the Generalized Continuum Hypothesis hold, but some subset of $\mathbb{N}$ is not a constructible set.
(2) $\mathbb{R}$ cannot be well-ordered.
(3) The Axiom of Choice holds but the Continuum Hypothesis fails.
(4) The Denumerable Axiom fails for some family of pairs of real functions.

Fundamental to these proofs was his definition of forcing, which ensured that in a certain definite sense the model would have only those properties that it was "forced" to have. He also pointed out that his proofs could be restated in first-order number theory to show that the consistency of **ZF** implies the consistency of both **ZF** and (i), where (i) was any of (1), (2), (3), and (4).

At the time, Cohen published only his result establishing the independence of the Continuum Hypothesis from **ZFC**. This model was sketched in two articles [Cohen 1963b and 1964] that Gödel submitted to the *Proceedings of the National Academy of Sciences*, where his own results on constructible sets had first appeared a quarter century earlier.[4]

The singular importance of Cohen's work lies not only in the fact that he established the independence of the Axiom of Choice from **ZF**, and of the Continuum Hypothesis from **ZFC**, but, above all, that his method of forcing permitted the relative deductive strength of many propositions to be determined rigorously. During the years since Cohen's discovery, an immense number of articles have been written which apply and develop the technique of forcing to obtain independence results.[5]

Even before Cohen's discovery reached print, several mathematicians associated with Stanford University and with the University of California at Berkeley—particularly Solomon Feferman and Azriel Levy, as well as Robert Solovay of Princeton—applied the method of forcing to a variety of problems. Many of their findings have been cited earlier in these pages. At

[3] With the later approach of Boolean-valued models, it was not necessary to begin with a countable model of **ZF** and the Axiom of Constructibility.

[4] Cohen published the remaining results in a book [1966] based on lectures that he gave at Harvard during the spring of 1965.

[5] Many of these results are discussed by Felgner [1971], Jech [1973; 1978], and Kunen [1980].

long last the intuitions of Russell, Zermelo, and others, as to which propositions require the Axiom of Choice, could be conclusively established. No longer was it necessary to say, as Sierpinski had done, only that all *known* proofs of a certain proposition used the Axiom. In particular, some of these results obtained by forcing bore on the earlier controversy about the Axiom. Thus, Levy showed that if **ZF** is consistent, then it is consistent to assume **ZF**, the Axiom, and the proposition that there is no set-theoretically definable well-ordering of the real numbers [1963]. By supposing the existence of a model of **ZF** which contains an inaccessible cardinal, Solovay obtained a model of **ZF**, the Principle of Dependent Choices, and the proposition that every subset of $\mathbb{R}$ is Lebesgue-measurable, has the Baire property, and, if uncountable, has a perfect subset. From the same assumption Solovay showed that, in a model of **ZFC** due to Levy, the Generalized Continuum Hypothesis holds and every *projective* subset of $\mathbb{R}$ is Lebesgue-measurable, has the Baire property, and, if uncountable, has a perfect subset.[6]

What was the effect of Cohen's discovery of forcing on the independence results previously obtained by the Fraenkel–Mostowski method? The usefulness of the earlier results was greatly enhanced in 1966 when Thomas Jech and Antonin Sochor demonstrated the Embedding Theorem, which in many cases permitted a Cohen-style model of **ZF** and a proposition *P* to be obtained from a Fraenkel–Mostowski model of **ZFU** and *P*.[7] Jech and Sochor belonged to a group of Czechoslovakian set theorists working at Prague on independence results. Prominent in this group was Petr Vopěnka, who had invented a technique related to Cohen's: ∇-models.[8] At Berkeley, independently of Vopěnka's research, Scott and Solovay developed an equivalent technique known as the method of Boolean-valued models. Unfortunately, their joint paper on the subject remained unpublished and became part of the mathematical folklore.[9] Another version of forcing, based on partial orders, was largely due to Shoenfield [1971].

In the wake of Cohen's discovery, it is not surprising that a new proof was found for the independence of the Axiom of Choice, as well as for its

[6] Solovay 1965 and 1970; *cf.* 1964. In his preprint, Cohen had suggested the problem of establishing the consistency of **ZF** with the proposition that every subset of $\mathbb{R}$ is measurable, and had later discussed the problem with Solovay. In 1964 or 1965 Solovay also showed that if **ZF** is consistent, then there is a model $\mathfrak{M}$ of **ZF** in which every subset of $\mathbb{R}$ is measurable, has the Baire property, and, if uncountable, has a perfect subset, but in which $\aleph_1$ is singular. He did not publish this result since he believed that the proper way to formulate the problem was by requiring the Principle of Dependent Choices to hold as well. It remains uncertain whether he knew then that Lebesgue measure is countably additive in the model $\mathfrak{M}$, as in fact it is (personal communication from Solovay, 13 March 1982).

[7] Jech and Sochor 1966. This theorem was extended by David Pincus [1972].

[8] See, in particular, Vopěnka 1967.

[9] Scott later wrote the preface to a book [Bell 1978] that superseded the unpublished paper. An earlier book on the Boolean-valued models of Scott and Solovay was written by Rosser [1969].

relative consistency. Around 1971 Kenneth Kunen came upon an independence proof that involved infinitary logic. Strengthening the constructible hierarchy, he let

$$M = \bigcup \{M_\alpha : \alpha \text{ is an ordinal}\},$$

where

$$M_0 = 0, \quad M_\beta = \bigcup_{\alpha < \beta} M_\alpha \quad \text{if } \beta \text{ is a limit ordinal,}$$

and $M_{\alpha+1}$ is the set of all those subsets of M_α definable from a countable sequence of elements of M_α by a formula with countably many first-order quantifiers and countable conjunctions. Then M was a model of **ZF**, but it could not be proved or refuted in **ZF** that M satisfies the Axiom of Choice. Assuming both **ZFC** and the existence of at least $\aleph_1$ measurable cardinals, Kunen deduced that the Axiom is false in M [1973].

The new proof for the relative consistency of the Axiom was likewise based on definability. A set A was said to be ordinal-definable if it was definable in first-order logic from finitely many ordinals, while A was called hereditarily ordinal-definable if the transitive closure of A was ordinal-definable. John Myhill and Dana Scott demonstrated that the class of all hereditarily ordinal-definable sets is an inner model of **ZF** satisfying the Axiom of Choice, indeed the largest inner model of **ZF** with the property that every non-empty class contains a definable element.[10]

There were limitations to the results that could be shown to be independent of **ZF** or consistent with it. In particular, it was impossible to establish the consistency relative to **ZF** of an axiom positing the existence of a large cardinal. Letting **IN** be the assumption that there exists a strongly inaccessible cardinal, one can deduce in **ZF** + **IN** that **ZF** is consistent. If one could prove in **ZF** that the consistency of **ZF** implies the consistency of **ZF** + **IN**, then it would follow that the consistency of **ZF** + **IN** can be shown in **ZF** + **IN**. But this would contradict Gödel's Second Incompleteness Theorem. On the other hand, for sufficiently simple statements of number theory, Shoenfield had established in 1961 that the corresponding sets of natural numbers are constructible. In particular, for Σ_2^1 or Π_2^1 predicates one could not violate the Axiom of Constructibility, much less the Axiom of Choice.[11] Later Shoenfield's result was extended by Richard

[10] Myhill and Scott 1971, 274. This property is reminiscent of Richard's 1907 attempt to deduce a restricted version of the Axiom; see 2.4. In 1946 Gödel introduced the notion of ordinal-definability while giving a lecture at Princeton. There he conjectured that this notion would permit an easier proof of the relative consistency of the Axiom than did the constructible sets; his lecture was first published in Davis 1965, 84–88.

[11] Shoenfield 1961, 135. A predicate whose arguments are natural numbers is Σ_n^1 (respectively, Π_n^1) if it can be written as a string of n function quantifiers, beginning with an existential (respectively, universal) quantifier and alternating, followed by some string of number quantifiers, then a quantifier-free recursive predicate.

Platek, who demonstrated that if a Π_4^1 predicate was deduced in **ZF** by using the Axiom of Choice, then it could be deduced in **ZF** alone [1969, 220]. Platek's result could not be strengthened any further, however, since Feferman and Levy had shown in the interim that there exists a Σ_4^1 predicate which is independent of **ZF** and which expresses a special case of the Axiom.[12]

Prior to 1962, even the Axiom's critics had not proposed a serious alternative to it. Although Church in 1927 and Specker in 1951 had considered a few alternatives, no one had gone beyond their preliminary investigations. Yet a decade later two Polish mathematicians, Jan Mycielski and Hugo Steinhaus, introduced such an alternative: the Axiom of Determinateness.[13] This axiom **AD** grew out of the theory of infinite positional games. It asserted that if S is a set of denumerable sequences of zeros and ones, and if two players alternately choose terms (each zero or one) and thereby form an infinite sequence, then the game is determined. That is, there is a strategy which will always ensure that the sequence is in S, or ensure that it is not in S. Steinhaus had already conjectured a very strong version of **AD** during 1925, but Banach and Mazur had used the Axiom of Choice to construct a game refuting his conjecture.[14] In 1964 Mycielski argued for **AD** partly on the grounds that it did not permit certain "unpleasant" consequences of the Axiom of Choice, such as the Banach–Tarski paradox. As he then showed, **AD** implies that every set of real numbers is Lebesgue-measurable. On the other hand, **AD** implies certain agreeable consequences of the Axiom of Choice such as proposition (3.6.3): the equivalence of continuity and sequential continuity for any real function. More generally, the Denumerable Axiom restricted to subsets of $\mathbb{R}$ follows from **AD**.[15] Nevertheless, Mycielski continued, the consistency of **AD** with respect to **ZF** was quite problematical.[16]

Recent research on **AD** involves large cardinals, partition properties, and descriptive set theory. The Axiom of Choice implies that a measurable cardinal must be very large, whereas Solovay deduced from **AD** that $\aleph_1$ is measurable. Similarly, by using the Axiom of Choice one can show that certain infinite partition relations are false, but their truth follows from **AD**.[17] As a result, interest has arisen in a restricted form of **AD**, which is not known

---

[12] Feferman and Levy 1963, 593.

[13] Mycielski and Steinhaus 1962, 1–3. Later it was often called the Axiom of Determinacy.

[14] Steinhaus 1965, 464–465. This original version of **AD** was self-contradictory; see Mycielski 1964, 218.

[15] Mycielski [1964, 219] also demonstrated that **AD** yields the regularity of $\aleph_1$, the countable additivity of Lebesgue measure, and the proposition that a countable union of sets of first category in $\mathbb{R}^n$ is of first category. Solovay observed that the Countable Union Theorem for subsets of $\mathbb{R}^n$ also follows from **AD**; see Mycielski 1971, 265.

[16] Mycielski 1964, 205–207; see also footnote 19 below. Later mathematicians, when using **AD**, have tended to assume the Principle of Dependent Choices as well. This appears to reflect Solovay's influence.

[17] See, for example, Jech 1973, 179–182, and Kleinberg 1977.

to contradict the Axiom of Choice: projective determinacy. In 1975 D. A. Martin established that Borel determinacy (the restriction of **AD** to the Borel sets of the Baire space of irrational numbers) can be deduced from the Axiom of Choice in **ZF** [1975]. The analogous but stronger assumption of projective determinacy (the restriction of **AD** to the projective sets) appeals to a number of mathematicians as a large cardinal axiom apparently consistent with the Axiom of Choice.[18] Finally, models of **ZF** + **AD** have been used to obtain an inner model of **ZF**, the Axiom of Choice, and various propositions P.[19]

[18] See Jech 1978, 559–563, and Moschovakis 1980, 9, 604–611.

[19] Fenstad [1971, 54] outlined an unpublished result of Solovay's in which P asserted the existence of a measurable cardinal.

Conclusion

The history of the Axiom of Choice is the history of how an assumption's status can change. At a given point in time, each assumption in mathematics forms part of a nexus of suppositions with varying degrees of explicitness. Throughout history, mathematicians have operated within conceptual frameworks in which certain assumptions were stated but in which, on the other hand, certain assumptions were tacit or even unconscious. Euclid, among others, presupposed properties of the continuum that were not recognized explicitly as necessary for geometry until the nineteenth century. During the late nineteenth century, Frege emphasized the need to make *all* one's assumptions explicit, and Hilbert's axiomatization of geometry furthered this process by using a *formal* axiomatic method as distinct from that of Euclid. Zermelo's Axiom of Choice is best viewed as a further attempt to formulate an implicit assumption explicitly.

At once there is a caveat to be added. There are many possible conceptual frameworks, but no assurance that the historical development had to take the form that it did. Yet, given an attempt to render mathematical assumptions explicit in so far as possible and given the recognition that infinite sets had to be studied in order for mathematics to develop, the emergence of the Axiom or one of its equivalents was all but inevitable.

In 1908 Zermelo argued that the Axiom had been applied implicitly by many researchers in diverse branches of mathematics before he stated it explicitly in 1904. His observation was essentially correct. In this book we have traced the prior evolution of arbitrary selections from the use of a finite number of choices, to an infinite number made by a stated or unstated rule, and finally to an infinite number of arbitrary choices for which no rule was possible.

The controversy generated by Zermelo's proof of the Well-Ordering Theorem had many sources, but two deserve special mention. For the German followers of Cantor, the proof was flawed by Burali-Forti's paradox. Zermelo ably circumvented this criticism by rejecting the assertion that the collection W of all ordinals is a set. However, a second and more troublesome source was found in the criticisms of Baire, Borel, Lebesgue, Peano, and Russell: The Axiom did not provide a *rule* by which to carry out the choices. Thus, if one could establish that there exists a mathematical object with a given property P only by defining a particular such object, then the Axiom was false. In this spirit many mathematicians rejected the Axiom. On the other hand, from the constructivist viewpoint espoused by Richard, the Axiom was true precisely because the notion of set was restricted to that of containing a definable element. The difficulties sensed by Baire, Borel, and Lebesgue stemmed from an attempt to blend an increasingly abstract form of analysis, as embodied in their own researches, with a constructivist philosophy of mathematics. Sierpiński pointed out just how arduous it would be to take such an approach consistently. Indeed, Borel and Lebesgue used the Axiom implicitly again and again, even after they had opposed it vigorously. Much of their research, including the theory of Borel sets and that of Lebesgue measure, would have collapsed without the Denumerable Axiom. It was not only the subtlety of the Axiom's forms that caught Borel and Lebesgue, but their own ambivalence toward the Axiom, toward Cantorian set theory, and even toward the general notion of real function.

The history of the Axiom of Choice lies at the crossroads where mathematics and philosophy meet. Repeatedly, mathematicians such as Borel (who had a low opinion of philosophers) were compelled to make quasi-philosophical judgments about the nature of mathematics. Three decades after the controversy began, Lebesgue continued to insist that the philosophy of mathematics must be created by mathematicians, not by philosophers. Yet one of the most regrettable aspects of the controversy was that, after the initial debate, mathematicians did not significantly deepen their philosophies of mathematics. The immense body of mathematical results concerning the Axiom's deductive strength and its relationship to other propositions—developed chiefly by the Warsaw school—did not lead to philosophical insights. In fact, the principal distinction had already been stated by Hadamard in 1905: The Axiom posits the *existence* of a choice function, not its *construction*. If one wished to restrict existence to constructions, then there was little more to be said and no dispute need arise. As a result, Brouwer and his followers took little part in the ensuing discussion. Only those who adopted a more ambivalent position toward modern mathematics and toward its constructivistic restrictions, such as Borel and Lebesgue, felt a need to continue the debate.

The Axiom epitomizes a fundamental transformation that took place in the late nineteenth and early twentieth centuries. The use of the actual infinite throughout mathematics, a by-product of set theory, made it evident that existence and construction were vastly different mathematical concepts. As Lebesgue and Hadamard both recognized early in the controversy over the Axiom, two diametrically opposed views of mathematics were at stake. The constructivists' restriction of mathematics (perhaps because it became an anomaly) acquired a name, whereas the viewpoint espoused by Hadamard and Zermelo became the established position and so found only the name: modern mathematics.

A different transformation deserves emphasis as well. The controversy over his proof of 1904, and particularly over the Axiom of Choice, led Zermelo to axiomatize set theory. This was the first step in treating set theory as a formal system. In time, it became necessary to specify the logic underlying this system as well. During the decades that followed Zermelo's axiomatization, the development of set theory came more and more to involve *models* of set theory with respect to *first-order* logic. In other words, there occurred a pronounced shift toward the first-order metamathematical and semantic study of set theory—culminating, with Cohen's results, in a fragmentation of the very foundations of set theory. The plaintive aside of Dana Scott, quoted at the beginning of the Epilogue, echoes the qualms of many mathematicians past and present: The Axiom of Choice is surely necessary, but if only there were some way to make it self-evident as well ....

Appendix 1

Five Letters on Set Theory

These letters are our translation of Baire *et alii* 1905, printed here with the kind permission of the *Société Mathématique de France*. They are discussed in 2.3. The original pagination is indicated in the margin and by two oblique lines in the text.

I. *Letter from Hadamard to Borel* 261

I have read with interest the arguments that you put forward (second issue of volume LX of *Mathematische Annalen*) against Zermelo's proof, found in the previous volume. However, I do not share your opinion on this matter. I do not agree, first of all, with the comparison that you make between the fact which Zermelo uses as his starting point and an argument which would enumerate the elements of the set one after another *transfinitely*. Indeed, there is a fundamental difference between the two cases: The latter argument requires a sequence of successive choices, *each of which depends on those made previously*; this is the reason why it is inadmissible to apply it transfinitely. I do not see how any analogy can be drawn, from the point of view which concerns us, between the choices in question and those used by Zermelo, which are *independent of each other*.

Moreover, you take exception to his procedure for a *non-denumerable* infinity of choices. But, for my part, I see no difference in this regard between the case of a non-denumerable infinity and that of a denumerable infinity. The difference//would be evident if the choices in question depended on 262 each other in some way, because then it would be necessary to consider the order in which one made them. To me the difference appears, once again, to vanish completely in the case of independent choices.

What is certain is that Zermelo provides no method to carry out *effectively* the operation which he mentions, and it remains doubtful that anyone will be able to supply such a method in the future. Undoubtedly, it would have been more interesting to resolve the problem in this manner. But the question posed in this way (the effective determination of the desired correspondence) is nonetheless completely distinct from the one that we are examining (does such a correspondence exist?). Between them lies all the difference, and it is fundamental, separating what Tannery[1] calls a *correspondence* that can be *defined* from a correspondence that can be *described.* Several important mathematical questions would completely change their meaning, and their solutions, if the first word were replaced by the second. You use correspondences, whose *existence* you establish without being able to *describe* them, in your important argument about [complex] series which can be continued along an arc across their circle of convergence. If only those entire series were considered whose law of formation can be described, the earlier view (*i.e.*, that entire series which can be continued along an arc across their circle of convergence are the exception) ought, in my opinion, to be regarded as the true one. Furthermore, this is purely a matter of taste since the notion of a correspondence "which can be described" is, to borrow your expression, "outside mathematics." It belongs to the field of psychology and concerns a property of our minds. To discover whether the correspondence used by Zermelo can be specified *in fact* is a question of this sort.

To render the existence of this correspondence possible, it appears sufficient to take *one* element from any given set, just as the following proposition A suffices for B:

263 A. *Given a number x, there exists a number y which is not//a value obtained from x in any algebraic equation with integer coefficients.*

B. *There exists a function y of x such that, for every x, y is not an algebraic number and is not a value obtained from x in any algebraic equation with integer coefficients.*

Undoubtedly, one can form such functions. But what I claim is that this fact is in no way necessary in order to assert the correctness of theorem B. I believe that many mathematicians would not take any more trouble than I do to verify this fact before using the theorem in question.

J. Hadamard

[1] *Revue générale des Sciences*, vol. VIII, 1897, p. 133ff.

II. *Letter from Baire to Hadamard*

Borel has communicated to me the letter in which you express your viewpoint in the great debate resulting from Zermelo's note. I beg your indulgence in presenting some thoughts that it suggested to me.

As you know, I share Borel's opinion in general, and if I depart from it, it is to go further than he does.

Let us suppose that one tries to apply Zermelo's method to the set M of sequences of positive integers. One takes from M a distinguished element m_1; there remains the set $M-\{m_1\}$, from which one takes a distinguished element m_2; and so on. Each of these successive choices, indeed, depends on those that precede it. But, so you say along with Zermelo, the choices are independent of each other because he permits as a starting point *some choice of a distinguished element in EVERY subset of M*. I do not find this satisfactory. To me it conceals the difficulty *by immersing it in a still greater difficulty.*

The expression *a given set* is used continually. Does it make sense? Not always, in my opinion. As soon as one speaks of the infinite (even the denumerable, and it is here that I am tempted to be more radical than Borel), the comparison, *conscious or unconscious*, with a bag of marbles passed from hand to hand must disappear completely. We are then, I believe, in the realm of *potentiality* [dans le *virtuel*].//That is to say, we establish conven- 264 tions that ultimately permit us, when an object is defined *by a new convention*, to assert certain properties of this object. But to hold that one can go farther than this does not seem legitimate to me. In particular, when a set is given (we agree to say, for example, that we are given the set of sequences of positive integers), *I consider it false to regard the subsets of this set as given.* I refuse, *a fortiori*, to attach any meaning to the act of supposing that a choice has been made in every subset of a set.

Zermelo says: "Let us suppose that to each subset of M there corresponds one of its elements." This supposition is, I grant, in no way contradictory. Hence all that it proves, as far as I am concerned, is that we do not perceive a contradiction in supposing that, in each set which is defined for us, the elements are positionally related to each other in exactly the same way as the elements of a well-ordered set. In order to say, then, that one has established that every set can be put in the form of a well-ordered set, the meaning of these words must be extended in an extraordinary way and, I would add, a fallacious one.

In the preceding paragraphs I have only managed to express my thinking very incompletely. I stated my viewpoint in the letter that Borel cited in his note. For me, progress in this matter would consist in delimiting the domain of the definable. And, despite appearances, in the last analysis everything must be reduced to the finite.

R. BAIRE

III. *Letter from Lebesgue to Borel*

You ask for my opinion about Zermelo's note (*Math. Annalen*, vol. LIX), about your objections to it (*Math. Annalen*, vol. LX), and about the letter from Hadamard that you communicated to me. Here is my reply. Forgive me for being so lengthy; I have tried to be clear.

First of all, I agree with you on the following point: Zermelo has very cleverly shown that we know how to resolve problem A:

265 // A. *To put a set M in the form of a well-ordered set,*

whenever we know how to resolve problem B:

B. *To assign to each set M′, formed from elements of M, a particular element m′ of M′.*

Unfortunately, problem B is not easy to resolve, so it seems, except for the sets that we know how to well-order. As a result, we do not have a general solution to problem A.

I strongly doubt that a general solution can be given to this problem, at least if one accepts (as Cantor does) that to define a set *M* is to name a property *P* which is possessed by certain elements of a previously defined set *N* and which characterizes, by definition, the elements of *M*. In fact, with this definition, we know nothing about the elements of *M* other than this: They possess all the *unknown* properties of the elements of *N* and they are the only ones that possess the *unknown* property *P*. Nothing about this permits two elements of *M* to be distinguished from each other, still less to be arranged as they would need to be in order to resolve A.

This objection, made *a priori* to any attempt to resolve A, obviously disappears if we particularize *N* or *P*. The objection disappears, for example, if *N* is the set of numbers. In general, all that one can hope to do is to indicate certain problems, such as B, whose resolution would entail that of A and which are possible in certain particular cases that are rarely encountered. In my opinion, this is why Zermelo's argument is interesting.

I believe that Hadamard is more faithful than you are to Zermelo's thought when he interprets this author's note as an attempt, not to resolve A effectively, but to demonstrate the existence of a solution. The question comes down to this, which is hardly new: *Can one prove the existence of a mathematical object without defining it?*

This is obviously a matter of convention. Nevertheless, I believe that we can only build solidly *by granting that it is impossible to demonstrate the existence of an object without defining it.* From this perspective, closely related to Kronecker's and Drach's, there is nothing to distinguish between A and problem C:

C. *Can every set be well-ordered?*

266 //I would have nothing more to say if the convention that I mentioned were universally adopted. But I must admit that one often uses, and that I

myself have often used, the word *existence* in other senses. For example, when Cantor's well-known argument is interpreted as saying that *there exists a non-denumerable infinity of numbers*, no means is given to name such an infinity. It is only shown, as you have said before me, that whenever one has a denumerable infinity of numbers, one can define a number not belonging to this infinity. (Here the word *define* always means *to name a property characterizing what is defined.*) This sort of existence can be used in an argument in the following fashion: A property is true if negating it leads to the assertion that all numbers can be arranged in a denumerable sequence. I believe that this kind of existence can enter an argument only in such a fashion.

Zermelo utilizes the *existence* of a *correspondence* between the subsets of M and certain of their elements. You see, even if the existence of these correspondences were not questionable, due to the way in which their existence had been proved, it would not be self-evident that one had the right to use their existence in the way that Zermelo did.

I come to the argument that you state in the following way: "It is possible to choose *ad libitum* a distinguished element m' from a particular set M; since this choice can be made for each of the sets M', it can be made for the set of these sets." From this argument the existence of the correspondences seems to follow.

First of all, when M' is given, is it self-evident that one can choose m'? It would be self-evident if M' existed in the almost Kroneckerian sense that I mentioned earlier, since to say that M' exists would then be to assert that one knew how to name certain of its elements. But let us extend the meaning of the word *exist*. The set Γ of correspondences between the subsets M' and the distinguished elements m' certainly *exists* for Hadamard and Zermelo; the latter even represents the number of its elements by a transfinite product. However, do we know how to choose an element from Γ? Obviously not, since this would give a determinate solution to problem B in the case of M.

It is true that I use the word *to choose* in the sense of *to name* and that perhaps it suffices for Zermelo's argument that//*to choose* mean *to think of*. 267
Yet it must be noted all the same that what one is thinking of is not stated and that Zermelo's argument still requires one to think *always of the same determinate correspondence*. Hadamard believes, it seems to me, that it is not necessary to prove that one can *determine* a unique element. In my opinion, this is the source of our differences of judgment.

So as to convey more clearly the difficulty that I see, I remind you that in my thesis I proved the existence (in a sense that is not Kroneckerian and is perhaps difficult to make precise) of sets that were measurable but were not Borel-measurable. Nevertheless, I continued to doubt that any such set could ever be named. Under these conditions, would I have the right to base an argument on this hypothesis—*I assume as chosen a set that is measurable but not Borel-measurable*—even though I doubt that anyone could ever name one?

Thus I already see a difficulty with the assertion that "in a determinate M' I can choose a determinate m'," since there exist sets (the set C for example, which can be regarded as a set M' coming from a more general set) in which it is perhaps impossible to choose an element. Then there is the difficulty that you pointed out concerning the infinity of choices. As a result, if we wish to regard Zermelo's argument as completely general, it must be granted that we are speaking about an infinity of choices whose power may be very large. Moreover, no law is given either for this infinity or for the choices. We do not know if it is possible to name a rule defining a set of choices having the power of the set of the M'. We do not know if it is possible, given an M', to name an m'.

In sum, when I scrutinize Zermelo's argument, I find it, like a number of other general arguments about sets, too little Kroneckerian to have meaning (only as an existence theorem giving a solution to C, of course).

You allude to the following argument: "To well-order a set, it suffices to choose one element from it, then another, and so on." Certainly this argument presents enormous difficulties which are even greater, at least in appearance, than Zermelo's. And I am tempted to believe, as Hadamard does, that pro-
268 gress//has been made by replacing an infinity of successive choices, which depend on each other, with an unordered infinity of independent choices. Perhaps this is only an illusion. Perhaps the apparent simplification resides only in the fact that an ordered infinity of choices must be replaced by an unordered infinity, but one of higher power. Consequently, the fact that one can reduce to the single difficulty, placed at the beginning of Zermelo's argument, all the difficulties of the simplistic argument that you cited merely shows that this single difficulty is very great. In any case, it does not seem to me that the difficulty vanishes just because it concerns an unordered set of independent choices. For example, if I believe that there exist functions $y(x)$ such that, whatever x may be, y is never a value obtained from x in any algebraic equation with integer coefficients, this is because I believe, as does Hadamard, that it is possible to construct such a function. But in my opinion this does not follow immediately from the existence, whatever x may be, of numbers y which are not a value obtained from x in any equation with integer coefficients.[2]

I agree completely with Hadamard when he states that to speak of an infinity of choices without giving a rule presents a difficulty that is just as great whether or not the infinity is denumerable. When one says, as in the argument that you criticize, "since this choice can be made for each of the sets M', it can be made for the set of these sets," one says nothing unless the terms being used are explained. To make a choice can be to write down or to name the element chosen. To make an infinity of choices cannot be to write down or to name the elements chosen, one by one; life is too short. Hence, one must say what it means to make them. By this, we understand in general that a rule

[2] While correcting the proofs, I will add that in fact the argument by which we ordinarily justify Hadamard's statement A (p. 262) justifies at the same time statement B. And, in my opinion, it is because it justifies B that it justifies A.

is given which defines the elements chosen. For me as for Hadamard, this rule is equally indispensable whether or not the infinity is denumerable.

All the same, perhaps I still agree with you on this point since, although I find no theoretical difference between//the two kinds of infinity, from the practical point of view I distinguish strongly between them. When I hear of a rule defining a transfinite infinity of choices, I am very suspicious because I have never seen such a rule, whereas I know of rules defining a denumerable infinity of choices. Still, this is only a question of habit. Upon reflection, I see difficulties which, in my opinion, are sometimes just as great in arguments involving only a denumerable infinity of choices as in arguments involving a transfinite number. For example, if I do not regard the classical argument as establishing the proposition that every non-denumerable set contains a subset whose power is that of Cantor's second number-class, I do not grant any greater validity to the argument showing that a set which is not finite has a denumerable subset. Although I seriously doubt that a set will ever be named which is neither finite nor infinite, it has not been proved to my satisfaction that such a set is impossible. But I have already spoken to you about these questions. 269

H. LEBESGUE

IV. *Letter from Hadamard to Borel*

The question appears quite clear to me now, after Lebesgue's letter. More and more plainly, it comes down to the distinction, made in Tannery's article, between what is *determined* and what can be *described*.

In this matter Lebesgue, Baire, and you have adopted Kronecker's viewpoint, which until now I believed to be peculiar to him. You answer in the negative the question posed by Lebesgue (above, p. 265): Can one prove the existence of a mathematical object without defining it? I answer it in the affirmative. I take as my own, in other words, the answer that Lebesgue himself gave regarding the set Γ (p. 266).

I grant that it is impossible for *us*, at least at present, to *name* an element in this set. That is the issue for you; it is not the issue for me.

There is only one point, it seems to me, where Lebesgue is inconsistent with himself. That is when he does or//does not allow himself to use the existence of an object, according to the way in which its existence was proved. For me, the *existence* about which he speaks is a fact like any other, or else it does not occur. 270

As for Baire, the question takes the same form. I would prefer not to base it as he does (p. 264), following Hilbert, on the *non-contradictory*, which still seems to me to depend on psychology and to take into account the properties of our brains. I do not understand very well how Zermelo could have *proved* that *we do not perceive* a contradiction, *etc.* This cannot be *proved* but only *ascertained*: One perceived it or one did not perceive it.

Leaving this point aside, it is clear that the principal question, that of knowing if a set can be well-ordered, does not mean the same thing to Baire (any more than to you or Lebesgue) that it does to me. I would say rather—Is a well-ordering possible?—and not even—Can *one* well-order a set?—for fear of having to think who this *one* might be. Baire would say: Can *we* well-order it? An altogether subjective question, to my way of thinking.

Consequently, there are two conceptions of mathematics, two mentalities, in evidence. After all that has been said up to this point, I do not see any reason for changing mine. I do not mean to impose it. At the most, I shall note in its favor the arguments that I stated in the *Revue générale des Sciences* (30 March 1905), to wit:

1. I believe that in essence the debate is the same as the one which arose between Riemann and his predecessors over the notion of function. The *rule* that Lebesgue demands appears to me to resemble closely the analytic expression on which Riemann's adversaries insisted so strongly.[3] And even an analytic expression that is not too unusual. Not only does the *cardinality* of the choices fail to alter the question, but, it seems to me, their

271 *uniqueness* does not alter it either. I do not see//how we have the right to say, "For each value of x there exists a number satisfying .... Let y be this number ...," whereas, since "the bride is too beautiful," we cannot say, "For each value of x there exists an infinity of numbers satisfying .... Let y be one of the numbers ...."

2. Tannery's arbitrary choices lead to numbers v which *we would be* incapable of defining. I do not think that these numbers fail to exist.

As for the arguments proposed by Bernstein (*Math. Annalen*, vol. LX, p. 187) and, consequently, his objections to Zermelo's proof, I do not find them convincing. All the same, my opinion on this matter is independent of the question that we have been discussing.

Bernstein begins with Burali-Forti's paradox (*Circolo Matematico di Palermo*, 1897) concerning the set W of *all* ordinal numbers. To circumvent the contradiction obtained by Burali-Forti, he supposes the ordinal number of W to be such that it is impossible to add one to it. In my opinion this supposition, as well as the arguments that Bernstein adduces in its favor, is unacceptable. In Cantor's theory the order established between the elements of W and the additional element (it is this order which Bernstein attacks) is merely a *convention* that one is always free to make and that the properties of W, whatever they may be, cannot alter.

---

[3] I believe it necessary to reiterate this point, which, if I were to express myself fully, appears to form the essence of the debate. From the invention of the infinitesimal calculus to the present, it seems to me, the essential progress in mathematics has resulted from successively annexing notions which, for the Greeks or the Renaissance geometers or the predecessors of Riemann, were "outside mathematics" because it was impossible to describe them.

The solution is different. It is the very existence of the set W that leads to a contradiction. In the definition of W, the general definition of the word *set* is incorrectly applied. We have the right to form a set only from previously existing objects, and it is easily seen that the definition of W supposes the contrary.

Same observation for *the set of all sets* (Hilbert, Heidelberg Congress).

Let us return to the original question. I submit in this regard, not an argument (since I believe that we shall rest eternally on our respective positions) but a consequence of your principles.

Cantor considered the set of all those functions on the interval (0, 1) that assume only the values zero and one. To my mind this set has//a clear meaning and its power is $2^{\aleph}$, as Cantor stated.[4] Likewise, the set of all functions of x makes sense to me, and I see clearly that its power is $\aleph^{\aleph}$. 272

What meaning does all of this have for you? It appears obvious to me that it cannot have any. For on each function you impose an additional condition which has no mathematical meaning—that of being *describable to us.*

Or rather, this is what it means: From your point of view, one should consider only those functions definable in a finite number of words. But, for this reason, the two sets formed above are *countable* and, indeed, so is every other possible set.

J. HADAMARD

V. *Letter from Borel to Hadamard*

... First of all, I would like to call your attention to an interesting remark that Lebesgue made at the meeting of the Society on 4 May: How can Zermelo be certain that in the different parts of his argument he is always speaking of *the same* choice of distinguished elements, since he characterizes them in no way *for himself.* (Here it is not a question that someone may contradict him but rather of his being intelligible to himself.)

As for your new objection, here is my response.

I prefer not to write alephs. Nevertheless, I willingly state arguments equivalent to those which you mention, without many illusions about their intrinsic value, *but intending them to suggest other more serious arguments.* To give you a practical example, I refer to Note III which I inserted at the end of my recent little book (*Leçons sur les fonctions de variables réelles*, edited by Maurice Fréchet). The argument used there was obviously suggested by Cantor's argument, which I reported in my first *Leçons sur la théorie des fonctions*, page 107.[5]

---

[4] [(Translator) Here $\aleph$ is the power $2^{\aleph_0}$ of the continuum, which is assumed to be an aleph.]

[5] In Notes I and II of this little book, I continually used arguments of the sort that you deny me the right to make. So I was constantly filled with scruples, and each of these two Notes ended with a very restrictive remark.

273 //The form that I adopted in Note III is not absolutely satisfactory, as I indicated at the bottom of the last page of my book. But the analogous argument which Lebesgue gives in his article in the *Journal de Jordan* (1905) is, I believe, completely irreproachable, in the sense that it leads to a precise result expressible in a finite number of words. Nevertheless, it originated from that of Cantor.

One may wonder what is the real value of these arguments that I do not regard as absolutely valid but that still lead ultimately to effective results. In fact, it seems that if they were completely devoid of value, they could not lead to anything, since they would be meaningless collections of words. This, I believe, would be too harsh. They have a value analogous to certain theories in mathematical physics, through which we do not claim to express reality but rather to have a guide that aids us, by analogy, in predicting new phenomena, which must then be verified. It would require considerable research to learn what is the real and precise sense that can be attributed to arguments of this sort. Such research would be useless, or at least it would require more effort than it would be worth. How these overly abstract arguments are related to the concrete becomes clear when the need is felt.

I would agree with you that it is self-contradictory to speak of the set of all sets, for, by the argument from page 107 cited above, we can form a set whose power is still greater. But I believe that this contradiction arises because sets that are not really defined have been introduced.

E. BOREL

Appendix 2

Deductive Relations Concerning the Axiom of Choice

This appendix consists of eleven tables. The first four of them concern the Denumerable Axiom and certain of its consequences. Tables 4, 5, and 6—dealing with the Principle of Dependent Choices, the existence of a well-ordering for $\mathbb{R}$, and the Partition Principle—should be considered together. The Boolean Prime Ideal Theorem is the subject of Tables 7 and 8, while Table 9 lists equivalents to the Axiom of Choice. Finally, Tables 10 and 11 give various propositions implying the Axiom and some alternatives to it.

In these tables, $P \to Q$ and $P \leftrightarrow Q$ mean that P implies Q is a theorem of **ZF**, and that the equivalence of P and Q can be proved in **ZF**, respectively. On the other hand, $P \twoheadrightarrow Q$ and $P \nrightarrow Q$ mean that, in **ZF**, P implies Q but Q does not imply P and that, in **ZF**, P does not imply Q, respectively.

Table 1. Deductive Relations Concerning the Proposition (1.3.3) that Every Dedekind-Finite Set is Finite

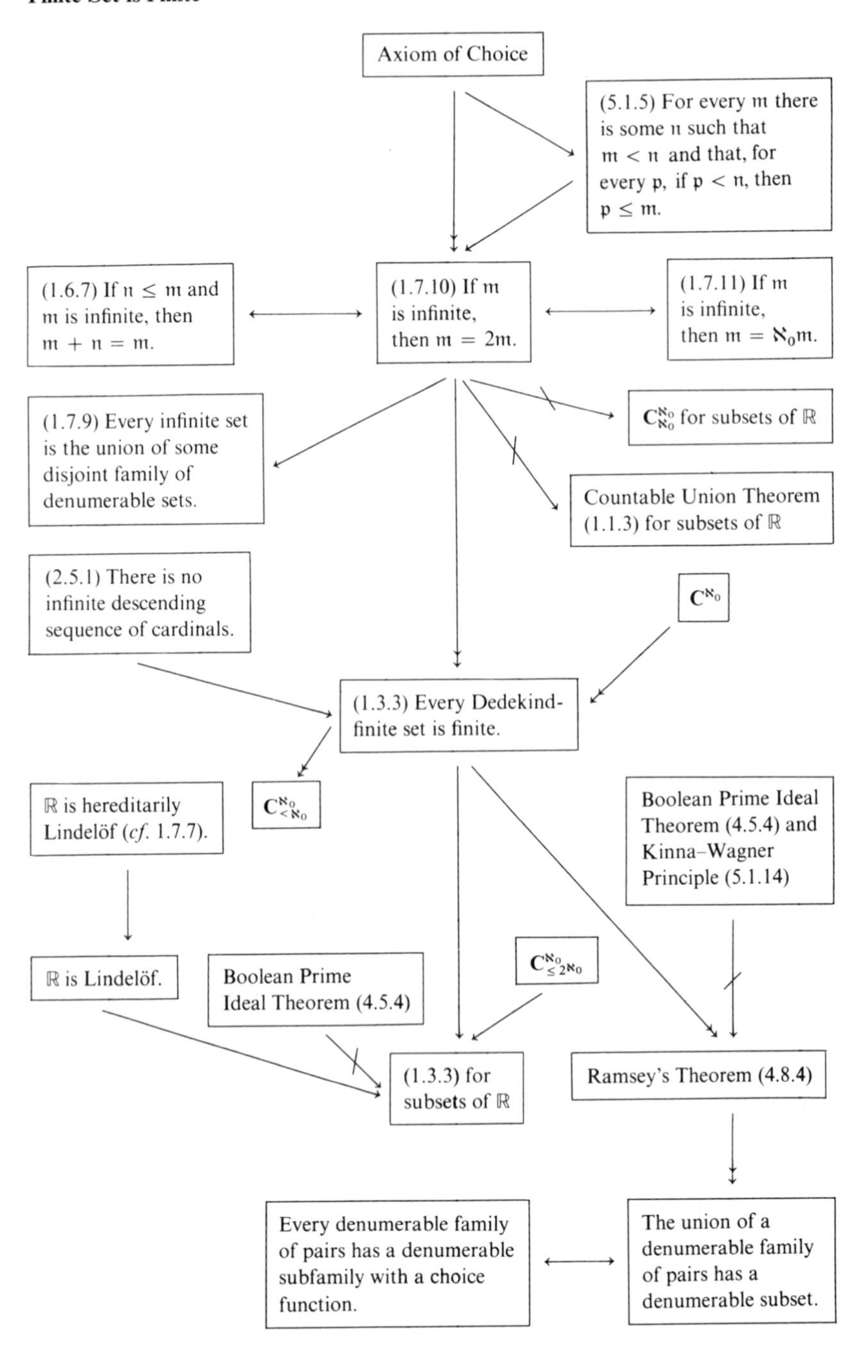

Table 2. Propositions Equivalent to (1.3.3)

(1.1.4)	Every infinite set has a denumerable subset.
(1.3.3)	Every Dedekind-finite set is finite.
(1.3.4)	If a set A is Dedekind-finite, then the power set of A is Dedekind-finite.
(1.3.5)	If A is Dedekind-finite and B is Dedekind-infinite, then $\overline{\overline{A}} \leq \overline{\overline{B}}$.
(1.3.6)	The union of a Dedekind-finite family of Dedekind-finite sets is Dedekind-finite.
(1.3.7)	If a class K contains the empty set and if $A \in K$ and $b \notin A$ imply $A \cup \{b\} \in K$, then K contains every Dedekind-finite set.
(2.7.3)	If A is a Dedekind-infinite family of non-empty sets, then the union of A is Dedekind-infinite.
(4.1.8)	If A is uncountable and B is denumerable, then $A—B$ has the same power as A.
(4.1.9)	If A is infinite and B is denumerable, then $A \cup B$ has the same power as A.
(4.1.10)	If $\aleph_0 < \overline{\overline{A}}$ and $\aleph_0 = \overline{\overline{B}}$, then $\aleph_0 < \overline{\overline{A-B}}$.
(4.3.15)	If $\aleph_0 \leq * \mathfrak{m}$, then $\aleph_0 \leq \mathfrak{m}$.
(4.3.16)	$\mathfrak{m} \leq \aleph_0$ or $\aleph_0 \leq \mathfrak{m}$.
(4.3.17)	If $\mathfrak{m} + \aleph_0 < \mathfrak{n} + \aleph_0$, then $\mathfrak{m} < \mathfrak{n}$.
(4.3.18)	If $\aleph_0 + \mathfrak{p} = \aleph_0 + \mathfrak{q}$, then $\mathfrak{p} = \mathfrak{q}$ or $\mathfrak{p}, \mathfrak{q} \leq \aleph_0$.

Table 3. Deductive Relations Concerning the Countable Union Theorem (1.1.3)

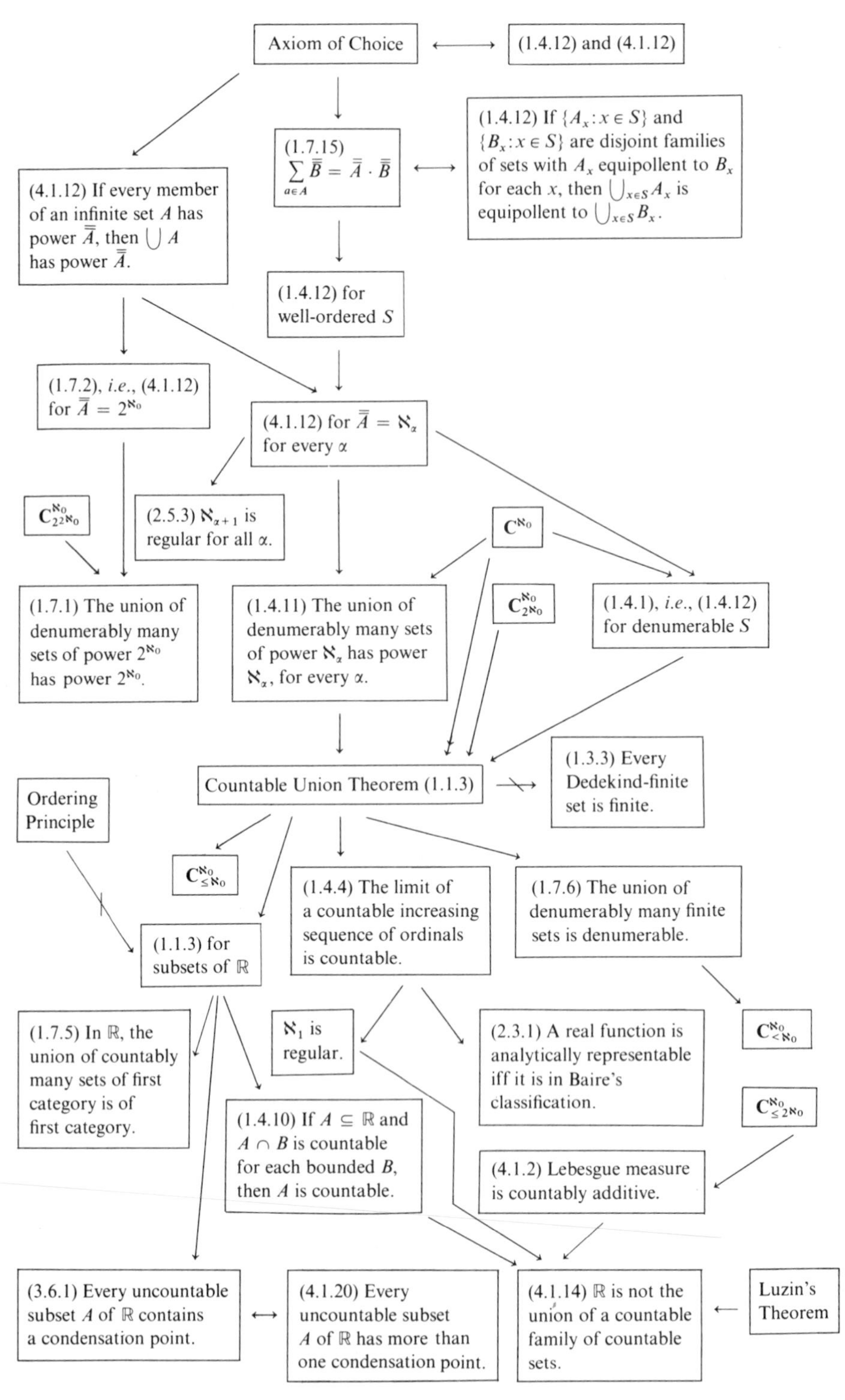

Table 4. Deductive Relations Concerning the Denumerable Axiom of Choice and the Principle of Dependent Choices

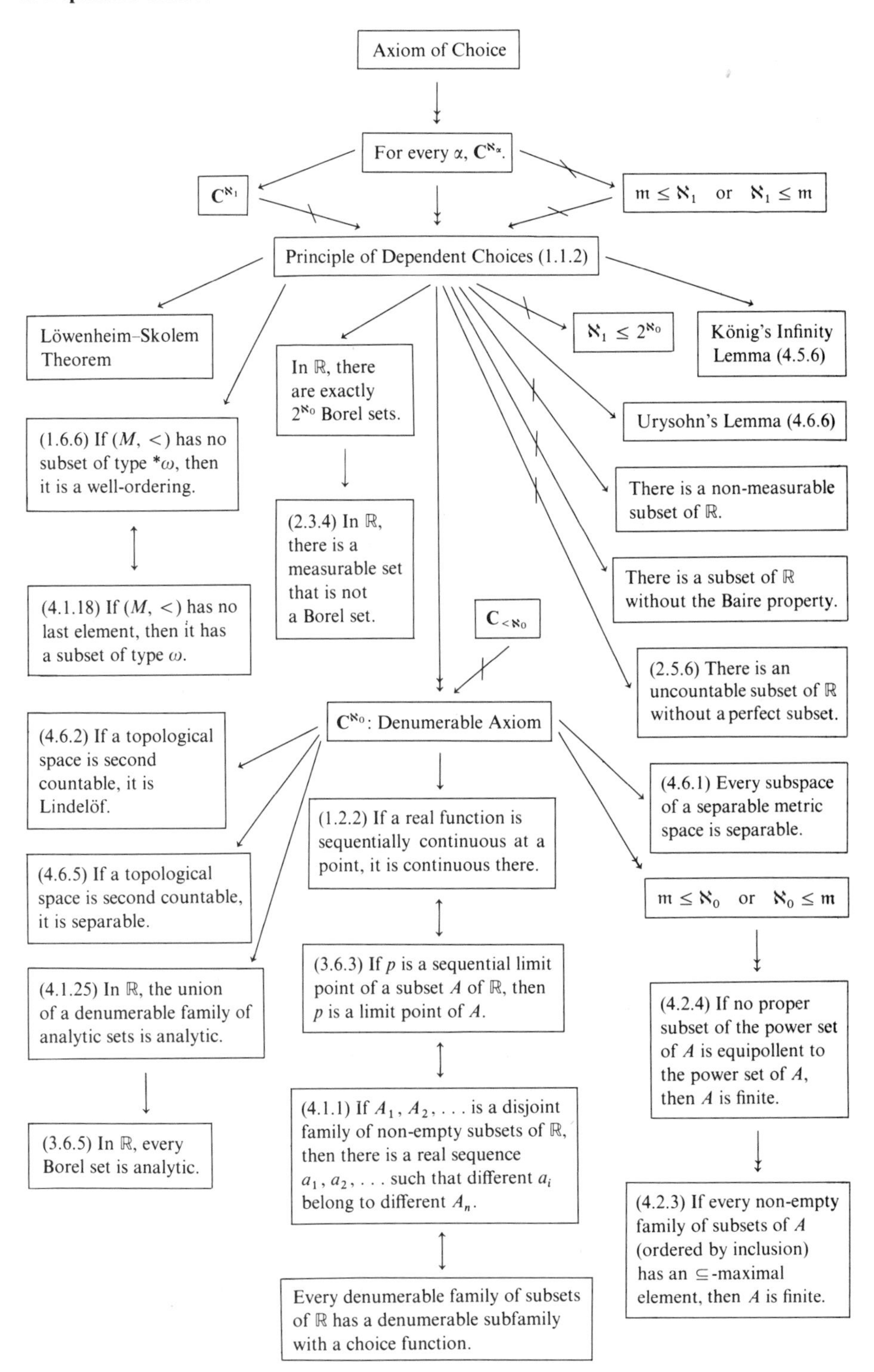

Table 5. Deductive Relations Concerning $\mathbf{C}^{\aleph_1}$, Non-Measurable Sets, and the Existence of a Well-Ordering for $\mathbb{R}$

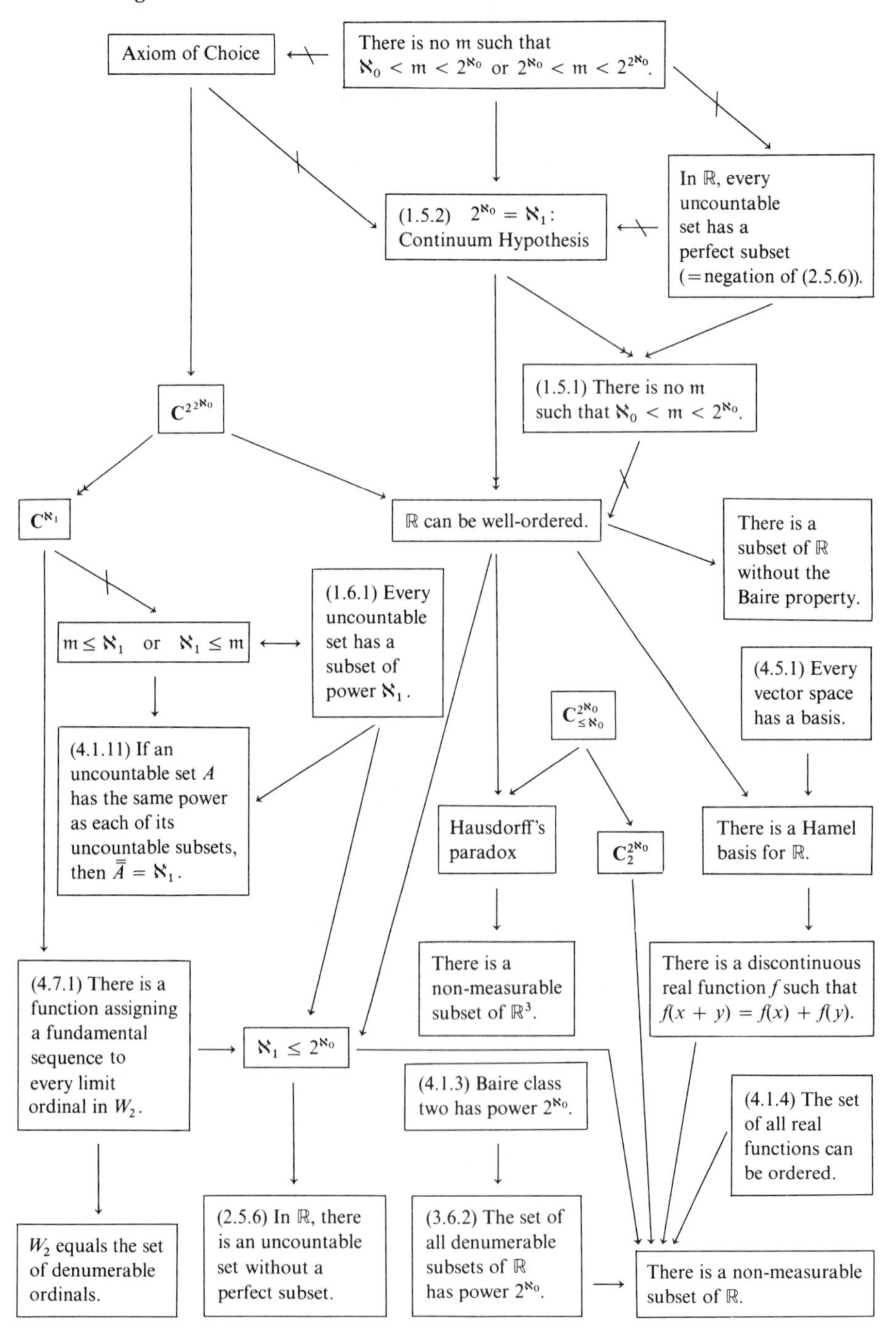

Table 6. Deductive Relations Concerning the Partition Principle (1.1.5)

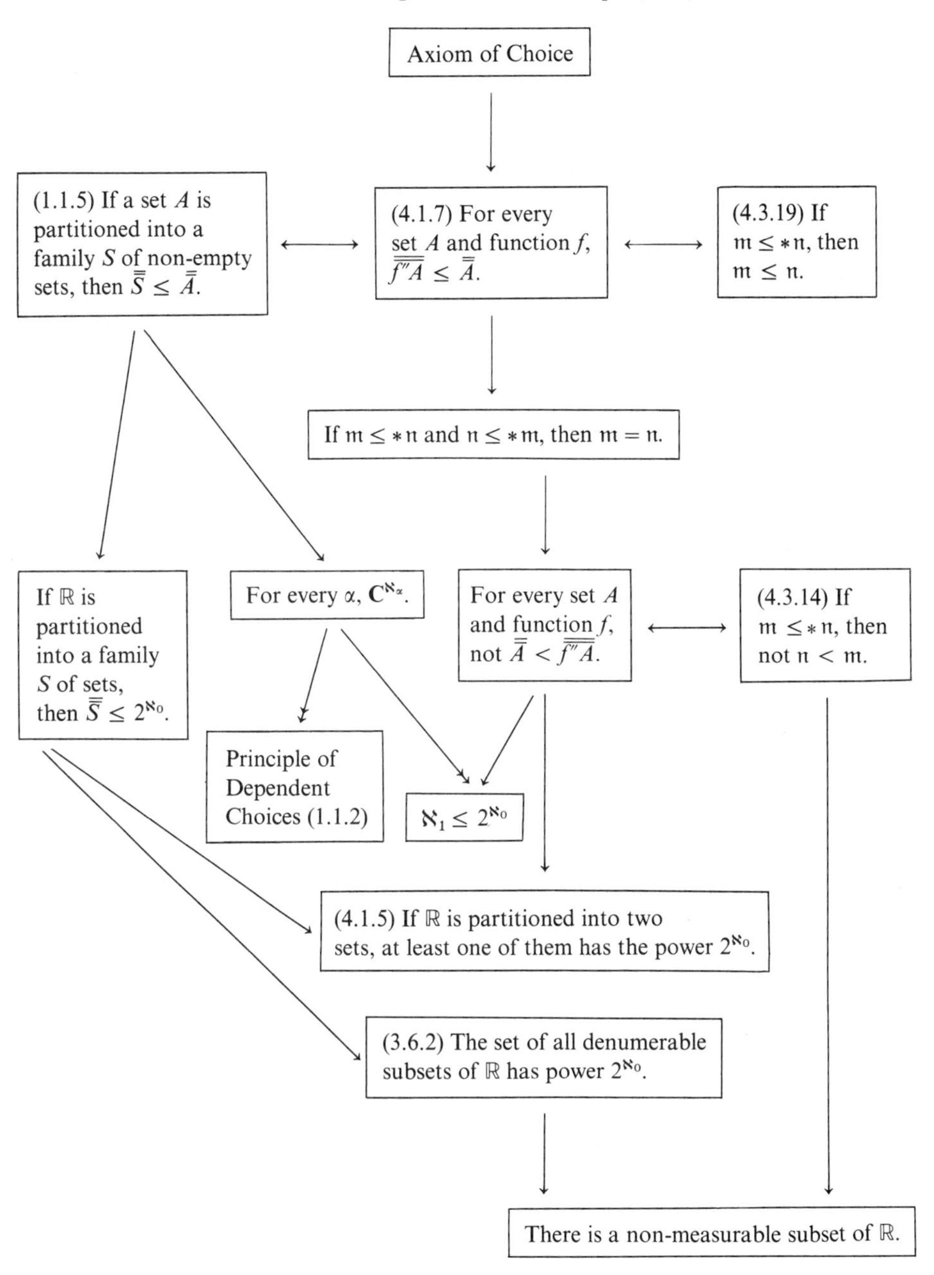

Table 7. Propositions Equivalent to the Boolean Prime Ideal Theorem

(4.5.4)	In a Boolean ring there exists at least one prime ideal.
(4.5.5)	On each infinite set there is a two-valued additive measure such that each singleton has measure zero.
(4.6.9)	(Ultrafilter Theorem) Every filter on a set can be extended to an ultrafilter.
(4.6.10)	The product of any family of compact Hausdorff spaces is a compact Hausdorff space.
(5.1.13(3))	If G is a graph such that every finite subgraph of G can be colored with 3 colors, then G itself can be colored with 3 colors.
Stone Representation Theorem for Boolean algebras (p. 230)	
Existence of Stone–Čech compactification of a Tychonoff space (p. 239)	
Completeness Theorem for propositional logic (p. 297)	
Completeness Theorem for first-order logic (p. 256)	
Compactness Theorem for first-order logic (pp. 257–258)	

Table 8. Deductive Relations Concerning the Boolean Prime Ideal Theorem, the Hahn–Banach Theorem, and the Ordering Principle

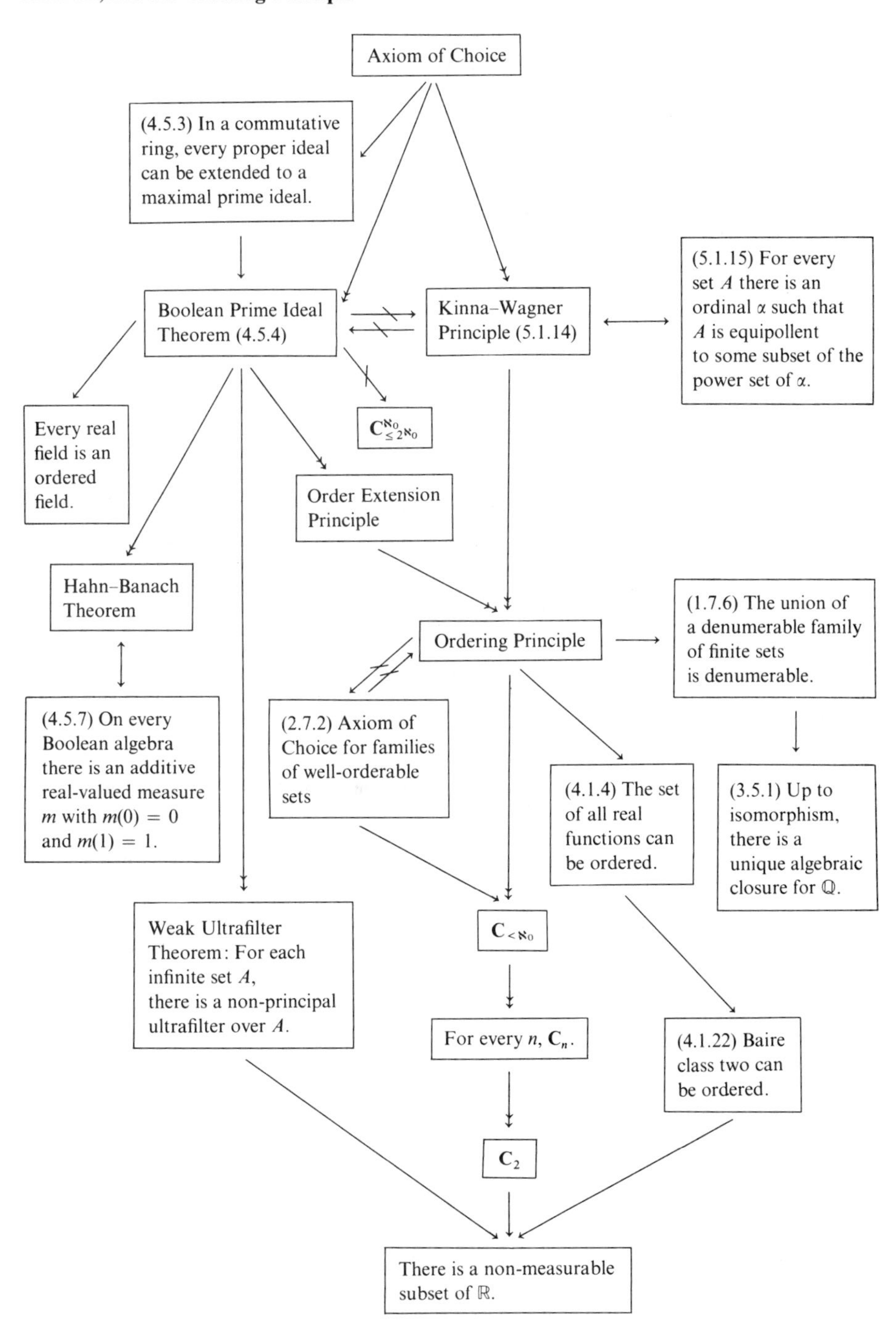

Table 9. Propositions Equivalent to the Axiom of Choice

A. Axiom of Choice and Related Principles

(1.1.1) (Axiom of Choice) Given any family T of non-empty sets, there is a function f such that $f(A) \in A$ for each A in T.

(5.1.16(n)) For every family T of non-empty sets there is a function f such that, for each A in T, $f(A)$ is a non-empty subset of A having at most n elements.

Zermelo's Principle: Given any set M, there is a function f such that for each non-empty subset A of M, $f(A) \in A$. (p. 90)

There is some n such that (5.1.16(n)).

B. Well-Ordering Theorem and Related Principles

(1.5.3) The cardinal numbers are well-ordered by magnitude.

(5.1.6) The power set of a well-ordered set can be well-ordered.

(5.1.7) Every set that can be ordered can also be well-ordered.

Well-Ordering Theorem: Every set can be well-ordered. (p. 39)

Aleph Theorem: Every infinite cardinal number is an aleph. (p. 45)

C. Multiplicative Axiom and Related Principles

(1.7.12) For every disjoint family T of non-empty sets, $\bigcup(T^\times) = \bigcup T$.

(1.7.13) For any disjoint families S and T, if $S^\times = T^\times$, then $S = T$.

(1.7.14) Suppose that T is a family of disjoint non-empty sets, that $A \subseteq \bigcup T$, and that no member of T contains more than one element of A. Then $A \subseteq M$ for some M in $T^\times$.

(2.7.1) (Multiplicative Axiom) If T is a disjoint family of non-empty sets, then $T^\times$ is non-empty.

D. Cardinal Equivalents

(1.1.6) (Trichotomy of Cardinals) $\mathfrak{m} < \mathfrak{n}$ or $\mathfrak{m} = \mathfrak{n}$ or $\mathfrak{m} > \mathfrak{n}$.

(1.6.8) If $\mathfrak{m}$ is infinite and $\mathfrak{n} \leq \mathfrak{m}$, then $\mathfrak{m}\mathfrak{n} = \mathfrak{m}$.

(3.2.1) (Zermelo–König Theorem) If for every t in T, $\mathfrak{m}_t < \mathfrak{n}_t$, then

$$\sum_{t \in T} \mathfrak{m}_t < \prod_{t \in T} \mathfrak{n}_t .$$

(4.1.16) For infinite $\mathfrak{m}$ and $\mathfrak{n}$, if $\mathfrak{m} < \mathfrak{p}$ and $\mathfrak{n} < \mathfrak{p}$, then $\mathfrak{m} + \mathfrak{n} \neq \mathfrak{p}$.

(4.1.17) If $S = \bigcup_{i=1}^{\infty} E_i$, where $E_1, E_2, \ldots$ is a sequence of disjoint sets, and if $\overline{\overline{T}} < \overline{\overline{S}}$, then $\overline{\overline{T}} < \overline{\overline{\bigcup_{i=1}^{n} E_i}}$ for some n.

(4.3.1) $\mathfrak{m}\mathfrak{n} = \mathfrak{m} + \mathfrak{n}$ for every infinite $\mathfrak{m}$ and $\mathfrak{n}$.

(4.3.2) $\mathfrak{m} = \mathfrak{m}^2$ for every infinite $\mathfrak{m}$.

(4.3.3) If $\mathfrak{m}^2 = \mathfrak{n}^2$, then $\mathfrak{m} = \mathfrak{n}$.

(4.3.4) If $\mathfrak{m} < \mathfrak{n}$ and $\mathfrak{p} < \mathfrak{q}$, then $\mathfrak{m} + \mathfrak{p} < \mathfrak{n} + \mathfrak{q}$.

(4.3.5) If $\mathfrak{m} < \mathfrak{n}$ and $\mathfrak{p} < \mathfrak{q}$, then $\mathfrak{m}\mathfrak{p} < \mathfrak{n}\mathfrak{q}$.

(4.3.6) If $\mathfrak{m} + \mathfrak{p} < \mathfrak{n} + \mathfrak{p}$, then $\mathfrak{m} < \mathfrak{n}$.

(4.3.7) If $\mathfrak{m}\mathfrak{p} < \mathfrak{n}\mathfrak{p}$, then $\mathfrak{m} < \mathfrak{n}$.

(4.3.8) If $\mathfrak{m} < \mathfrak{n}$, then $\mathfrak{n} = \mathfrak{m} + \mathfrak{p}$ for a unique $\mathfrak{p}$.

(4.3.9) $\mathfrak{m} + \mathfrak{p} = \mathfrak{m} + \mathfrak{q}$ implies either that $\mathfrak{p} = \mathfrak{q}$ or that $\mathfrak{p} \leq \mathfrak{m}$ and $\mathfrak{q} \leq \mathfrak{m}$.

(4.3.10) If $\mathfrak{m} + \mathfrak{m} < \mathfrak{m} + \mathfrak{n}$, then $\mathfrak{m} < \mathfrak{n}$.

(4.3.11) If $\mathfrak{p}^{\mathfrak{m}} < \mathfrak{q}^{\mathfrak{m}}$, then $\mathfrak{p} < \mathfrak{q}$.

(4.3.12) If $\mathfrak{m} \neq 0$ and $\mathfrak{m}^{\mathfrak{p}} < \mathfrak{m}^{\mathfrak{q}}$, then $\mathfrak{p} < \mathfrak{q}$.

(4.3.13) $\mathfrak{m} \leq * \mathfrak{n}$ or $\mathfrak{n} \leq * \mathfrak{m}$.

(4.3.26) If $\mathfrak{m}$, $\mathfrak{n}$, and $\mathfrak{p}$ are infinite, and if $\mathfrak{n}$ and $\mathfrak{p}$ are equal or consecutive, then $\mathfrak{m}\mathfrak{n}$ and $\mathfrak{m}\mathfrak{p}$ are equal or consecutive.

(4.3.27) For every set A there is a set B such that if $X \subseteq B$ and not $\overline{\overline{A}} \leq \overline{\overline{X}}$, then $X \in B$.

(4.3.28) If $\mathfrak{m}$ is infinite, where $\mathfrak{n} \leq \mathfrak{m}$ and $\overline{\overline{M}} = \mathfrak{m}$, then $\{X \subseteq M\colon \text{not } \mathfrak{n} < \overline{\overline{X}}\}$ is equipollent to $\mathfrak{m}^{\mathfrak{n}}$.

(4.3.29) For every set A there is a set M equipollent to $\{X \subseteq M\colon \text{not } \overline{\overline{A}} \leq \overline{\overline{X}}\}$.

(5.1.1) For infinite $\mathfrak{m}$ and $\mathfrak{n}$, if $\mathfrak{m} < \mathfrak{p}$ and $\mathfrak{n} < \mathfrak{p}$, then $\mathfrak{m}\mathfrak{n} \neq \mathfrak{p}$.

(5.1.4) For every $\mathfrak{m}$ there is some $\mathfrak{n}$ such that (i) $\mathfrak{m} < \mathfrak{n}$ and (ii′) for every $\mathfrak{p}$, if $\mathfrak{m} < \mathfrak{p}$, then $\mathfrak{n} \leq \mathfrak{p}$.

If $\mathfrak{p} < \mathfrak{m}, \mathfrak{n} < 2^{\mathfrak{p}}$, then either $\mathfrak{m} \leq \mathfrak{n}$ or $\mathfrak{n} \leq \mathfrak{m}$. (p. 296)

E. Maximal Principles

(3.4.1) Suppose that M is an infinite set and that S is a non-empty family of subsets of M. If S is extendable by limits, then S contains an $\subseteq$-maximal subset.

(3.4.2) (Hausdorff's Maximal Principle) Every partially ordered set M has an ordered subset A that is $\subseteq$-maximal among ordered subsets of M.

(3.4.3) Every partially ordered set M has an ordered subset A that is $\subseteq$-maximal among ordered subsets of M and also includes B, a given subset of M.

(4.4.1) If M is a non-empty family of sets and is extendable above by limits, then M contains an $\subseteq$-maximal set.

(4.4.2) If M is a non-empty family of sets and is extendable below by limits, then M contains an $\subseteq$-minimal set.

(4.4.3) Suppose that S consists of those subsets of a set E which have a given property P, that E is in S, and that, for every non-empty subset T of S well-ordered by $\supseteq$, $\bigcap T$ is in S. Then S contains an $\subseteq$-minimal set.

(4.4.4) Suppose that S consists of those subsets of a set E which have a given property P and each of which includes a given set A, that A is in S, and that, for every non-empty subset T of S well-ordered by $\subseteq$, $\bigcup T$ is in S. Then S contains an $\subseteq$-maximal set.

(4.4.5) Suppose that G is a non-empty family of sets and that, for every subfamily H of G ordered by inclusion, there is some A in G such that $A \subseteq B$ for every B in H. Then G contains an $\subseteq$-minimal set.

(4.4.6) (Zorn's Lemma) If A is a family of sets such that the union of every chain $B \subseteq A$ is in A, then A contains an $\subseteq$-maximal set.

(4.4.7) If E is a set partially ordered by a relation S and if every subset of E ordered by S has an S-upper bound in E, then E contains an S-maximal element.

(4.4.8) If a non-empty family M of subsets of a set A has finite character, then M contains an $\subseteq$-maximal set.

(4.4.9) Every partially ordered set A has an ordered subset B which contains all its upper bounds (if any) in A.

(4.4.10) For every set M and every set S of propositional functions

$$P_s(x_1, x_2, \ldots, x_{n_s}),$$

there is an $\subseteq$-maximal subset A of M such that for every P_s in S and for every $x_1, x_2, \ldots, x_{n_s}$ in A, $P_s(x_1, x_2, \ldots, x_{n_s})$ is true.

(5.1.2) Every family of sets has an $\subseteq$-maximal subfamily whose members are disjoint.

Kurepa's Principle: Every family of sets has an $\subseteq$-maximal subfamily F such that every pair of sets A, B in F is in the relation S, where S is either overlap ($A—B \neq 0$ and $B—A \neq 0$ and $A \cap B \neq 0$), non-overlap ($A—B = 0$ or $B—A = 0$ or $A \cap B = 0$), non-disjointness ($A \cap B \neq 0$), or incomparability (neither $A \subseteq B$ nor $B \subseteq A$). (p. 295)

F. Algebraic Equivalents

(5.1.8) If $\mathfrak{A}$ is an algebra and if b is an element in some basis for $\mathfrak{A}$, then there is an $\subseteq$-maximal subalgebra of $\mathfrak{A}$ not containing b.

(5.1.9) If $\mathfrak{A}$ is an algebra with a finite basis and if $\mathfrak{B}$ is a proper subalgebra of $\mathfrak{A}$, then there is an $\subseteq$-maximal proper subalgebra of $\mathfrak{A}$ which includes $\mathfrak{B}$.

(5.1.10) Every lattice with a unit and at least one other element has a maximal ideal.

(5.1.11) If V is a real vector space, then for every subspace S there is a subspace S' such that $S \cap S' = \{0\}$ and $S \cup S'$ generates V.

(5.1.12) If a subset A of a vector space V generates V, then A includes a basis.

G. Equivalents in Logic

(4.8.5) (Downward Löwenheim–Skolem Theorem) Every model of an infinite set A of first-order sentences has a submodel whose power is at most that of A.

Löwenheim–Skolem–Tarski Theorem: If a countable set of first-order sentences has an infinite model, then it has a model of each infinite cardinality. (pp. 257–258)

H. Topological Equivalents

(4.6.8) (Tychonoff's Compactness Theorem) The product of any family of compact topological spaces is compact.

Weak Tychonoff Theorem: The product of any number of copies of a given compact topological space is compact. (p. 239)

I. Equivalents Involving Functions and Relations

(1.5.6) If A and B are uncountable sets, then there is a function $f: A \to B$ which is either one–one or onto.

(2.7.4) Every function f includes a one–one function with the same range as f.

(2.7.5) Every relation S includes a function with the same domain as S.

Principle of $\aleph_\alpha$-Dependent Choices for every α (p. 299)

J. Equivalents Involving Finiteness

(4.2.7) A set A is finite if and only if, for each well-ordering S of A, the relation converse to S also well-orders A.

(4.2.8) A set A is finite if and only if A has at most one element or can be partitioned into two sets, each of smaller power than A.

(4.2.9) A set A if finite if and only if A has at most one element or is of smaller power than the Cartesian product $A \times A$.

K. Pairs of Propositions Jointly Equivalent to the Axiom of Choice

(1.4.12) If $\{A_x: x \in T\}$ and $\{B_x: x \in T\}$ are disjoint families of sets with A_x equipollent to B_x for each x, then $\bigcup_{x \in T} A_x$ is equipollent to $\bigcup_{x \in T} B_x$; and

(4.1.12) If every member of an infinite set A is equipollent to A, then $\bigcup A$ is equipollent to A.

The Order Extension Principle together with each of the following:

(4.1.19) (Cofinality Principle) Every ordered set is cofinal with some well-ordered set.

If a family K of sets has an $\subseteq$-maximal subfamily F such that every pair of sets in F is in the relation S, then K has an $\subseteq$-maximal subfamily G such that every pair of sets in G is in the relation T—provided that S and T are any distinct relations among disjointness, non-disjointness, overlap, non-overlap, comparability, and incomparability. (pp. 295–296)

Table 10. Propositions Implying the Axiom of Choice but Stronger Than It

(4.3.20)	(Generalized Continuum Hypothesis) For every infinite cardinal $\mathfrak{m}$ there is no $\mathfrak{n}$ with $\mathfrak{m} < \mathfrak{n} < 2^{\mathfrak{m}}$.
(4.3.30)	(Axiom of Inaccessible Sets) For every set A there is a set B such that (a) $A \in B$; (b) if $X \in B$ and $Y \subseteq X$, then $Y \in B$; (c) if $X \in B$, then the power set of X is a member of B; and (d) if $X \subseteq B$ and $\overline{\overline{X}} < \overline{\overline{B}}$, then $X \in B$.
(4.8.2)	(Axiom of Global Choice) For every class M of non-empty sets there is a function that assigns to each A in M an element of A.
(4.9.1)	A class M is a proper class if and only if M is equipollent to the class of all sets.
(4.9.2)	The class of all sets can be well-ordered.
Hilbert's ε-axiom (p. 255)	
Axiom of Constructibility (p. 280)	

Table 11. Alternatives to the Axiom of Choice

(4.7.2)	There is no function assigning a fundamental sequence to every limit ordinal in W_2, but for each α in W_2 there is a function assigning a fundamental sequence to every limit ordinal less than α.
(4.7.3)	For some α in W_2 there is no function assigning a fundamental sequence to every limit ordinal less than α.
(5.1.19)	$\mathbb{R}$ is a countable union of countable sets.
(5.1.20)	The power set of any set having cardinality $\mathfrak{m}$ is the union of countably many sets of cardinality $\mathfrak{m}$.
Axiom of Determinateness (p. 306)	

Journal Abbreviations Used in the Bibliography

AAMS	Abstracts of Papers Presented to the American Mathematical Society
AC	Acta Mathematica
ACL	Actas Academie Nacional de Ciencas Exactas, Fis., y Naturales, Lima
ADM	Archiv der Mathematik
AFSP	Archiv für systematische Philosophie
AH	Archive for History of Exact Sciences
AJM	American Journal of Mathematics
AL	Accademia nazionale dei Lincei, Memorie, Classe di scienze fisiche, matematiche e naturale
ALG	Archiv für mathematische Logik und Grundlagenforschung
ALR	Accademia nazionale dei Lincei, Rendiconti, Classe di scienze fisiche, matematiche e naturale
ALS	Acta Litterarum ac Scientiarum Regiae Universitatis Hungaricae Francisco–Josephinae, sectio scientiarum mathematicarum
AM	Annals of Mathematics
AMH	Acta Mathematica Academiae Scientiarum Hungaricae
AML	Annals of Mathematical Logic
AMM	American Mathematical Monthly
AMST	American Mathematical Society Translations

AP	Bulletin de l'Académie Polonaise des Sciences, Série des Sciences Math., Astron., et Phys.
AR	Bulletin Scientifique de la République Populaire Roumaine, Section de Sciences Math. et Phys.
AS	Annals of Science
ASB	Memorie della R. Accademia delle Scienze del'Istituto di Bologna
ASC	Bulletin de l'Académie des Sciences de Cracovie, Classe des Sciences Math., Série A
ASP	Annales de la Société Polonaise de Mathématiques
ASPD	Annales de la Société Polonaise de Mathématiques (Dodatek)
AT	Accademia delle Scienze di Torino, Classe di Scienze Fisiche, Matematiche, e Naturale (Atti)
BAMS	Bulletin of the American Mathematical Society
BMRS	Biographical Memoirs of the Royal Society
BSM	Bulletin des sciences mathématiques et astronomiques
BUM	Bolletino della Unione Matematica Italiana
CMC	Commentationes Mathematicae Universitatis Carolinae
CMH	Commentarii Mathematici Helvetici
CMP	Circolo Matematico di Palermo, Rendiconti
CMS	Commentarii Mathematici Universitatis Sancti Pauli
COL	Colloquium Mathematicum
CPS	Proceedings of the Cambridge Philosophical Society
CRP	Comptes Rendus Hebdomadaires des Séances de l'Académie des Sciences, Paris
CRR	Comptes Rendus des Séances de l'Institut des Sciences de Roumanie
CRV	Comptes Rendus des Séances de la Société des Sciences et des Lettres de Varsovie, Classe III
CSG	Commentationes societatis regiae scientiarum Gottingensis recentiores
DAN	Doklady Akademii Nauk SSSR
DM	Duke Mathematical Journal
DMV	Jahresbericht der Deutschen Mathematiker-Vereinigung
DMVA	Jahresbericht der Deut. Math.-Verein. (Angelegenheiten)
DT	Deutsche Mathematik
EM	Enseignement mathématique
ENS	Annales Scientifiques de l'Ecole Normale Supérieure

ERK	Erkenntnis
FM	Fundamenta Mathematicae
FV	Finska Vetenskaps-Societeten, Förhandlingar, A. Matematik och Naturvetenskaper
GC	Atti della Accademia Gioenia di Scienze Naturale in Catania
GM	Giornale di matematiche di Battaglini
HB	Hibbert Journal
HM	Historia Mathematica
HPL	History and Philosophy of Logic
IAN	Izvestiya Akademii Nauk SSSR, Seriya Matematicheskaya
IJM	Israel Journal of Mathematics
IL	Istituto Lombardo di Scienze e Lettere, Rendiconti
IM	Indagationes Mathematicae
IST	Istoriko-Matematicheskie Issledovaniya
JEP	Journal de l'Ecole Polytechnique
JHP	Journal of the History of Philosophy
JHS	Japanese Studies in the History of Science
JJM	Japanese Journal of Mathematics
JM	Journal für die reine und angewandte Mathematik (Crelle)
JPA	Journal de mathématiques pures et appliquées
JPL	Journal of Philosophical Logic
JPP	Journal of Philosophy, Psychology, and Scientific Methods
JSL	Journal of Symbolic Logic
LMS	Proceedings of the London Mathematical Society
MA	Mathematische Annalen
MC	Matematiche (Catania)
MCJ	Mathematica (Cluj)
MD	Mind
MF	Rendiconti del Seminario matematico e fisico di Milano
MG	Mathematics Magazine
MGZ	Mathematical Gazette
MI	The Mathematical Intelligencer
MJO	Mathematical Journal of Okayama University
MM	Messenger of Mathematics
MMP	Monatshefte für Mathematik und Physik
MN	Mathematische Nachrichten

MNB	Mathematisch-naturwissenschaftliche Blätter
MPA	Annali di matematica pura ed applicata
MR	Mathematical Reviews
MS	Matematicheskii Sbornik
MSH	Abhandlungen aus dem mathematischen Seminar der Hamburgischen Universität
MZ	Mathematische Zeitschrift
NA	Koninklijke Nederlandse Akademie van Wetenschappen te Amsterdam, Proceedings of the section of mathematical sciences
NAL	Deutsche Akademie der Naturforscher, Nova Acta Leopoldina
NAMS	Notices of the American Mathematical Society
NDJ	Notre Dame Journal of Formal Logic
NG	Nachrichten, Akademie der Wissenschaften, Göttingen
NMF	Norsk Matematisk Forening, Skrifter
NP	Nauka Polska
NT	Nature
PAS	Proceedings of the National Academy of Sciences (U.S.A.)
PAW	Sitzungsberichte der Preussischen Akademie der Wissenschaften, Physik.-math. Klasse
PJM	Pacific Journal of Mathematics
PM	Philosophical Magazine
PMD	Publicationes Mathematicae (Decebren)
PP	Polish Perspectives
PR	Polish Review
QM	Quarterly Journal of Pure and Applied Mathematics
RDM	Rivista di matematica
RE	Rete
RG	Revue générale des sciences pures et appliquées
RHS	Revue d'histoire des sciences et de leurs applications
RM	Revue du mois
RMM	Revue de métaphysique et de morale
RPL	Revue philosophique de Louvain
SA	Scientific American
SC	Scientia
SCM	Scripta Mathematica
SHMT	Séminaire d'histoire des mathématiques, Toulouse
SI	Science

SM	Studia Mathematica
SMD	Soviet Mathematics, Doklady
SMF	Bulletin de la Société Mathématique de France
SMFC	Bulletin de la Société Mathématique de France (Comptes Rendus)
SMR	Rendiconti del Seminario Matematico della R. Università di Roma
SP	Annali della Scuola Normale Superiore de Pisa
SR	Science Progress
SW	Königlich Sächsischen Gesellschaft der Wissenschaften zu Leipzig, Math.-Phys. Klasse, Sitzungsberichte
TAMS	Transactions of the American Mathematical Society
TM	Tôhoku Mathematical Journal
UBA	Contribuciones científicas de la Facultad de Ciencias Exactas y Naturales de la Universidad de Buenos Aires
UCM	University of California Publications in Mathematics; Seminar Reports (Los Angeles)
UMA	Revista de la Unión Matematica Argentina
UZ	Ucheneye Zapiski, Fiziko-matematicheskie Nauki
VS	Videnskaps-selskapets Skrifter, I. Mat.-Nat. Klasse
ZML	Zeitschrift für mathematische Logik und Grundlagen der Mathematik
ZP	Zeitschrift für Philosophie und philosophische Kritik

Bibliography

Ackerman, Wilhelm

1924 "Begründung des 'tertium non datur' mittels der Hilbertschen Theorie der Widerspruchsfreiheit," **MA 93**, 1–36.

1937 "Die Widerspruchsfreiheit der allgemeinen Mengenlehre," **MA 114**, 305–315.

See Hilbert, David, and Ackermann, Wilhelm.

Addison, John

1959 "Some Consequences of the Axiom of Constructibility," **FM 46**, 337–357.

Addison, John; Henkin, Leon; and Tarski, Alfred

1965 (eds.) *The Theory of Models: Proceedings of the 1963 International Symposium at Berkeley* (Amsterdam: North-Holland).

Alexander, J. W.

1939 "Ordered Sets, Complexes, and the Problem of Compactification," **PAS 25**, 296–298.

Alexandroff, Paul

1916 "Sur la puissance des ensembles mesurables B," **CRP 162**, 323–325.

1969 (ed.) *Problemy Gilberta: Sbornik* (Moscow: Nauk).

1971 (ed.) *Die Hilbertschen Probleme* (Leipzig: Akad. Verlagsgesellschaft); German translation of [1969].

Alexandroff, Paul, and Urysohn, Paul

1929 "Mémoire sur les espaces topologiques compacts," **NA 14**, 1–96.

Arboleda, L. C.

1979 "Les débuts de l'école soviétique topologique: notes sur les lettres de Paul S. Alexandroff et Paul S. Urysohn à Maurice Fréchet," **AH 20**, 73–89.

Aristotle
1941 *The Basic Works of Aristotle*, ed. by Richard McKeon (New York: Random House).

Artin, Emil, and Schreier, Otto
1927 "Algebraische Konstruction reeller Körper," **MSH 5**, 85–99.

Arzelà, Cesare
1889 "Funzioni di linee," **ALR (4) 5**, 342–348.
1895 "Sulle funzioni di linee," **ASB (5) 5**, 225–244.
1895a "Sull'integrabilità delle equazioni differenziali ordinarie," **ASB (5) 5**, 257–270.
1901 *Lezioni di calcolo infinitesimale* (Florence: Le Monnier), vol. I, part 1a.

Ascoli, Giulio
1884 "Le curve limite di una varietà data di curve," **AL (3) 18**, 521–586.

Baer, Reinhold
1928 "Über ein Vollständigkeitsaxiom in der Mengenlehre," **MZ 27**, 536–539.

Bagnera, Giuseppe
1915 *Corso di analisi infinitesimale* (Palermo).

Baire, René
1898 "Sur les fonctions discontinues qui se rattachent aux fonctions continues," **CRP 129**, 1621–1623.
1899 "Sur les fonctions de variables réelles," **MPA (3) 3**, 1–123.

Baire, René; Borel, Emile; Hadamard, Jacques; and Lebesgue, Henri
1905 "Cinq lettres sur la théorie des ensembles," **SMF 33**, 261–273; translated as Appendix 1 of this book.

Banach, Stefan
1923 "Sur le problème de la mesure," **FM 4**, 7–33.
1929 "Sur les fonctionelles linéaires. II," **SM 1**, 223–239.
1930 "Über additive Massfunktionen in abstrakten Mengen," **FM 15**, 97–101.

Banach, Stefan, and Kuratowski, Kazimierz
1929 "Sur une généralisation du problème de la mesure," **FM 14**, 127–131.

Banach, Stefan, and Tarski, Alfred
1924 "Sur la décomposition des ensembles de points en parties respectivement congruentes," **FM 6**, 244–277.

Banaschewski, Bernhard
1953 "Über den Satz von Zorn," **MN 10**, 181–186.

Bar-Hillel, Yehoshua
See Fraenkel, Abraham, and Bar-Hillel, Yehoshua.
See Fraenkel, Abraham; Bar-Hillel, Yehoshua; and Levy, Azriel.

Bar-Hillel, Yehoshua, *et alii*
1961 (eds.) *Essays on the Foundations of Mathematics* (Jerusalem: Magnes Press).

Bartle, R. G.
1955 "Nets and Filters in Topology," **AMM 62**, 551–557.

Barwise, Jon
1974 "Axioms for Abstract Model Theory," **ML 7**, 221–265.
1977 (ed.) *Handbook of Mathematical Logic* (Amsterdam: North-Holland).
1980 "Infinitary Logics," in E. Agazzi (ed.), *Modern Logic—A Survey* (Dordrecht: Reidel), 93–112.

Bell, J. L.
1978 *Boolean-Valued Models and Independence Proofs in Set Theory* (Oxford: Clarendon).

Bell, J. L., and Fremlin, D. H.
1972 "A Geometric Form of the Axiom of Choice," **FM 77**, 167–170.

Bendixson, Ivar
1883 "Quelques théorèmes de la théorie des ensembles de points," **AC 2**, 415–429.

Bernacerraf, Paul, and Putnam, Hilary
1964 *Philosophy of Mathematics: Selected Readings* (Englewood Cliffs, N.J.: Prentice-Hall).

Bernays, Paul
1937 "A System of Axiomatic Set Theory, Part I," **JSL 2**, 65–77.
1941 "A System of Axiomatic Set Theory, Part II," **JSL 6**, 1–17.
1942 "A System of Axiomatic Set Theory, Part III," **JSL 7**, 65–89.
1976 *Sets and Classes*, ed. by G. H. Müller (Amsterdam: North-Holland).
See Hilbert, David, and Bernays, Paul.

Bernstein, Felix
1901 *Untersuchungen aus der Mengenlehre* (Ph.D. dissertation: Göttingen; printed at Halle); reprinted with some alterations as [1905c].
1904 "Bemerkung zur Mengenlehre," **NG**, 557–560.
1905 "Über die Reihe der transfiniten Ordnungszahlen," **MA 60**, 187–193.
1905a "Zum Kontinuumproblem," **MA 60**, 463–464.
1905b "Zur Mengenlehre," **DMV 14**, 198–199.
1905c "Untersuchungen aus der Mengenlehre," **MA 61**, 117–155; *cf.* [1901].
1908 "Zur Theorie der trigonometrischen Reihe," **SW 60**, 325–338.
1919 "Die Mengenlehre Georg Cantors und der Finitismus," **DMV 28**, 63–78.

Beth, Evert
1959 *The Foundations of Mathematics* (Amsterdam: North-Holland).

Bettazzi, Rodolfo
1892 "Sui punti di discontinuità delle funzioni di variabile reale," **CMP 6**, 173–195.
1896 "Sulla catena di un ente in un gruppo," **AT 31**, 446–456.
1896a "Gruppi finiti ed infiniti di enti," **AT 31**, 506–512.
1897 "Sulla definizione del gruppo finito," **AT 32**, 352–355.

Birkhoff, Garrett
1933 "On the Combination of Subalgebras," **CPS 29**, 441–464.
1937 "Moore-Smith Convergence in General Topology," **AM (2) 38**, 39–56.
1940 *Lattice Theory*, Colloquium Publications, vol. 25 (New York: American Mathematical Society).
1967 Third edition of [1940].

Bishop, Errett
1967 *Foundations of Constructive Analysis* (New York: McGraw-Hill).

Blass, Andreas
1977 "Ramsey's Theorem in the Hierarchy of Choice Principles," **JSL 42**, 387–390.
1977a "A Model without Ultrafilters," **AP** 25, 329–331.
1981 "The Model of Set Theory Generated by Countably Many Generic Reals," **JSL 46**, 732–752.

Bleicher, M. H.
1964 "Some Theorems on Vector Spaces and the Axiom of Choice," **FM 54**, 95–107.
1965 "Multiple Choice Axioms and Axioms of Choice for Finite Sets," **FM 57**, 247–252.

Blumberg, Henry
1918 "Non-Measurable Functions Connected with Functional Equations," **BAMS 24**, 217,220.

Bochner, Salomon
1928 "Fortsetzung Riemannscher Flächen," **MA 98**, 406–421.

Bolzano, Bernard
1817 *Rein analytischer Beweis des Lehrsatzes, dass zwischen je zwei Werthen, die ein entgegengesetztes Resultat gewaehren, wenigstens eine reelle Wurzel der Gleichung liege* (Prague: Gottlieb Haase); translated as [1980].
1851 *Paradoxien des Unendlichen*, ed. by F. Přihonský (Leipzig: C. H. Reclam); translated as [1950].
1950 *Paradoxes of the Infinite*, trans. by D. A. Steele (London: Routledge).
1980 Translation of [1817] by J. B. Russ, **HM 7**, 156–185.

Borel, Emile
1895 "Sur quelques points de la théorie des fonctions," **ENS 12**, 9–55.
1898 *Leçons sur la théorie des fonctions* (Paris: Gauthier–Villars).
1905 "Quelques remarques sur les principes de la théorie des ensembles," **MA 60**, 194–195.
1905a *Leçons sur les fonctions de variables réelles et les développements en series de polynomes*, ed. by Maurice Fréchet (Paris: Gauthier-Villars).
1908 "Sur les principes de la théorie des ensembles," *Atti del IV Congresso Internazionale dei Matematici* (*Roma*) **2**, 15–17; reprinted in [1914] and [1950].
1908a "Les 'paradoxes' de la théorie des ensembles," **ENS (3) 25**, 443–448; reprinted in [1914] and [1950].
1909 "La théorie des ensembles et les progrès récents de la théorie des fonctions," **RG 20**, 315–324.
1912 "La philosophie mathématique et l'infini," **RM 14**, 219–227; reprinted in [1914] and [1950].

1912a "Le calcul des intégrales définies," **JPA (6) 8**, 159–210.
1912b "Sur les théorèmes fondamentaux de la théorie des fonctions de variables réelles," **CRP 154**, 413–415.
1914 Second edition of [1898].
1914a "L'infini mathématique et la réalité," **RM 18**, 71–83; reprinted in [1914] and [1950].
1946 "L'axiome du choix et la mesure des ensembles," **CRP 222**, 309–310.
1947 "Les paradoxes de l'axiome du choix," **CRP 224**, 1537–1538; reprinted in [1950].
1947a "Sur les difficultés des définitions asymptotiques," **CRP 224**, 1597–1599.
1950 Fourth edition of [1898].
1972 *Oeuvres*, 4 vols. (Paris: C.N.R.S.).
See Baire, René; Borel, Emile; Hadamard, Jacques; and Lebesgue, Henri.

Bourbaki, Nicolas
1939 *Eléments de mathématique. Première partie: Les structures fondamentales de l'analyse. Livre I. Théorie des ensembles* (Paris: Hermann).
1940 *Eléments de mathématique. II. Première partie: Les structures fondamentales de l'analyse. Livre III. Topologie générale* (Paris: Hermann).
1950 "Sur le théorème de Zorn," **ADM 2**, 434–437.
1954 *Eléments de mathématique. Livre I. Théorie des ensembles. Chapitre I et II* (Paris: Hermann).
1969 *Eléments d'histoire de mathématiques*, second edition (Paris: Hermann).

Boyer, Carl
1968 *A History of Mathematics* (New York: Wiley).

Brouwer, L. E. J.
1907 *Over de Grondslagen der Wiskunde* (Amsterdam: Maas & van Suchtelen); translated in [1975, 11–101].
1911 "On the Structure of Perfect Sets of Points (second communication)," **NA 14**, 137–147.
1917 "Addenda en corrigenda over de grondslagen der wiskunde," **NA** (*Verslagen*) **25**, 1418–1423; translated in [1975, 145–149].
1975 *Collected Works*, ed. by A. Heyting, vol. I (Amsterdam: North-Holland).

Brown, Arlen, and Pearcy, Carl
1977 *Introduction to Operator Theory. I* (New York: Springer).

Bunn, Robert
1975 *Infinite Sets and Numbers* (Ph.D. dissertation: University of British Columbia).
1980 "Developments in the Foundations of Mathematics, 1870–1910," in Grattan-Guinness 1980, 220–255.

Burali-Forti, Cesare
1894 "Sulle classi ordinate e i numeri transfiniti," **CMP 8**, 169–179.
1896 "Le classi finite," **AT 32**, 34–52.
1896a "Sopra un teorema del Sig. G. Cantor," **AT 32**, 229–237.
1897 "Una questione sui numeri transfiniti," **CMP 11**, 154–164; translated in van Heijenoort 1967, 104–111.
1897a "Sulle classe ben ordinate," **CMP 11**, 260; translated in van Heijenoort 1967, 111–112.

Burkhardt, Johann
1938 "Zur Neubegründung der Mengenlehre," **DMV 48**, 146–165.
1939 "Zur Neubegründung der Mengenlehre. Folge," **DMV 49**, 146–155.

Burstin, Celestyn
1929 "Ein Beitrag zur Theorie der Ordnung der linearen Systeme," **TM 31**, 296–299.

Butterfield, Herbert
1965 *The Whig Interpretation of History* (New York: Norton).

Campbell, Paul J.
1978 "The Origin of 'Zorn's Lemma'," **HM 5**, 77–89.

Cantor, Georg
1872 "Über die Ausdehnung eines Satzes aus der Theorie der trignometrischen Reihen," **MA 5**, 123–132.
1874 "Über eine Eigenschaft des Inbegriffes aller reellen algebraischen Zahlen," **JM 77**, 258–262.
1878 "Ein Beitrag zur Mannigfaltigkeitslehre," **JM 84**, 242–258.
1880 "Über unendliche, lineare Punktmannichfaltigkeiten. II," **MA 17**, 355–358.
1882 "Über unendliche, lineare Punktmannichfaltigkeiten. III," **MA 20**, 113–121.
1883 "Über unendliche, lineare Punktmannichfaltigkeiten. IV," **MA 21**, 51–58.
1883a "Über unendliche, lineare Punktmannichfaltigkeiten. V," **MA 21**, 545–591.
1883b "Sur divers théorèmes de la théorie des ensembles de points situés dans un espace continu à *n* dimensions. Première communication," **AC 2**, 409–414.
1884 "Über unendliche, lineare Punktmannichfaltigkeiten. VI," **MA 23**, 453–488.
1884a "De la puissance des ensembles parfaits de points," **AC 4**, 381–392.
1885 "Über verschiedene Theoreme aus der Theorie der Punktmengen in einem *n*-fach ausgedehnten stetigen Raume G_n. Zweite Mitteilung," **AC 7**, 105–124.
1887 "Mitteilungen zur Lehre vom Transfiniten," **ZP 91**, 81–125, 252–270.
1888 "Mitteilungen zur Lehre vom Transfiniten," **ZP 92**, 240–265.
1891 "Über eine elementare Frage der Mannigfaltigkeitslehre," **DMV 1**, 75–78.
1895 "Beiträge zur Begründung der transfiniten Mengenlehre. I," **MA 46**, 481–512; translated in [1915].
1897 "Beiträge zur Begrundung der transfiniten Mengenlehre. II," **MA 49**, 207–246; translated in [1915].
1915 *Contributions to the Founding of the Theory of Transfinite Numbers*, trans. and ed. by P. Jourdain (Chicago: Open Court); reprinted (Dover).
1932 *Gesammelte Abhandlungen mathematischen und philosophischen Inhalts*, ed. by E. Zermelo (Berlin: Springer); reprinted (Hildesheim: Olms, 1962).

Carathéodory, Constantin
1906 "Über die starken Maxima und Minima bei einfachen Integralen," **MA 62**, 449–503.

Cartan, Henri
1937 "Théorie des filtres," **CRP 205**, 595–598.
1937a "Filtres et ultrafiltres," **CRP 205**, 777–779.

Cassina, Ugo
1936 "Sul principio della scelta ed alcuni problemi dell'infinito," **MF 10**, 53–81.

Cassinet, J.

1980 "L'axiome multiplicatif, forme dominante de l'axiome du choix chez les mathématiciens brittaniques de 1900 à 1913," **SHMT 1.1**, 1–57.

1980a "Rodolfo Bettazzi, un des premiers à mettre clairement en évidence l'utilisation d'un principe de choix dans les démonstrations," **SHMT 1.1**, 1–5.

Cassinet, J., and Guillemot, M.

1980 "Préhistoire de l'axiome du choix," **SHMT 1.2**, C1–C21.

Cauchy, Augustin

1821 *Cours d'analyse de l'Ecole Royale Polytechnique* (Paris: Imprimerie Royale); reprinted in *Oeuvres complètes d'Augustin Cauchy* (Paris: Gauthier-Villars, 1897), series 2, vol. 3.

Cavaillès, Jean

1947 *Transfini et continu* (Paris: Hermann); reprinted in [1962].

1962 *Philosophie mathématique* (Paris: Hermann).

See Noether, Emmy, and Cavaillès, Jean.

Čech, Eduard

1937 "On Bicompact Spaces," **AM (2) 38**, 823–844.

Chang, Chen-Chung

1960 "Maximal *n*-Disjointed Sets and the Axiom of Choice," **FM 49**, 11–14.

Church, Alonzo

1927 "Alternatives to Zermelo's Assumption," **TAMS 29**, 178–208.

1968 "Paul J. Cohen and the Continuum Problem," *Proceedings of the International Congress of Mathematicians* (*Moscow–1966*), 15–20.

Chwistek, Leon

1924 "The Theory of Constructive Types. Part II. Cardinal Arithmetic," **ASP 3**, 92–141.

1924a "Sur les fondements de la logique moderne," *Atti del V Congresso Internaz. di Filos., Napoli*, 24–28.

1926 "Über die Hypothesen der Mengenlehre," **MZ 25**, 439–473.

Cipolla, Michele

1913 "Sul postulato di Zermelo e la teoria dei limiti delle funzioni," **GC (5) 6**, 1–13.

1924 "Sui fondamenti logici della matematica secondo le recenti vedute di Hilbert," **MPA (4) 1**, 19–29.

Cohen, Paul J.

1963 "A Minimal Model for Set Theory," **BAMS 69**, 537–540.

1963a "The Independence of the Axiom of Choice" (mimeographed).

1963b "The Independence of the Continuum Hypothesis. I," **PAS 50**, 1143–1148.

1964 "The Independence of the Continuum Hypothesis. II," **PAS 51**, 105–110.

1965 "Independence Results in Set Theory," in Addison *et alii* 1965, 39–54.

1966 *Set Theory and the Continuum Hypothesis* (New York: Benjamin).

Cohen, Paul J., and Hersh, Reuben

1967 "Non-Cantorian Set Theory," **SA 217** (December), 104–116.

Collins, George E.
1954 "Distributivity and an Axiom of Choice," **JSL 19**, 275–277.

Couturat, Louis
1896 *De l'infini mathématique* (Paris: Alcan); reprinted (New York: Franklin, 1969) and (Paris: Blanchard, 1973).
1900 "Les mathématiques au Congrès de Philosophie," **EM 2**, 397–410.
1905 *Les principes des mathématiques* (Paris: Alcan).
1906 "Pour la logistique (Réponse à M. Poincaré)," **RMM 14**, 208–250.

Crossley, J. N.
1975 (ed.) *Algebra and Logic*, Lecture Notes in Mathematics, No. 450 (Berlin: Springer).

Dauben, Joseph W.
1979 *Georg Cantor: His Mathematics and Philosophy of the Infinite* (Cambridge: Harvard University Press).

Dawson, John W., and Howard, Paul E.
1976 "Factorials of Infinite Cardinals," **FM 93**, 185–195.

Davis, Martin
1965 (ed.) *The Undecidable* (Hewlett, N.Y.: Raven Press).

Dedekind, Richard
1857 "Abriss einer Theorie der höhern Congruenzen in Bezug auf einen reellen Primzahl-Modulus," **JM 54**, 1–26.
1877 "Sur la théorie des nombres entiers algébriques," **BSM 11**, 278–288.
1888 *Was sind und was sollen die Zahlen?* (Braunschweig: Vieweg).
1893 Second edition of [1888]; translated by W. W. Beman in *Essays on the Theory of Numbers* (Chicago: Open Court, 1901; reprinted by Dover, 1963).
1932 *Gesammelte mathematische Werke*, ed. by Robert Fricke, Emmy Noether, and Öystein Ore, 3 vols. (Braunschweig: Vieweg).
See Lejeune Dirichlet, Peter, and Dedekind, Richard.

de la Vallée-Poussin, Charles
1916 *Intégrales de Lebesgue, fonctions d'ensemble, classes de Baire* (Paris: Gauthier-Villars).
1934 Second edition of [1916].

Denjoy, Arnaud
1946 *L'énumération transfinie. Livre I. La notion du rang* (Paris: Gauthier-Villars).

de Renna e Souza, Celso
1965 "A Note on Continuous Games, the Notion of Strategy and Zermelo's Axiom," **NDJ 6**, 183–189.

de Swart, H.
See Gielen, W.; de Swart, H.; and Veldman, W.

Devlin, Keith
1977 *The Axiom of Constructibility: A Guide for the Mathematician*, Lecture Notes in Mathematics, No. 617 (Berlin: Springer).

Dieudonné, Jean
1970 "The Work of Nicolas Bourbaki," **AMM 77**, 134–145.
1976 Preface to [Dugac 1976].

Dingler, Hugo
1911 *Über die Bedeutung der Burali-Fortischen Antinomie für die Wohlordnungssätze der Mengenlehre* (Munich: Ackermann).
1913 "Über die logischen Paradoxien der Mengenlehre und eine paradoxienfreie Mengendefinition," **DMV 22**, 307–315.

Dold, A., and Eckmann, B.
1974 *Victoria Symposium on Non-Standard Analysis, University of Victoria, 1972*, Lecture Notes in Mathematics, No. 369 (Berlin: Springer).

Doss, Raouf
1945 "Note on Two Theorems of Mostowski," **JSL 10**, 13–15.

du Bois-Reymond, Paul
1882 *Die allgemeine Functionenlehre I* (Tübingen: Laupp); *cf.* [1887].
1887 *Théorie générale des fonctions. Première partie*, French translation of [1882] by G. Milhaud and A. Girot (Nice: Imprimerie Niçoise).

Dugac, Pierre
1973 "Eléments d'analyse de Karl Weierstrass," **AH 10**, 41–176.
1976 *Richard Dedekind et les fondements des mathématiques* (*avec de nombreux textes inédits*) (Paris: Vrin).
1976a "Notes et documents sur la vie et l'oeuvre de René Baire," **AH 15**, 297–383.

Dummett, Michael
1977 *Elements of Intuitionism* (Oxford: Clarendon).

Easton, W. B.
1964 *Powers of Regular Cardinals* (Ph.D. dissertation: Princeton University).
1964a "Proper Classes of Generic Sets," **NAMS 11**, 205.
1970 "Powers of Regular Cardinals," **AML 1**, 139–178.

Ellentuck, Erik
1965 "The Universal Properties of Dedekind-Finite Cardinals," **AM 82**, 225–248.

Faedo, Sandro
1940 "Il principio di Zermelo per gli spazi astratti," **SP (2) 9**, 263–276.

Feferman, Solomon
1964 "Independence of the Axiom of Choice from the Axiom of Dependent Choices," **JSL 29**, 226.
1965 "Some Applications of the Notions of Forcing and Generic Sets (Summary)," in Addison *et alii* 1965, 89–95.
1965a "Some Applications of the Notions of Forcing and Generic Sets," **FM 56**, 325–345.

Feferman, Solomon, and Levy, Azriel
1963 "Independence Results in Set Theory by Cohen's Method. II," **NAMS 10**, 593.

Felgner, Ulrich
1969 "Die Existenz wohlgeordneter, konfinaler Teilmengen in Ketten und das Auswahlaxiom," **MZ 111**, 221–232; correction, **MZ 115**, 392.
1971 *Models of ZF Set Theory*, Lecture Notes in Mathematics, No. 223 (Berlin: Springer).
1971a "Comparison of the Axioms of Local and Universal Choice," **FM 71**, 43–62.

Felgner, Ulrich, and Jech, Thomas J.
1973 "Variants of the Axiom of Choice in Set Theory with Atoms," **FM 79**, 79–85.

Fenstad, Jens E.
1971 "The Axiom of Determinateness," *Proceedings of the Second Scandinavian Logic Symposium*, ed. by J. E. Fenstad (Amsterdam: North-Holland), 41–62.

Finsler, Paul
1926 "Über die Grundlegung der Mengenlehre. Erster Teil. Die Mengen und ihre Axiome," **MZ 25**, 683–712; reprinted in [1975].
1975 *Aufsätze zur Mengenlehre* (Darmstadt: Wissenschaftliche Buchgemeinschaft).

Fort, M. K.
1948 "A Specialization of Zorn's Lemma," **DM 15**, 763–765.

Fraenkel, Abraham
1919 *Einleitung in die Mengenlehre* (Berlin: Springer).
1921 "Über die Zermelosche Begründung der Mengenlehre," **DMVA 30**, 97–98.
1922 "Zu den Grundlagen der Cantor–Zermeloschen Mengenlehre," **MA 86**, 230–237.
1922a "Über den Begriff 'definit' und die Unabhängigkeit des Auswahlaxioms," **PAW**, 253–257; translated in van Heijenoort 1967, 284–289.
1922b "Axiomatische Begründung der transfiniten Kardinalzahlen. I," **MZ 13**, 153–188.
1922c "Zu den Grundlagen der Mengenlehre," **DMVA 31**, 101–102.
1925 "Untersuchungen über die Grundlagen der Mengenlehre," **MZ 22**, 250–273.
1927 *Zehn Vorlesungen über die Grundlegung der Mengenlehre* (Leipzig: Teubner).
1928 "Gelöste und ungelöste Probleme in Umkreis des Auswahlprinzips," *Atti di Congresso Internazionale dei Matematici* (*Bologna*) **2**, 255–259.
1928a "Über die Ordnungsfähigkeit beliebiger Mengen," **PAW**, 90–91.
1931 "Sur une atténuation essentielle de l'axiome du choix," **CRP 192**, 1072.
1932 "Über die Axiome der Teilmengenbildung," *Verhandlungen des Internationalen Mathematiker-Kongresses, Zurich 1932* **2**, 341–342.
1935 "Sur l'axiome du choix," **EM 34**, 32–51.
1937 "Über eine abgeschwächte Fassung des Auswahlaxioms," **JSL 2**, 1–25.
1952 "L'axiome du choix," **RPL 50**, 429–459.
1967 *Lebenskreise* (Stuttgart: Deutsche Verlags-Anstalt).

Fraenkel, Abraham, and Bar-Hillel, Yehoshua
1958 *Foundations of Set Theory* (Amsterdam: North-Holland).

Fraenkel, Abraham; Bar-Hillel, Yehoshua; and Levy, Azriel
1973 Second edition of [Fraenkel and Bar-Hillel 1958].

Frascella, William J.
1965 "A Generalization of Sierpiński's Theorem on Steiner Triples and the Axiom of Choice," **NDJ 6**, 163–179.

Fréchet, Maurice
1904 "Généralisation d'un théorème de Weierstrass," **CRP 139**, 848–850.
1906 "Sur quelques points du calcul fonctionnel" (dissertation), **CMP 22**, 1–72.
1913 "Pri la funkcia ekvacio $f(x + y) = f(x) + f(y)$," **EM 15**, 390–393.
1934 *L'arithmétique de l'infini*, Actualités scientifiques et industrielles, vol. 144 (Paris: Hermann).

Frege, Gottlob
1884 *Die Grundlagen der Arithmetik, ein logisch-mathematische Untersuchung über den Begriff der Zahl* (Breslau).
1893 *Grundgesetze der Arithmetik, begriffschriftlich abgeleitet*, vol. 1 (Jena).
1895 "Kritische Beleuchtung einiger Punkte in E. Schröders Vorlesungen über die Algebra der Logik," **AFSP 1**, 433–456; translated in [1960, 86–106].
1903 Vol. 2 of [1893].
1960 *Translations from the Philosophical Writings of Gottlob Frege*, ed. by P. Geach and M. Black, second edition (Oxford: Blackwell).
1980 *Philosophical and Mathematical Correspondence*, ed. by G. Gabriel *et alii*, trans. by H. Kaal (Chicago: University of Chicago Press).

Fremlin, D. H.
See Bell, J. L., and Fremlin, D. H.

Frewer, Magdalene
1977 *Das wissenschaftliche Werk Felix Bernsteins* (dissertation: Göttingen).

Gaifman, Haim
1975 "Global and Local Choice Functions," **IJM 22**, 257–265.

Galilei, Galileo
1974 *Two New Sciences*, trans. by Stillman Drake (Madison: University of Wisconsin Press).

Garciadiego, Alejandro R.
See Moore, Gregory H., and Garciadiego, Alejandro R.

Gauntt, Robert J.
1968 "Some Restricted Versions of the Axiom of Choice," **NAMS 15**, 351.
1970 "Axiom of Choice for Finite Sets—A Solution to a Problem of Mostowski," **NAMS 17**, 454.
1970a Correction to [Truss 1970], **NAMS 17**, 694.

Gauss, Carl Friedrich
1801 *Disquisitiones Arithmeticae* (Leipzig: Fleischer).
1832 "Theoria residuorum biquadraticum. Commentatio secunda," **CSG 7**; reprinted in [1876, 94–148].

1876 *Carl Friedrich Gauss Werke* (Göttingen: Königlichen Gesellschaft der Wissenschaften), vol. 2.
1966 Translation of [1801] by A. A. Clarke (New Haven: Yale University Press).

Gauthier, Yvon
1977 *Fondements des mathématiques: Introduction à une philosophie constructiviste* (Montréal: Université de Montréal).

Gentzen, Gerhard
1936 "Die Widerspruchsfreiheit der reinen Zahlentheorie," **MA 112**, 493–565; translated in [1969, 132–213].
1969 *The Collected Papers of Gerhard Gentzen*, ed. by M. E. Szabo (Amsterdam: North-Holland).

Gielen, W.; de Swart, H.; and Veldman, W.
1981 "The Continuum Hypothesis in Intuitionism," **JSL 46**, 121–136.

Gitik, Mordechai
1980 "All Uncountable Cardinals Can Be Singular," **IJM 35**, 61–88.

Gödel, Kurt
1930 "Die Vollständigkeit der Axiome des logischen Funktionenkalkuls," **MMP 37**, 349–360; translated in van Heijenoort 1967, 582–591.
1931 "Über formal unentscheidbare Sätze der Principia Mathematica und verwandter Systeme. I," **MMP 38**, 173–198; translated in van Heijenoort 1967, 596–616.
1938 "The Consistency of the Axiom of Choice and of the Generalized Continuum-Hypothesis," **PAS 24**, 556–557.
1939 "Consistency-Proof for the Generalized Continuum-Hypothesis," **PAS 25**, 220–224.
1940 *The Consistency of the Axiom of Choice and of the Generalized Continuum-Hypothesis with the Axioms of Set Theory*, Annals of Mathematics Studies, No. 3 (Princeton: Princeton University Press).
1944 "Russell's Mathematical Logic," in Schilpp 1944, 123–153 (reprinted in Bernacerraf and Putnam 1964, 211–232, and in Pears 1972, 192–226).
1947 "What Is Cantor's Continuum Problem?" **AMM 54**, 515–525; *cf.* [1964].
1964 Revised version of [1947], in Bernacerraf and Putnam 1964, 258–273.
1965 "Remarks before the Princeton Bicentennial Conference on Problems in Mathematics," in Davis 1965, 84–88.
See Hahn *et alii.*

Gonseth, Ferdinand
1926 *Les fondements des mathématiques* (Paris: Blanchard).
1941 (ed.) *Les entretiens de Zurich, 6–9 décembre 1938* (Zurich: Leeman).

Gottschalk, W. H.
1952 "The Extremum Law," **PAMS 3**, 631.

Grattan-Guinness, Ivor
1970 "An Unpublished Paper by Georg Cantor: Principien einer Theorie der Ordnungstypen. Erste Mittheilung," **AC 124**, 65–107.
1971 "Towards a Biography of Georg Cantor," **AS 27**, 345–392.

1971a "The Correspondence between Georg Cantor and Philip Jourdain," **DMV 73**, 111–130.

1972 "Bertrand Russell on His Paradox and the Multiplicative Axiom. An Unpublished Letter to Philip Jourdain," **JPL 1**, 103–110.

1974 "The Rediscovery of the Cantor–Dedekind Correspondence," **DMV 76**, 104–139.

1975 "Preliminary Notes on the Historical Significance of Quantification and of the Axioms of Choice in the Development of Mathematical Analysis," **HM 2**, 475–487.

1976 Review of [Dugac 1976], **AS 33**, 589–593.

1977 *Dear Russell–Dear Jourdain* (London: Duckworth).

1978 "How Bertrand Russell Discovered His Paradox," **HM 5**, 127–137.

1979 "In Memoriam Kurt Gödel: His 1931 Correspondence with Zermelo on His Incompletability Theorem," **HM 6**, 294–304.

1980 (ed.) *From the Calculus to Set Theory: 1630–1910. An Introductory History* (London: Duckworth).

1981 "On the Development of Logics between the Two World Wars," **AMM 88**, 495–509.

Grelling, Kurt

1910 *Die Axiome der Arithmetik mit besonderer Berücksichtigung der Beziehungen zur Mengenlehre* (Ph.D. dissertation: Göttingen).

Guillemot, M.

1980 "Baire et l'axiome du choix," **SHMT 1.1**, 1–38.

See Cassinet, J., and Guillemot, M.

Hadamard, Jacques

1904 "Le troisième Congrès international des Mathématiciens," **RG 15**, 961–962.

1905 "La théorie des ensembles," **RG 16**, 241–242.

1905a "Les principes des mathématiques et le problème des ensembles," **RG 16**, 541–543; *cf.* Richard 1905.

1926 "Les fondements des mathématiques," **BSM (2) 51**, 66–73 (=preface to [Gonseth 1926]).

See Baire, René; Borel, Emile; Hadamard, Jacques; and Lebesgue, Henri.

Hahn, Hans

1921 *Theorie der reellen Funktionen* (Berlin: Springer).

1980 *Empiricism, Logic, and Mathematics: Philosophical Papers*, ed. by Brian McGuinness (Dordrecht: Reidel).

See Schoenflies, Arthur, and Hahn, Hans.

Hahn, Hans; Carnap, Rudolf; Gödel, Kurt; Heyting, Arend; Reidemeister, Kurt; Scholz, Heinrich; von Neumann, John

1931 "Diskussion zur Grundlegung der Mathematik," **ERK 2**, 135–151; translated in part in Hahn 1980, 31–38.

Hájek, Petr

1966 "The Consistency of the Church's Alternatives," **AP 14**, 423–430.

Halpern, James D.

1961 "The Independence of the Axiom of Choice from the Boolean Prime Ideal Theorem," **NAMS 8**, 279–280.

1964 "The Independence of the Axiom of Choice from the Boolean Prime Ideal Theorem," **FM 55**, 57–66.
1966 "Bases in Vector Spaces and the Axiom of Choice," **PAMS 17**, 670–673.
1972 "On a Question of Tarski and a Maximal Theorem of Kurepa," **PJM 41**, 111–121.

Halpern, James D., and Howard, Paul E.
1970 "Cardinals $\mathfrak{m}$ such that $2\mathfrak{m} = \mathfrak{m}$," **PAMS 26**, 487–490.
1974 "Cardinal Addition and the Axiom of Choice," **BAMS 80**, 584–586.

Halpern, James D., and Levy, Azriel
1964 "The Ordering Theorem Does Not Imply the Axiom of Choice," **NAMS 11**, 56.
1971 "The Boolean Prime Ideal Theorem Does Not Imply the Axiom of Choice," in Scott 1971, 83–134.

Hamel, Georg
1905 "Eine Basis aller Zahlen und die unstetigen Lösungen der Funktionalgleichung: $f(x + y) = f(x) + f(y)$," **MA 60**, 459–462.

Hannequin, Arthur
1895 *Essai critique sur l'hypothèse des atomes dans la science contemporaine* (Paris: Masson).

Hardy, G. H.
1903 "A Theorem Concerning the Infinite Cardinal Numbers," **QM 35**, 87–94; reprinted in [1979, 427–434].
1906 "The Continuum and the Second Number Class," **LMS (2) 4**, 10–17; reprinted in [1979, 438–445].
1979 *Collected Papers of G. H. Hardy*, ed. by I. W. Busbridge and R. A. Rankin (Oxford: Clarendon), vol. VII.

Harper, Judith, and Rubin, Jean
1976 "Variations of Zorn's Lemma, Principles of Cofinality, and Hausdorff's Maximal Principle. Part I: Set Forms," **NDJ 17**, 565–588.

Hartogs, Friedrich
1915 "Über das Problem der Wohlordnung," **MA 76**, 436–443.

Harward, A. E.
1905 "On the Transfinite Numbers," **PM (6) 10**, 439–460.

Hausdorff, Felix
1904 "Der Potenzbegriff in der Mengenlehre," **DMV 13**, 569–571.
1906 "Untersuchungen über Ordnungstypen," **SW 58**, 106–169.
1907 "Untersuchungen über Ordnungstypen," **SW 59**, 84–159.
1907a "Über dichte Ordnungstypen," **DMV 16**, 541–546.
1908 "Grundzüge einer Theorie der geordneten Mengen," **MA 65**, 435–505.
1909 "Die Graduierung nach dem Endverlauf," **SW 61**, 297–334.
1914 *Grundzüge der Mengenlehre* (Leipzig: de Gruyter); reprinted (New York: Chelsea, 1965).
1914a "Bemerkung über den Inhalt von Punktmengen," **MA 75**, 428–433.

1916 "Die Mächtigkeit der Borelschen Mengen," **MA 77**, 430–437.
1927 Second edition of [1914].
1932 "Zur Theorie der linearen metrischen Räume," **JM 167**, 294–311.
1937 Third edition of [1914]; translated (New York: Chelsea, 1957).

Heine, Eduard
1872 "Die Elemente der Functionenlehre," **JM 74**, 172–188.

Henkin, Leon
1947 *The Completeness of Formal Systems* (Ph.D. dissertation: Princeton).
1954 "Metamathematical Theorems Equivalent to the Prime Ideal Theorem for Boolean Algebra," **BAMS 60**, 387–388.
1954a "Boolean Representation through Propositional Calculus," **FM 41**, 89–96.
See Addison, John; Henkin, Leon; and Tarski, Alfred.

Henkin, Leon, and Mostowski, Andrzej
1959 Review of [Malcev 1941], **JSL 24**, 55–57.

Herbrand, Jacques
1930 *Recherches sur la théorie de la démonstration* (Ph.D. dissertation: Paris); translated in part in van Heijenoort 1967, 525–581.

Hersh, Reuben
See Cohen, Paul J., and Hersh, Reuben.

Hessenberg, Gerhard
1906 *Grundbegriffe der Mengenlehre* (Göttingen: Vandenhoeck & Ruprecht); also in *Abhandlungen der Fries'schen Schule* N.S. **1**, Heft 4.
1908 "Willkürliche Schöpfungen des Verstandes?" **DMV 17**, 145–162.
1909 "Kettentheorie und Wohlordnung," **JM 135**, 81–133.

Heyting, Arend
1955 *Les fondements des mathématiques. Intuitionnisme. Théorie de la démonstration* (Paris: Gauthier-Villars).
1956 *Intuitionism: An Introduction* (Amsterdam: North-Holland).
1966 Second edition of [1956].
See Hahn *et alii.*

Hilbert, David
1897 "Die Theorie der algebraischen Zahlkörper," **DMV 4**, 175–546.
1899 *Grundlagen der Geometrie* (Leipzig: Teubner); translated by E. J. Townsend (Chicago: Open Court, 1902).
1900 "Mathematische Probleme. Vortrag, gehalten auf dem internationalem Mathematiker-Kongress zu Paris. 1900," **NG**, 253–297; translated in **BAMS (2) 8** (1902), 437–479.
1904 "Über die Grundlagen der Logik und der Arithmetik," *Verhandlungen des dritten internationalen Mathematiker-Kongresses* (Leipzig: Teubner, 1905), 174–185; translated in van Heijenoort 1967, 129–138.
1922 "Neubegründung der Mathematik. Erste Mitteilung," **MSH 1**, 157–177.
1923 "Die logischen Grundlagen der Mathematik," **MA 88**, 151–165.
1926 "Über das Unendliche," **MA 95**, 161–190; translated in van Heijenoort 1967, 367–392.

1928 "Die Grundlagen der Mathematik," **MSH 6**, 65–92; translated in van Heijenoort 1967, 464–479.
1930 "Probleme der Grundlegung der Mathematik," **MA 102**, 1–9.
1932–1935 *Gesammelte Abhandlungen*, 3 vols. (Berlin: Springer).

Hilbert, David, and Ackermann, Wilhelm
1928 *Grundzüge der theoretischen Logik* (Berlin: Springer).

Hilbert, David, and Bernays, Paul
1939 *Grundlagen der Mathematik. II* (Berlin: Springer).

Hobson, Ernest
1905 "On the General Theory of Transfinite Numbers and Order Types," **LMS (2) 3**, 170–188.
1906 "On the Arithmetic Continuum," **LMS (2) 4**, 21–28.
1907 *The Theory of Functions of a Real Variable and the Theory of Fourier's Series* (Cambridge: Cambridge University Press).
1921 Second edition of [1907], vol. I.
1926 Second edition of [1907], vol. II.
1927 Third edition of [1907], vol. I.

Hochstadt, Harry
1980 "Eduard Helly, Father of the Hahn–Banach Theorem," **MI 2**, 123–125.

Höft, Hartmut, and Howard, Paul E.
1973 "A Graph-Theoretic Equivalent to the Axiom of Choice," **ZML 19**, 191.

Howard, Paul E.
See Dawson, John W., and Howard, Paul E.
See Halpern, James D., and Howard, Paul E.
See Höft, Hartmut, and Howard, Paul E.

Huntington, Edward V.
1905 "The Continuum as a Type of Order: An Exposition of the Modern Theory. V–VI. With an Appendix on the Transfinite Numbers," **AM (2) 7**, 15–43; reprinted in [1917].
1917 *The Continuum and Other Types of Serial Order* (Cambridge: Harvard University Press); reprinted (New York: Dover, 1955).

Inagaki, Takeshi
1952 "Sur deux théorèmes concernant un ensemble partiellement ordonné," **MJO 1**, 167–176.

Jacobsthal, Ernst
1909 "Über den Aufbau der transfiniten Arithmetik," **MA 66**, 145–194.

Jaegermann, M.
1965 "The Axiom of Choice and Two Definitions of Continuity," **AP 13**, 699–704.

Janiszewski, Zygmunt
1910 "Sur la géometrie de lignes cantoriennes," **CRP 151**, 198–201.
1912 "Sur les continus irréductibles entre deux points," **JEP (2) 16**, 79–170.
1918 " O Potrzebach Matematyki w Polsce," **NP 1**, 11–20.

Jech, Thomas J.
1966 "On Ordering of Cardinalities," **AP 14**, 293–296.
1966a "On Cardinals and Their Successors," **AP 14**, 533–537.
1966b "Interdependence of Weakened Forms of the Axiom of Choice," **CMC 7**, 359–371.
1966c *Axiom výběru* (Ph.D. dissertation: Prague).
1968 "ω_1 Can Be Measurable," **IJM 6**, 363–367.
1971 "On Models for Set Theory without AC," in Scott 1971, 135–141.
1973 *The Axiom of Choice* (Amsterdam: North-Holland).
1974 (ed.) *Axiomatic Set Theory*, Proceedings of Symposia in Pure Mathematics, vol. 13, Part 2 (Providence: American Mathematical Society).
1977 "About the Axiom of Choice," in Barwise 1977, 345–370.
1978 *Set Theory* (New York: Academic Press).
1982 "On Hereditarily Countable Sets," **JSL 47**, 43–47.
See Felgner, Ulrich, and Jech, Thomas J.

Jech, Thomas J., and Sochor, Antonin
1966 "Applications of the θ-Model," **AP 14**, 351–355.

Jensen, Ronald B.
1966 "Independence of the Axiom of Dependent Choices from the Countable Axiom of Choice," **JSL 31**, 294.
1967 "Consistency Results for ZF," **NAMS 14**, 137.

Jordan, Camille
1892 "Remarques sur les intégrales définies," **JPA (4) 8**, 69–99.
1893 *Cours d'analyse de l'Ecole Polytechnique: Tome premier. Calcul différentiel*, second edition (Paris: Gauthier-Villars).

Jourdain, Philip E. B.
1904 "On the Transfinite Cardinal Numbers of Well-Ordered Aggregates," **PM (6) 7**, 61–75.
1904a "On the Transfinite Cardinal Numbers of Number-Classes in General," **PM (6) 7**, 294–303.
1905 "On Transfinite Cardinal Numbers of the Exponential Form," **PM (6) 9**, 42–56.
1905a "On a Proof that Every Aggregate Can Be Well-Ordered," **MA 60**, 465–470.
1905b "The Definition of a Series Similarly Ordered to the Series of All Ordinal Numbers," **MM 35**, 56–58.
1906 "On the Question of the Existence of Transfinite Numbers," **LMS (2) 4**, 266–283.
1906a "The Multiplication of an Infinity of Order Types," **MM 36**, 13–16.
1906b "On Sets of Intervals in a Simply-Ordered Series," **MM 36**, 61–69.
1906c Review of [Young 1906], **MGZ 3**, 373–375.
1907 "On the Comparison of Aggregates," **QM 38**, 352–367.
1908 "The Multiplication of Alephs," **MA 65**, 506–512.
1908a "On Infinite Sums and Products of Cardinal Numbers," **QM 39**, 375–384.
1910 "A Theorem in the General Theory of Ordered Aggregates," **QM 41**, 214–219.
1918 "A Proof that Any Transfinite Aggregate Can Be Well-Ordered," **NT 101**, 84.
1918a "Démonstration du théoreme d'après lequel tout ensemble peut être bien ordonné," **CRP 166**, 520–523.
1918b "A Proof that Any Aggregate Can Be Well-Ordered," **MD** (N.S.) **27**, 386–388.

1918c "A Proof that Any Aggregate Can Be Well-Ordered," **NT 101**, 304.
1918d "Démonstration du théorème d'après lequel tout ensemble peut être bien ordonné," **CRP 166**, 984–986.
1918e "Problems of Arrangement of an Infinite Class," **SR 13**, 299–304.
1919 "A Proof that Any Aggregate Can Be Well-Ordered," **NT 103**, 45.
1919a "A Proof that Any Aggregate Can Be Well-Ordered," **MD** (N.S.) **28**, 382–384.
1921 "A Proof that Every Aggregate Can Be Well-Ordered," **AC 43**, 239–261.

Kamke, Erich
1937 "Zur Definition der affinen Abbildung," **DMV 36**, 145–156.

Kelley, John
1950 "The Tychonoff Product Theorem Implies the Axiom of Choice," **FM 37**, 75–76.

Kennedy, Hubert C.
1975 "Nine Letters from Giuseppe Peano to Bertrand Russell," **JHP 13**, 205–220.
1980 *Peano: Life and Works of Giuseppe Peano* (Dordrecht: Reidel).
See Peano, Giuseppe.

Keyser, Cassius J.
1901 "Theorems Concerning Positive Definitions of Finite Assemblage and Infinite Assemblage," **BAMS (2) 7**, 218–226.
1904 "The Axiom of Infinity: A New Presupposition of Thought," **HB 2**, 532–552; reprinted in [1916].
1905 "Some Outstanding Problems for Philosophy," **JPP 2**, 207–213.
1905a "The Axiom of Infinity," **HB 3**, 380–383.
1916 *The Human Worth of Rigorous Thinking* (New York: Columbia University Press).

Kinna, W., and Wagner, K.
1955 "Über eine Abschwächung des Auswahlpostulates," **FM 42**, 75–82.

Kleene, Stephen
1976 "The Work of Kurt Gödel," **JSL 41**, 761–778.
1978 "An Addendum to 'The Work of Kurt Gödel'," **JSL 43**, 613.

Kleene, Stephen, and Vesley, Richard
1965 *The Foundations of Intuitionistic Mathematics, Especially in Relation to Recursive Functions* (Amsterdam: North-Holland).

Kleinberg, Eugene
1969 "The Independence of Ramsey's Theorem," **JSL 34**, 205–206.
1977 *Infinitary Combinatorics and the Axiom of Determinateness*, Lecture Notes in Mathematics, No. 162 (Berlin: Springer).

Klimovsky, Gregorio
1949 "Un teorema equivalente al de Zorn," **UMA 14**, 47–48.
1956 "Tres enunciados equivalentes al teorema de Zorn," **UBA 2.1**, 1–29.
1958 "El teorema de Zorn y la existencia de filtros e ideales maximales en los reticulados distributivos," **UMA 18**, 160–164.
1962 "El axioma de elección y la existencia de subgrupos conmutativos maximales," **UMA 20**, 267–287.

Kline, Morris
1972 *Mathematical Thought from Ancient to Modern Times* (New York: Oxford University Press).

Kneser, Helmuth
1950 "Eine direkte Ableitung des Zornschen Lemmas aus dem Auswahlaxiom," **MZ 53**, 110–113.

Kondô, Motokiti
1939 "Sur l'uniformisation des complémentaires analytiques et les ensembles projectifs de la seconde classe," **JJM 15**, 197–230.

König, Dénes
1908 "Zur Theorie der Mächtigkeiten," **CMP 26**, 339–342.
1916 "Über Graphen und ihre Anwendung auf Determinantstheorie und Mengenlehre," **MA 77**, 453–465.
1926 "Sur les correspondances multivoques des ensembles," **FM 8**, 114–134.
1927 "Über eine Schlussweise aus dem Endlichen ins Unendliche: Punktmengen. Kartenfärben. Verwandtschaftsbeziehungen. Schachspiel," **ALS 3**, 121–130.

König, Julius
1904 "Zum Kontinuum-Problem," *Verhandlungen des dritten internat. Math.-Kongress* (Leipzig: Teubner, 1905), 144–147.
1905 "Über die Grundlagen der Mengenlehre und das Kontinuumproblem," **MA 61**, 156–160; translated in van Heijenoort 1967, 145–149.
1905a "Zum Kontinuumproblem," **MA 60**, 177–180, 462.
1906 "Über die Grundlagen der Mengenlehre und das Kontinuum-Problem (Zweite Mitteilung)," **MA 63**, 217–221.
1906a "Sur la théorie des ensembles," **CRP 143**, 110–112.
1914 *Neue Grundlagen der Logik, Arithmetik, und Mengenlehre*, ed. by D. König (Leipzig: Veit).

Korselt, Alwin
1911 "Über einen Beweis des Äquivalenzsatzes," **MA 70**, 294–296.

Kowalewski, Gerhard
1950 *Bestand und Wandel* (Munich: Oldenbourg).

Kreisel, Georg
1960 "La prédicativité," **SMF 88**, 371–391.
1962 "The Axiom of Choice and the Class of Hyperarithmetic Functions," **IM 24**, 307–319.
1980 "Kurt Gödel," **BMRS 26** (December), 149–224.

Krull, Wolfgang
1923 "Algebraische Theorie der Ringe. I," **MA 88**, 80–111.
1929 "Die Idealtheorie in Ringen ohne Endlichkeitsbedingungen," **MA 101**, 729–744.

Kruse, Arthur H.
1962 "Some Observations on the Axiom of Choice," **ZML 8**, 125–146.
1963 "A Problem on the Axiom of Choice," **ZML 9**, 207–218.

Kunen, Kenneth
1973 "A Model for the Negation of the Axiom of Choice," in Mathias and Rogers 1973, 489–494.
1980 *Set Theory: An Introduction to Independence Proofs* (Amsterdam: North-Holland).

Kuratowski, Kazimierz
1921 "Sur la notion de l'ordre dans la théorie des ensembles," **FM 2**, 161–171.
1922 "Une méthode d'élimination des nombres transfinis des raisonnements mathématiques," **FM 3**, 76–108.
1923 "Sur l'existence effective des fonctions représentables analytiquement de toute classe de Baire," **CRP 176**, 229–232.
1924 "Sur l'état actuel de l'axiomatique de la théorie des ensembles," **ASP 3**, 146–147.
1933 *Topologie I*, Monografje Matematyczne, vol. 3 (Warsaw: Garasiński).
1975 "Birth of the Polish School of Mathematics," *Studies in Topology*, ed. by Nick Stavrakas and Keith Allen (New York: Academic Press), xix–xxii.
1980 *A Half Century of Polish Mathematics: Remembrances and Reflections* (Warsaw: Polish Scientific Publishers).
See Banach, Stefan, and Kuratowski, Kazimierz.

Kuratowski, Kazimierz, and Mostowski, Andrzej
1968 *Set Theory* (Amsterdam: North-Holland).
1976 Second edition of [1968].

Kuratowski, Kazimierz, and Sierpiński, Wacław
1926 "Sur un problème de M. Fréchet concernant les dimensions des ensembles linéaires," **FM 8**, 193–200.

Kurepa, Djuro
1952 "Sur la relation d'inclusion et l'axiome de choix de Zermelo," **SMF 80**, 225–232.
1953 "Über das Auswahlaxiom," **MA 126**, 381–384.

Kuzawa, Sister Mary Grace
1968 *Modern Mathematics: The Genesis of a School in Poland* (New Haven: College & University Press).
1970 "Fundamenta Mathematicae: An Examination of Its Founding and Significance," **AMM 77**, 485–492.

Lakatos, Imre
1967 (ed.) *Problems in the Philosophy of Mathematics* (Amsterdam: North-Holland).

Läuchli, Hans
1962 "Auswahlaxiom in der Algebra," **CMH 37**, 1–18.
1964 "The Independence of the Ordering Principle from a Restricted Axiom of Choice," **FM 54**, 31–43.
1971 "Coloring Infinite Graphs and the Boolean Prime Ideal Theorem," **IJM 9**, 422–429.

Lebesgue, Henri
1902 "Intégrale, longueur, aire," **MPA (3) 7**, 231–359.
1904 *Leçons sur l'intégration et la recherche des fonctions primitives* (Paris: Gauthier-Villars).

1905 "Sur les fonctions représentables analytiquement," **JPA 60**, 139–216.
1907 "Contributions à l'étude des correspondances de M. Zermelo," **SMF 35**, 202–212.
1907a "Sur les transformations ponctuelles, transformant les plans en plans, qu'on peut définir par des procédés analytiques," **AT 42**, 532–539.
1917 "Sur certaines démonstrations d'existence," **SMF 45**, 132–144.
1918 "Remarques sur les théories de la mesure et de l'intégration," **ENS (3) 35**, 191–250.
1921 "Sur les correspondances entre les points de deux espaces," **FM 2**, 256–285.
1922 *Notice sur les travaux scientifiques* (Toulouse: E. Privat).
1922a "A propos d'une nouvelle revue mathématique: Fundamenta Mathematicae," **BSM (2) 46**, 35–48.
1928 Second edition of [1904].
1941 "Les controverses sur la théorie des ensembles et la question des fondements," in Gonseth 1941, 109–122.
1971 "A propos de quelques travaux mathématiques récents," **EM (2) 17**, 1–48.
1972–1973 *Oeuvres scientifiques*, 5 vols. (Geneva: Kundig).
See Baire, René; Borel, Emile; Hadamard, Jacques; and Lebesgue, Henri.

Lejeune Dirichlet, Peter, and Dedekind, Richard
1863 *Vorlesungen über Zahlentheorie* (Braunschweig: Vieweg).
1871 Second edition of [1863].
1879 Third edition of [1863].
1893 Fourth edition of [1863].

Lennes, N. J.
1922 "On the Foundations of the Theory of Sets," **BAMS 28**, 300.

Levi, Beppo
1900 "Sulla teoria delle funzioni e degli insiemi," **ALR (5) 9**, 72–79.
1902 "Intorno alla teoria degli aggregati," **IL (2) 35**, 863–868.
1918 "Riflessioni sopra alcuni principii della teoria degli aggregati e delle funzioni," *Scritti matematici offerti ad Enrico d'Ovidio* (Turin: Bocca), 305–324.
1923 "Sui procedimenti transfiniti (Auszug aus einem Briefe an Herrn Hilbert)," **MA 90**, 164–173.
1934 "La nozione di 'dominio deduttivo' e la sua importanza in taluni argomenti relativi ai fondamenti dell'analisi," **FM 23**, 63–74.

Levy, Azriel
1958 "The Independence of Various Definitions of Finiteness," **FM 46**, 1–13.
1961 "Comparing the Axioms of Local and Universal Choice," in Bar-Hillel *et alii* 1961, 83–90.
1962 "Axioms of Multiple Choice," **FM 50**, 475–483.
1963 "Independence Results in Set Theory by Cohen's Method. I," **NAMS 10**, 592–593.
1963a "Remarks on a Paper by J. Mycielski," **AMH 14**, 125–130.
1964 "The Interdependence of Certain Consequences of the Axiom of Choice," **FM 54**, 135–157.
1965 "The Fraenkel–Mostowski Method for Independence Proofs in Set Theory," in Addison *et alii* 1965, 221–228.
1968 "On Von Neumann's Axiom System for Set Theory," **AMM 75**, 762–763.
See Feferman, Solomon, and Levy, Azriel.
See Fraenkel, Abraham; Bar-Hillel, Yehoshua; and Levy, Azriel.
See Halpern, James D., and Levy, Azriel.

Lévy, Paul
1950 "Axiome de Zermelo et nombres transfinis," **ENS (3) 67**, 15–49.

Lindelöf, Ernst
1903 "Sur quelques points de la théorie des ensembles," **CRP 137**, 697–700.
1905 "Remarques sur un théorème fondamental de la théorie des ensembles," **AC 29**, 183–190.

Lindenbaum, Adolf, and Mostowski, Andrzej
1938 "Über die Unabhängigkeit des Auswahlaxioms und einiger seiner Folgerungen," **CRV 31**, 27–32.

Lindenbaum, Adolf, and Tarski, Alfred
1926 "Communication sur les recherches de la théorie des ensembles," **CRV 19**, 299–330.

Łoś, Jerzy, and Ryll-Nardzewski, Czesław
1951 "On the Application of Tychonoff's Theorem in Mathematical Proofs," **FM 38**, 233–237.
1955 "Effectiveness of the Representation Theory for Boolean Algebras," **FM 41**, 49–56.

Löwenheim, Leopold
1915 "Über Möglichkeiten im Relativkalkul," **MA 76**, 447–470; translated in van Heijenoort 1967, 228–251.

Luxemburg, W. A. J.
1969 "Reduced Powers of the Real Number System and Equivalents of the Hahn–Banach Extension Theorem," *Applications of Model Theory to Algebra, Analysis, and Probability*, ed. by W. A. J. Luxemburg (New York: Holt, Rinehart, and Winston), 123–137.

Luzin, Nikolai
1912 "Sur les propriétés des fonctions mesurables," **CRP 154**, 1688–1690.
1914 "Sur un problème de M. Baire," **CRP 158**, 1258–1261.
1917 "Sur la classification de M. Baire," **CRP 164**, 91–94.
1927 "Sur une question concernant la propriété de M. Baire," **FM 9**, 116–118.
1927a "Sur les ensembles analytiques," **FM 10**, 1–95.
1930 *Leçons sur les ensembles analytiques et leurs applications* (Paris: Gauthier-Villars); reprinted (New York: Chelsea, 1972).
1930a "Sur le problème de M. J. Hadamard d'uniformisation des ensembles," **CRP 190**, 349–351.
1930b "Sur le problème de M. Jacques Hadamard d'uniformisation des ensembles," **MCJ 4**, 54–65.
1947 "O chastyakh natural'nogo ryada," **IAN (5) 11**, 403–410.

Luzin, Nikolai, and Novikov, P. S.
1935 "Choix effectif d'un point dans un complémentaire analytique arbitraire, donné par un crible," **FM 25**, 559–560.

Luzin, Nikolai, and Sierpiński, Wacław
1917 "Sur une propriété du continu," **CRP 165**, 498–500.
1917a "Sur une décomposition d'un intervalle en une infinité non dénombrable d'ensembles non mesurables," **CRP 165**, 422–424.
1918 "Sur quelques propriétés des ensembles (A)," **ASC**, 35–48.
1923 "Sur un ensemble non mesurable B," **JPA (9) 2**, 53–72.

Lyubekii, V. A.
1970 "The Existence of a Non-Measurable Set of Type A_2 Implies the Existence of an Uncountable Set of Type *CA* which Does Not Include a Perfect Subset," **SMD 11**, 1513–1515.

Mahlo, Paul
1911 "Über lineare transfinite Mengen," **SW 63**, 187–225.

Malcev, Anatolii
1936 "Untersuchungen aus dem Gebiete der mathematischen Logik," **MS 43**, 323–336; translated in [1971, 1–14].
1941 "On a General Method for Obtaining Local Theorems in Group Theory" (Russian), **UZ 1**, 3–9; translated in [1971, 15–21].
1971 *The Metamathematics of Algebraic Systems. Collected Papers: 1936–1967*, ed. by B. F. Wells III (Amsterdam: North-Holland).

Marczewski, Edward
See Szpilrajn, Edward.

Martin, D. A.
1975 "Borel Determinacy," **AM 102**, 363–371.

Mathias, A. R. D., and Rogers, H.
1973 (eds.) *Cambridge Summer School in Mathematical Logic*, Lecture Notes in Mathematics, No. 337 (Berlin: Springer).

May, Kenneth O.
1971 Review of [Grattan-Guinness 1970], **MR 41**, 948–949.

Mazurkiewicz, Stefan
1910 "Sur la théorie des ensembles," **CRP 151**, 296–298.

McCoy, Neal
1938 "Subrings of Infinite Direct Sums," **DM 4**, 486–494.

Medvedev, Fedor Andreyevich
1965 *Razvitie teorii mnozhestv v XIX veke* (Moscow: Nauka).
1966 "Ranniya isotoriya teorii ekvivalentnosti," **IST 17**, 229–246.
1976 *Frantsuzskaya shkola teorii funktsii i mnozhestv na rubezhe XIX–XX bb.* (Moscow: Nauka).

Mehrtens, Herbert
1979 *Die Entstehung der Verbandstheorie* (Hildesheim: Gerstenberg).

Mendelson, Elliott
1956 "The Independence of a Weak Axiom of Choice," **JSL 21**, 350–366.
1956a "Some Proofs of Independence in Axiomatic Set Theory," **JSL 21**, 291–303.
1958 "The Axiom of Fundierung and the Axiom of Choice," **ALG 4**, 67–70.

Merzbach, Julius
1925 *Bemerkungen zur Axiomatik der Mengenlehre* (Ph.D. dissertation: Marburg).

Meschkowski, Herbert
1967 *Probleme des Unendlichen. Werk und Leben Georg Cantors* (Braunschweig: Vieweg).

Mirimanoff, Dimitry
1917 "Les antinomies de Russell et de Burali-Forti et le problème fondamental de la théorie des ensembles," **EM 19**, 37–52.
1917a "Remarques sur la théorie des ensembles et les antinomies cantoriennes. I," **EM 19**, 209–217.

Mollerup, Johannes
1907 "Sur les sous-ensembles bien ordonnées du continu," **CMP 23**, 351–357.

Monna, A. F.
1972 "The Concept of Function in the 19th and 20th Centuries in particular with regard to the Discussions between Baire, Borel, and Lebesgue," **AH 9**, 57–84.

Mooij, J. J. A.
1966 *La philosophie des mathématiques de Henri Poincaré* (Paris: Gauthier-Villars).

Moore, E. H., and Smith, H. L.
1922 "A General Theory of Limits," **AJM 44**, 102–121.

Moore, Gregory H.
1976 "Ernst Zermelo, A. E. Harward, and the Axiomatization of Set Theory," **HM 3**, 206–209.
1978 "The Origins of Zermelo's Axiomatization of Set Theory," **JPL 7**, 307–329.
1980 "Beyond First-Order Logic: The Historical Interplay between Mathematical Logic and Axiomatic Set Theory," **HPL 1**, 95–137.

Moore, Gregory H., and Garciadiego, Alejandro R.
1981 "Burali-Forti's Paradox: A Reappraisal of Its Origins," **HM 8**, 319–350.

Moore, R. L.
1932 *Foundations of Point Set Theory*, Colloquium Publications, vol. 13 (New York: American Mathematical Society).
1962 Second edition of [1932].

Morris, Douglass B.
1969 "Choice and Cofinal Well-Ordered Subsets," **NAMS 16**, 1088.

Moschovakis, Yiannis N.
1980 *Descriptive Set Theory* (Amsterdam: North-Holland).

Moss, Barbara
1979 "Beppo Levi and the Axiom of Choice," **HM 6**, 54–56.

Moss, J. M. B.
1972 "Some B. Russell's Sprouts (1903–1908)," Lecture Notes in Mathematics, No. 255 (Berlin: Springer), 211–250.

Mostowski, Andrzej
1938 "O niezalezności definicji skończoności w systemie logiki" ("On the Independence of the Definitions of Finiteness in a System of Logic," Ph.D. dissertation), **ASPD 16**, 1–54; *cf.* Tarski, **JSL 3** (1938), 115–116.
1938a "Über den Begriff einer endlichen Menge," **CRV 31**, 13–20.
1939 "Über die Unabhängigkeit des Wohlordnungssatzes vom Ordnungsprinzip," **FM 32**, 201–252.
1945 "Axiom of Choice for Finite Sets," **FM 33**, 137–168.
1948 "On the Principle of Dependent Choices," **FM 35**, 127–130.
1950 "Some Impredicative Definitions in the Axiomatic Set Theory," **FM 37**, 111–124.
1958 "On a Problem of W. Kinna and K. Wagner," **COL 6**, 207–208.
1967 "Recent Results in Set Theory," in Lakatos 1967, 82–108.
See Henkin, Leon, and Mostowski, Andrzej.
See Kuratowski, Kazimierz, and Mostowski, Andrzej.
See Lindenbaum, Adolf, and Mostowski, Andrzej.

Mrowka, S.
1956 "On the Ideals' Extension Theorem and Its Equivalence to the Axiom of Choice," **FM 43**, 46–49.
1958 "Two Remarks on My Paper 'On the Ideals' Extension Theorem and Its Equivalence to the Axiom of Choice'," **FM 46**, 127–130.

Murata, Tamotsu
1958 "French Empiricism. I—One of the Studies of the Foundations of Mathematics," **CMS 6**, 92–114.
1966 "On the Meaning of 'Virtualité' in the History of Set Theory," **JHS 5**, 119–139.

Murphey, Murray G.
1961 *The Development of Peirce's Philosophy* (Cambridge: Harvard University Press).

Mycielski, Jan
1961 "Some Remarks and Problems on the Colouring of Infinite Graphs and the Theorem of Kuratowski," **AMH 12**, 125–129.
1964 "On the Axiom of Determinateness," **FM 53**, 205–224.
1964a "Two Remarks on Tychonoff's Product Theorem," **AP 12**, 439–441.
1966 "On the Axiom of Determinateness (II)," **FM 59**, 203–212.
1971 "On Some Consequences of the Axiom of Determinateness," in Scott 1971, 265–266.

Mycielski, Jan, and Steinhaus, Hugo
1962 "A Mathematical Axiom Contradicting the Axiom of Choice," **AP 10**, 1–3.

Myhill, John, and Scott, Dana
1971 "Ordinal Definability," in Scott 1971, 271–278.

Neumann, Bernhard H.
1937 "Some Remarks on Infinite Groups," **LMS 12**, 120–127.

Noether, Emmy
1916 "Die allgemeinsten Bereiche aus ganzen transzendenten Zahlen," **MA 77**, 103–128.
1927 "Abstrakter Aufbau der Idealtheorie in algebraischen Zahl- und Funktionenkörpern," **MA 96**, 26–61.

Noether, Emmy, and Cavaillès, Jean
1937 (eds.) *Briefwechsel Cantor–Dedekind* (Paris: Hermann); French translation in Cavaillès 1962.

Novak, Ilse L.
1950 "A Construction for Models of Consistent Systems," **FM 37**, 87–110.

Novikoff, Albert, and Barone, Jack
1977 "The Borel Law of Large Numbers, the Borel Zero–One Law, and the Work of Van Vleck," **HM 4**, 43–65.

Obreanu, Filip
1949 "Zorn's Theorem" (Rumanian), **AR1**, 687–692.

Pasch, Moritz
1882 *Vorlesungen über neuere Geometrie* (Leipzig: Teubner).

Peano, Giuseppe
1890 "Démonstration de l'intégrabilité des équations différentielles ordinaires," **MA 37**, 182–228.
1895–1908 *Formulaire de mathématiques*, 5 vols. (Turin: Bocca).
1902 "Confronto col Formulario," **RDM 8**, 7–11.
1906 "Additione," **RDM 8**, 143–157; translated in [1973, 206–218].
1906a "Super theorema de Cantor–Bernstein," **CMP 21**, 360–366.
1957–1959 *Opera scelte*, 3 vols., ed. by U. Cassina (Rome: Cremonese).
1973 *Selected Works of Giuseppe Peano*, ed. and trans. by H. C. Kennedy (Toronto: University of Toronto Press).

Pearcy, Carl
See Brown, Arlen, and Pearcy, Carl.

Pears, David F.
1972 (ed.) *Bertrand Russell* (New York).

Peirce, C. S.
1881 "On the Logic of Number," **AJM 4**, 85–95.
1885 "On the Algebra of Logic: A Contribution to the Philosophy of Notation," **AJM 7**, 180–202.
1889 Definition of "finite," *Century Dictionary* (New York: Century Co.), 2225.

1900 "Infinitesimals," **SI** (N.S.) **11**, 430–433.
1933 *Collected Papers*, ed. by Charles Hartshorne and Paul Weiss (Cambridge: Harvard University Press), vol. IV.
1976 *The New Elements of Mathematics*, ed. by Carolyn Eisele (The Hague: Mouton), vol. 3, part 2.

Pelc, Andrzej
1978 "On Some Weak Forms of the Axiom of Choice in Set Theory," **AP 26**, 585–589.

Phillips, Esther R.
1978 "Nicolai Nicolaevich Luzin and the Moscow School of the Theory of Functions," **HM 5**, 275–305.

Pieri, Mario
1906 "Sur la compatibilité des axiomes de l'arithmétique," **RMM 14**, 196–207.

Pincherle, Salvatore
1880 "Saggio di una introduzione alla teoria delle funzioni analitiche secondo i principii del Prof. C. Weierstrass," **GM 18**, 178–254, 317–357.

Pincus, David
1968 "Comparison of Independence Results in Mostowski's System G and in Zermelo–Fraenkel Set Theory," **NAMS 15**, 234.
1971 "Support Structures for the Axiom of Choice," **JSL 36**, 28–38.
1972 "Zermelo–Fraenkel Consistency Results by Fraenkel–Mostowski Methods," **JSL 37**, 721–743.
1972a "Independence of the Prime Ideal Theorem from the Hahn–Banach Theorem," **BAMS 78**, 766–770.
1974 "Cardinal Representatives," **IJM 18**, 321–344.
1974a "The Strength of the Hahn–Banach Theorem," in Dold and Eckmann 1974, 203–248.
1977 "Adding Dependent Choice," **AML 11**, 105–145.

Platek, Richard
1969 "Eliminating the Continuum Hypothesis," **JSL 34**, 219–225.

Poincaré, Henri
1905 "Les mathématiques et la logique," **RMM 13**, 815–835.
1906 "Les mathématiques et la logique," **RMM 14**, 17–34.
1906a "Les mathématiques et la logique," **RMM 14**, 294–317.
1906b "A propos de la logistique," **RMM 14**, 866–868.
1909 "Réflexions sur les deux notes précédentes," **AC 32,** 195–200.
1909a "La logique de l'infini," **RMM 17**, 461–482.
1912 "La logique de l'infini," **SC 12**, 1–11.

Putnam, Hilary
See Bernacerraf, Paul, and Putnam, Hilary.

Quine, W. V.
1937 "New Foundations for Mathematical Logic," **AMM 44**, 70–80.
1966 *The Ways of Paradox* (New York: Random House).

Ramsey, Frank
1925 "The Foundations of Mathematics," **LMS 2 (25)**, 338–384.
1929 "On a Problem of Formal Logic," **LMS (2) 30**, 264–286.
1931 *The Foundations of Mathematics and Other Logical Essays* (London: Kegan Paul).

Rang, Bernhard, and Thomas, Wolfgang
1981 "Zermelo's Discovery of the 'Russell Paradox'," **HM 8**, 15–22.

Rav, Yehuda
1980 "Birkhoff's Subdirect Product Representation Theorem for Rings Implies the Axiom of Choice," **AAMS 1**, 587.

Richard, Jules
1905 "Lettre à Monsieur le rédacteur de la Revue générale des Sciences," **AC 30**, 295–296; translated in van Heijenoort 1967, 142–144 (*cf.* Hadamard 1905a).
1907 "Sur un paradoxe de la théorie des ensembles et sur l'axiome Zermelo," **EM 9**, 94–98.
1929 "Sur l'axiome de Zermelo," **BSM (2) 53**, 106–109.

Rosser, J. Barkley
1937 "Gödel Theorems for Non-Constructive Logics," **JSL 2**, 129–137.
1969 *Simplified Independence Proofs* (New York: Academic Press).

Rubin, Herman
1958 "On a Problem of Kurepa Concerning the Axiom of Choice," **NAMS 5**, 378.
1960 "Two Propositions Equivalent to the Axiom of Choice Only under Both the Axioms of Extensionality and Regularity," **NAMS 7**, 381.

Rubin, Herman, and Rubin, Jean
1960 "Some New Forms of the Axiom of Choice," **NAMS 7**, 380–381.
1963 *Equivalents of the Axiom of Choice* (Amsterdam: North-Holland).

Rubin, Herman, and Scott, Dana
1954 "Some Topological Theorems Equivalent to the Boolean Prime Ideal Theorem," **BAMS 60**, 389.

Rubin, Jean
See Harper, Judith, and Rubin, Jean.
See Rubin, Herman, and Rubin, Jean.

Russ, S. B.
1980 "A Translation of Bolzano's Paper on the Intermediate Value Theorem," **HM 7**, 156–185.

Russell, Bertrand
1896 Review of [Hannequin 1895], **MD** (N.S.) **5**, 410–417.
1897 *An Essay on the Foundations of Geometry* (Cambridge: Cambridge University Press).
1901 "Sur la logique des relations . . . ," **RDM 7**, 115–148.
1902 "Théorie générale des séries bien ordonnées," **RDM 8**, 12–43.
1903 *The Principles of Mathematics* (Cambridge: Cambridge University Press).

1904 "The Axiom of Infinity," **HB 2**, 809–812; reprinted in [1973, 256–259].
1906 "On Some Difficulties in the Theory of Transfinite Numbers and Order Types," **LMS (2) 4**, 29–53; reprinted in [1973, 135–164].
1906a "Les paradoxes de la logique," **RMM 14**, 627–650; translated in [1973, 190–214].
1908 "Mathematical Logic as Based on the Theory of Types," **AJM 30**, 222–262; reprinted in van Heijenoort 1967, 150–182.
1911 "Sur les axiomes de l'infini et du transfini," **SMFC 39**, 22–35; translated in Grattan-Guinness 1977, 161–174.
1919 *Introduction to Mathematical Philosophy* (London: Allen and Unwin).
1927 *The Analysis of Matter* (London: Allen and Unwin).
1931 Review of [Ramsey 1931], **MD** (N.S.) **40**, 476–482.
1937 Second edition of [1903] (London: Allen and Unwin).
1959 *My Philosophical Development* (London: Allen and Unwin).
1973 *Essays in Analysis*, ed. by D. Lackey (London: Allen and Unwin).
1982–? *Collected Papers of Bertrand Russell*, ed. by K. Blackwell, A. Brink, N. Griffin, R. Rempel, and J. Slater, 28 vols. (London: Allen and Unwin).

Russell, Bertrand, and Whitehead, Alfred North
1910 *Principia Mathematica* (Cambridge: Cambridge University Press), vol. 1.
1912 Vol. 2 of [1910].
1913 Vol. 3 of [1910].
1925 Second edition of [1910].
1927 Second edition of [1912].
1927a Second edition of [1913].

Ryll-Nardzewski, Czesław
See Łoś, Jerzy, and Ryll-Nardzewski, Czesław.

Sageev, Gershon
1973 "An Independence Result Concerning the Axiom of Choice," **NAMS 20**, A-22.
1975 "An Independence Result Concerning the Axiom of Choice," **AML 8**, 1–184.

Scheeffer, Ludwig
1884 "Zur Theorie der stetigen Funktionen einer reellen Veränderlichen," **AC 5**, 279–296.

Schilpp, Paul Arthur
1944 (ed.) *The Philosophy of Bertrand Russell* (New York: Tudor).

Schmidt, Jürgen
1953 "Einige grundlegende Begriffe und Sätze aus der Theorie der Hullenoperatoren," *Bericht über die Mathematiker-Tagung in Berlin von 14. bis 18. Januar 1953* (Berlin: Deutscher Verlag der Wissenschaften), 21–48.
1962 "Einige algebraische Äquivalente zum Auswahlaxiom," **FM 50**, 485–496.

Schoenflies, Arthur
1900 "Die Entwickelung der Lehre von den Punktmannigfaltigkeiten," **DMV 8** (ii), 1–251.
1905 "Über wohlgeordnete Mengen," **MA 60**, 181–186.
1908 "Die Entwickelung der Lehre von den Punktmannigfaltigkeiten. Zweiten Teil," **DMV**, Ergänzungsband **2**, 1–331.

1911 "Über die Stellung der Definition in der Axiomatik," **DMV 20**, 222–255.
1921 "Zur Axiomatik der Mengenlehre," **MA 83**, 173–200.
1922 "Zur Erinnerung an Georg Cantor," **DMV 31**, 90–106.
1922a "Bemerkung zur Axiomatik der Grössen und Mengen," **MA 85**, 60–64.
1927 "Die Krisis in Cantors mathematischen Schaffen," **AC 50**, 1–23.

Schoenflies, Arthur, and Hahn, Hans
1913 *Entwickelung der Mengenlehre und ihrer Anwendungen, Erster Hälfte* (Leipzig: Teubner).

Schreier, Otto
1927 "Die Untergruppen der freien Gruppen," **MSH 5**, 161–183.
See Artin, Emil, and Schreier, Otto.

Schröder, Ernst
1890 *Vorlesungen über die Algebra der Logik* (*exacte Logik*) (Leipzig: Teubner), vol. 1; reprinted (New York: Chelsea, 1966).
1898 "Ueber zwei Definitionen der Endlichkeit und G. Cantor'sche Sätze," **NAL 71**, 303–362.
1900 "Sur une extension de l'idée d'ordre," *First International Congress of Philosophy* (*Paris*) **3**, 235–240.

Scorza-Dragoni, Giuseppe
1930 "Sull'approssimazione dell'integrale di Lebesgue mediante integrale di Riemann," **MPA (4) 7**, 61–70.
1936 "Sul principio di approssimazione nella teoria degli insiemi e sulla quasi-continuità delle funzioni misurabili," **SMR 1**, 53–58.

Scott, Dana
1954 "The Theorem on Maximal Ideals in Lattices and the Axiom of Choice," **BAMS 60**, 83.
1954a "Prime Ideal Theorems for Rings, Lattices, and Boolean Algebras," **BAMS 60**, 390.
1955 "Definitions by Abstraction in Axiomatic Set Theory," **BAMS 61**, 442.
1961 "Measurable Cardinals and Constructible Sets," **AP 7**, 145–149.
1971 (ed.) *Axiomatic Set Theory*, Proceedings of Symposia in Pure Mathematics, vol. 13, part 1 (Providence: American Mathematical Society).
1974 "Axiomatizing Set Theory," in Jech 1974, 207–214.
See Myhill, John, and Scott, Dana.
See Rubin, Herman, and Scott, Dana.

Shelah, Saharon
1980 "Going to Canossa," **AAMS 1**, 630.

Shepherdson, John C.
1951 "Inner Models of Set Theory," **JSL 16**, 161–190.
1952 "Inner Models of Set Theory, Part II," **JSL 17**, 225–237.
1953 "Inner Models of Set Theory, Part III," **JSL 18**, 145–167.

Shoenfield, Joseph R.
1955 "The Independence of the Axiom of Choice," **JSL 20**, 202.
1961 "The Problem of Predicativity," in Bar-Hillel *et alii* 1961, 132–139.
1971 "Unramified Forcing," in Scott 1971, 357–381.

Siegmund-Schultze, Reinhard

1978 *Der Strukturwandel in der Mathematik vom 19- zum 20-Jahrhundert, untersucht am Beispiel der Entstehung der ersten Begriffsbildungen der Funktionalanalysis* (Ph.D. dissertation: Halle).

Sierpiński, Wacław

1916 "Sur le rôle de l'axiome de M. Zermelo dans l'analyse moderne," **CRP 163**, 688–691.

1917 "Sur la démonstration du théorème de Cantor–Bendixson et sur l'énumération des points séparés d'un ensemble," **FV 59**, no. 17.

1917a "Sur quelques problèmes qui impliquent des fonctions non-mesurables," **CRP 164**, 882–884.

1918 "L'axiome de M. Zermelo et son rôle dans la théorie des ensembles et l'analyse," **ASC**, 97–152.

1920 "Une démonstration du théorème sur la structure des ensembles de points," **FM 1**, 1–6.

1920a "Sur l'équation fonctionelle $f(x + y) = f(x) + f(y)$," **FM 1**, 116–122.

1920b "Sur la décomposition des ensembles de points en parties homogènes," **FM 1**, 28–34.

1920c "Sur un problème de M. Lebesgue," **FM 1**, 152–158.

1920d "Sur la question de la mesurabilité de la base de M. Hamel," **FM 1**, 105–111.

1921 "Les exemples effectifs et l'axiome du choix," **FM 2**, 112–118.

1922 Aksioma Zermelo i ee rol v teorii mnozhest i analize," **MS 31**, 94–128 (Russian translation of [1918]).

1922a "Sur l'égalite $2\mathfrak{m} = 2\mathfrak{n}$ pour les nombres cardinaux," **FM 3**, 1–6.

1923 "Sur l'invariance topologique de la propriété de Baire," **FM 23**, 319–323.

1924 "Un exemple effectif d'un ensemble mesurable (B) de classe α," **FM 6**, 39–44.

1928 "Sur une décomposition d'ensembles," **MMP 35**, 239–243.

1928a *Leçons sur les nombres transfinis* (Paris: Gauthier-Villars).

1928b *Zarys teorii mnogości*, part I, third edition (Warsaw).

1928c "Sur un ensemble non dénombrable dont to te image continue est de première catégorie," **ASC**, 455–458.

1934 *Introduction to General Topology*, trans. by Cecilia Krieger (Toronto: University of Toronto Press).

1934a *Hypothèse du continu*, Monografie Matematyczne, vol. 4 (Warsaw: Garasiński).

1937 "Sur une décomposition du segment en plus que $2^{\aleph_0}$ ensembles non mesurables et presque disjoints," **FM 28**, 111–114.

1937a "Sur deux propositions, dont l'ensemble équivaut à l'hypothèse du continu," **FM 29**, 31–33.

1938 "Fonctions additives non complètement additives et fonctions non mesurables," **FM 30**, 96–99.

1941 "L'axiome du choix et l'hypothèse du continu," in Gonseth 1941, 125–143.

1945 "Sur le paradoxe de MM. Banach et Tarski," **FM 33**, 229–234.

1945a "Sur le paradoxe de la sphère," **FM 33**, 235–244.

1946 "Sur une proposition équivalente à l'axiome du choix," **ACL 9**, 111–112.

1947 "Sur la différence de deux nombres cardinaux," **FM 34**, 119–126.

1947a "L'hypothèse généralisée du continu et l'axiome du choix," **FM 34**, 1–5.

1947b "Sur une proposition qui entraîne l'existence des ensembles non mesurables," **FM 34**, 157–162.

1948 "Sur une proposition de A. Lindenbaum équivalente à l'axiome du choix," **CRV 40**, 1–3.

1954 "Sur une proposition équivalente à l'existence d'un ensemble de nombres réels de puissance $\aleph_1$," **AP 2**, 53–54.
1955 "L'axiome du choix pour les ensembles finis," **MC 10**, 92–99.
1959 "The Warsaw School of Mathematics and the Present State of Mathematics in Poland," **PR 4**, 51–63.
1963 "Polish School of Mathematics," **PP 6 (8)**, 25–35.
1965 "Sur un théorème équivalent à l'axiome du choix," **NDJ 6**, 161–162.
1975 *Oeuvres choisies*, vol. II: *Théorie des ensembles et ses applications, travaux des années 1908–1929* (Warsaw: PWN).
1976 Vol. III of [1975]: *Théorie des ensembles et ses applications, travaux des années 1930–1966* (Warsaw: PWN).
See Kuratowski, Kazimierz, and Sierpiński, Wacław.

Sierpiński, Wacław, and Tarski, Alfred
1930 "Sur une propriété charactéristique des nombres inaccessibles," **FM 15**, 292–300; reprinted in Sierpiński 1976, 29–35.

Sinaceur, Mohamed A.
1979 "La méthode mathématique de Dedekind," **RHS 32**, 107–142.

Skolem, Thoralf
1920 "Logisch-kombinatorische Untersuchungen über die Erfüllbarkeit oder Beweisbarkeit mathematischer Sätze nebst einem Theoreme über dichte Mengen," **VS** (1920, issue 4), 1–36; translated in van Heijenoort 1967, 252–263.
1923 "Einige Bemerkungen zur axiomatischen Begründung der Mengenlehre," *Den femte skandinaviska matematikerkongressen, Redogörelse* (Helsinki: Akademiska Bokhandeln), 217–232; translated in van Heijenoort 1967, 290–301.
1923a "Begründung der elementaren Arithmetik durch die rekurrierende Denkweise ohne Anwendung scheinbarer Veränderlichen mit unendlichem Ausdehnungsbereich," **VS** (1923, issue 6), 1–38; translated in van Heijenoort 1967, 302–333.
1929 "Über einige Grundlagenfragen der Mathematik," **VS** (1929, issue 4), 1–49.
1930 "Einige Bemerkungen zu der Abhandlung von E. Zermelo: 'Über die Definitheit in der Axiomatik'," **FM 15**, 337–341.
1933 "Über die Unmöglichkeit einer vollständigen Charakterisierung der Zahlenreihe mittels eines endlichen Axiomensystems," **NMF (2)** no. 10, 73–82.
1934 "Über die Nicht-charakterisierbarkeit der Zahlenreihe mittels endlich oder abzählbar unendlich vieler Aussagen mit ausschliesslich Zahlenvariablen," **FM 23**, 150–161.
1941 "Sur la portée du théorème de Löwenheim–Skolem," in Gonseth 1941, 25–52.
1970 *Selected Works in Logic*, ed. by J. E. Fenstad (Oslo: Universitetsforlaget).

Sobociński, Bolesław
1960 "A Simple Formula Equivalent to the Axiom of Choice," **NDJ 1**, 115–117.
1964 "A Theorem of Sierpiński on Triads and the Axiom of Choice," **NDJ 5**, 51–58.
1965 "A Note on Certain Set-Theoretical Formulas," **NDJ 6**, 157–160.

Sochor, Antonin
See Jech, Thomas, and Sochor, Antonin.

Solovay, Robert M.
1964 "The Measure Problem," **JSL 29**, 227–228.
1965 "The Measure Problem," **NAMS 12**, 217.
1970 "A Model of Set Theory in which Every Set of Reals is Lebesgue Measurable," **AM 92**, 1–56.

Specker, Ernst
1953 "The Axiom of Choice in Quine's 'New Foundations for Mathematical Logic'," **PAS 39**, 972–975.
1954 "Verallgemeinerte Kontinuumshypothese und Auswahlaxiom," **ADM 5**, 332–337.
1957 "Zur Axiomatik der Mengenlehre (Fundierungs- und Auswahlaxiom)," **ZML 3**, 173–210.

Steinhaus, Hugo
1965 "Games, an Informal Talk," **AMM 72**, 457–468.
See Mycielski, Jan, and Steinhaus, Hugo.

Steinitz, Ernst
1910 "Algebraische Theorie der Körper," **JM 137**, 167–309.

Stone, Marshall
1934 "Boolean Algebras and Their Application to Topology," **PAS 20**, 197–202.
1936 "The Theory of Representations for Boolean Algebras," **TAMS 40**, 37–111.
1937 "Applications of the Theory of Boolean Rings to General Topology," **TAMS 41**, 375–481.
1938 "The Representation of Boolean Algebras," **BAMS 44**, 807–816.

Sudan, Gabriel
1939 "Sur une note de A. Tarski," **CRR 3**, 7–8.

Suslin, Mikhail
1917 "Sur une définition des ensembles mesurables B sans nombres transfinis," **CRP 164**, 88–91.

Szele, T.
1950 "On Zorn's Lemma," **PMD 1**, 254–256.

Szmielew, Wanda
1947 "On Choices from Finite Sets," **FM 34**, 75–80.

Szpilrajn, Edward
1930 "Sur l'extension de l'ordre partiel," **FM 16**, 386–389.

Tambs Lyche, Ralph
1924 "Sur l'équation fonctionelle d'Abel," **FM 5**, 331–333.

Tannery, Jules
1886 *Introduction à la théorie des fonctions d'une variable* (Paris: Hermann).
1897 "De l'infini mathématique," **RG 8**, 129–140.

Tarski, Alfred
1924 "Sur quelques théorèmes qui équivalent à l'axiome du choix," **FM 5**, 147–154.
1924a "Sur les ensembles finis," **FM 6**, 45–95.
1925 "Quelques théorèmes sur les alephs," **FM 7**, 1–14.
1927 "Remarque concernant l'arithmétique des nombres cardinaux," **ASP 5**, 101.
1929 "Geschichtliche Entwickelung und gegenwärtiger Zustand der Gleichmächtigkeitstheorie und der Kardinalarithmetik," **ASPD 7**, 48–54.
1930 "Une contribution à la théorie de la mesure," **FM 15**, 42–50.
1934 Editor's note to [Skolem 1934], **FM 23**, 161.
1938 "Ein Überdeckungssatz für endliche Mengen nebst einigen Bemerkungen über die Definitionen der Endlichkeit," **FM 30**, 156–163.
1938a "Über unerreichbare Kardinalzahlen," **FM 30**, 68–89.
1938b "Eine äquivalente Formulierung des Auswahlaxioms," **FM 30**, 197–201.
1939 "On Well-Ordered Subsets of Any Set," **FM 32**, 176–183.
1939a "Ideale in vollständigen Mengenkörpern. I," **FM 32**, 45–63.
1948 "Axiomatic and Algebraic Aspects of Two Theorems on Sums of Cardinals," **FM 35**, 79–104.
1950 "Some Notions and Methods on the Borderline of Algebra and Metamathematics," *Proceedings of the International Congress of Mathematics; Cambridge, Massachusetts, August 30–September 6, 1950*, 705–720.
1954 "Theorems on the Existence of Successors of Cardinals, and the Axiom of Choice," **IM 16**, 26–32.
1954a "Prime Ideal Theorems for Boolean Algebras and the Axiom of Choice," **BAMS 60**, 390–391.
1955 "The Notion of Rank in Axiomatic Set Theory and Some of Its Applications," **BAMS 61**, 443.
1965 "On the Existence of Large Sets of Dedekind Cardinals," **NAMS 12**, 719.
See Addison, John; Henkin, Leon; and Tarski, Alfred.
See Banach, Stefan, and Tarski, Alfred.
See Lindenbaum, Adolf, and Tarski, Alfred.
See Sierpiński, Wacław, and Tarski, Alfred.

Teichmüller, Oswald
1936 "Operatoren in Wachsschen Raum," **JM 174**, 73–124.
1939 "Braucht der Algebraiker das Auswahlaxiom?" **DT 4**, 567–577.

Thomas, Wolfgang
See Rang, Bernhard, and Thomas, Wolfgang.

Tonelli, Leonida
1913 "Sul valore di un certo ragionamento," **AT 49**, 4–14.
1921 *Fondamenti di calcolo delle variazioni*, vol. 1 (Bologna: Zanichelli).
1923 "Sulla nozione di integrale," **MPA (4) 1**, 105–145.

Troelstra, A. S.
1977 "Aspects of Constructive Mathematics," in Barwise 1977, 973–1052.

Truss, John K.
1970 "Finite Versions of the Axiom of Choice," **NAMS 17**, 577.
1973 "Finite Axioms of Choice," **AML 6**, 147–176.
1973a "On Successors in Cardinal Arithmetic," **FM 78**, 7–21.

Tukey, John
1940 *Convergence and Uniformity in Topology*, Annals of Mathematics Studies, No. 2 (Princeton: Princeton University Press).

Tychonoff, Andrei
1930 "Über die topologische Erweiterung von Räumen," **MA 102**, 544–561.

Ulam, Stanisław
1929 "Concerning Functions of Sets," **FM 14**, 231–233.
1930 "Zur Masstheorie in der allgemeinen Mengenlehre," **FM 16**, 140–150.
1958 "John von Neumann, 1903–1957," **BAMS 64** (May Supplement), 1–49.
1974 *Sets, Numbers, and Universes: Selected Works*, ed. by W. A. Beyer *et alii* (Cambridge: MIT Press).

Urysohn, Paul
1925 "Über die Mächtigkeit der zusammenhängenden Mengen," **MA 94**, 262–295.
1925a "Zum Metrisationproblem," **MA 94**, 309–315.
See Alexandroff, Paul, and Urysohn, Paul.

van der Waerden, Bartel
1930 *Moderne Algebra*, vol. 1 (Berlin: Springer).
1937 Second edition of [1930].
1950 Third edition of [1930].

van Heijenoort, Jean
1967 *From Frege to Gödel. A Source Book in Mathematical Logic, 1879–1931* (Cambridge: Harvard University Press).

Van Vleck, Edward
1908 "On Non-Measurable Sets of Points, with an Example," **TAMS 9**, 237–244.

Vaughan, Herbert E.
1952 "Well-Ordered Subsets and Maximal Members of Ordered Sets," **PJM 2**, 407–412.

Vaught, Robert
1952 "On the Equivalence of the Axiom of Choice and a Maximal Principle," **BAMS 58**, 66.
1954 "Applications of the Löwenheim–Skolem–Tarski Theorem to Problems of Completeness and Decidability," **IM 16**, 467–472.
1956 "On the Axiom of Choice and Metamathematical Theorems," **BAMS 62**, 262–263.
1974 "Model Theory before 1945," *Proceedings of the Tarski Symposium*, ed. by Leon Henkin, vol. 25 of Proceedings of Symposia in Pure Mathematics (New York: American Mathematical Society), 153–172.

Veblen, Oswald
1908 "Continuous Increasing Functions of Finite and Transfinite Ordinals," **TAMS 9**, 280–292.
1908a "On the Well-Ordered Subsets of the Continuum," **CMP 25**, 235–236, 397.

Veldman, W.
See Gielen, W.; de Swart, H.; and Veldman, W.

Vesley, Richard
See Kleene, Stephen, and Vesley, Richard.

Vieler, Heinrich
1926 *Untersuchungen über Unabhängigkeit und Tragweite der Axiome der Mengenlehre in der Axiomatik Zermelos und Fraenkels* (Ph.D. dissertation: Marburg; printed at Göttingen).

Viola, Tullio
1931 "Riflessioni intorno ad alcune applicazioni del postulata della scelta di E. Zermelo e del principio di approssimazione di B. Levi nella teoria degli aggregati," **BUM 10**, 287–294.
1932 "Sul principio di approssimazione di B. Levi nella teoria della misura degli aggregati e in quella dell'integrale di Lebesgue," **BUM 11**, 74–78.
1934 "Ricerche assiomatiche sulle teorie delle funzioni d'insieme e dell'integrale di Lebesgue," **FM 23**, 75–10X.

Vitali, Giuseppe
1905 *Sul problema della misura dei gruppi di punti di una retta* (Bologna: Tip. Gamberini e Parmeggiani).

Vivanti, Giulio
1908 "Sopra alcune recenti obiezioni alla teoria dei numeri trasfiniti," **CMP 25**, 205–208.

von Neumann, John
1923 "Zur Einführung der transfiniten Zahlen," **ALS 1**, 199–208; translated in van Heijenoort 1967, 346–354.
1925 "Eine Axiomatisierung der Mengenlehre," **JM 154**, 219–240; translated in van Heijenoort 1967, 393–413.
1928 "Die Axiomatisierung der Mengenlehre," **MZ 27**, 669–752.
1928a "Über die Definition durch transfinite Induktion und verwandte Fragen der allgemeinen Mengenlehre," **MA 99**, 373–391.
1929 "Über eine Widerspruchfreiheitsfrage in der axiomatischen Mengenlehre," **JM 160**, 227–241.
1929a "Zur allgemeinen Theorie des Masses," **FM 13**, 73–116.
1961 *Collected Works*, vol. I: *Logic, Theory of Sets, and Quantum Mechanics*, ed. by A. H. Taub (Oxford: Pergamon).
See Hahn *et alii*.

Vopěnka, Petr
1967 "General Theory of ∇-Models," **CMC 8**, 145–170.

Wagner, K.
See Kinna, W., and Wagner, K.

Wallace, A. D.
1944 "A Substitute for the Axiom of Choice," **BAMS 50**, 278.

Wallman, Henry
1938 "Lattices and Topological Spaces," **AM (2) 39**, 112–126.

Wang, Hao
1949 "On Zermelo's and Von Neumann's Axioms for Set Theory," **PAS 35**, 150–155.
1970 Introduction to [Skolem 1970].
1978 "Kurt Gödel's Intellectual Development," **MI 1**, 182–185.
1981 "Some Facts about Kurt Gödel," **JSL 46**, 653–659.

Ward, L. E.
1962 "A Weak Tychonoff Theorem and the Axiom of Choice," **PAMS 13**, 757–758.

Warner, Seth
1965 *Modern Algebra*, vol. II (Englewood Cliffs, N.J.: Prentice-Hall).

Weber, Heinrich
1893 "Die allgemeinen Grundlagen der Galois'schen Gleichungstheorie," **MA 43**, 521–549.

Weyl, Hermann
1910 "Über die Definitionen der mathematischen Grundbegriffe," **MNB 7**, 93–95, 109–113.
1917 *Das Kontinuum* (Leipzig: Veit).
1968 *Gesammelte Abhandlungen*, 4 vols., ed. by K. Chandrasekharan (Berlin: Springer).

Whitehead, Alfred North
1902 "On Cardinal Numbers," **AJM 24**, 367–394.
1904 "Theorems on Cardinal Numbers," **AJM 26**, 31–32.
See Russell, Bertrand, and Whitehead, Alfred North.

Willard, Stephen
1970 *General Topology* (Reading, Mass.: Addison-Wesley).

Wilson, Edwin
1908 "Logic and the Continuum," **BAMS 14**, 432–443.

Witt, Ernst
1951 "Beweisstudien zum Satz von M. Zorn," **MN 4**, 434–438.

Wrinch, Dorothy
1923 "On Mediate Cardinals," **AJM 45**, 87–92.

Yesenin-Volpin, A. S.
1954 "Nedokazyemost gipotezy Suslina bez pomoshchi aksiomy vybora v systeme aksiom Bernaysa–Mostovskogo," **DAN 96**, 9–12; translated as [1963].
1963 "The Unprovability of Suslin's Hypothesis without the Aid of the Axiom of Choice in the Bernays–Mostowski Axiom System," **AMST (2) 23**, 83–88.

Young, Grace C.
See Young, William H., and Young, Grace C.

Young, William H.
1902 "Sets of Intervals on the Straight Line," **LMS 35**, 245–268.
1903 "Overlapping Intervals," **LMS 35**, 384–388.

Young, William H., and Young, Grace C.
1906 *The Theory of Sets of Points* (Cambridge: Cambridge University Press); reprinted (New York: Chelsea, 1972).

Zermelo, Ernst
1901 "Addition transfiniter Cardinalzahlen," **NG**, 34–38.
1904 "Beweis, dass jede Menge wohlgeordnet werden kann (Aus einem an Herrn Hilbert gerichteten Briefe)," **MA 59**, 514–516; translated in van Heijenoort 1967, 139–141.
1908 "Neuer Beweis für die Möglichkeit einer Wohlordung," **MA 65**, 107–128; translated in van Heijenoort 1967, 183–198.
1908a "Untersuchungen über die Grundlagen der Mengenlehre. I," **MA 65**, 261–281; translated in van Heijenoort 1967, 199–215.
1909 "Sur les ensembles finis et le principe de l'induction complète," **AC 32**, 185–193.
1909a "Über die Grundlagen der Arithmetik," *Atti del IV Congresso Internazionale dei Matematici, Roma* **2**, 8–11.
1914 "Über ganze transzendente Zahlen," **MA 75**, 434–442.
1929 "Über den Begriff der Definitheit in der Axiomatik," **FM 14**, 339–344.
1930 "Über Grenzzahlen und Mengenbereiche: Neue Untersuchungen über die Grundlagen der Mengenlehre," **FM 16**, 29–47.
1932 "Über Stufen der Quantifikation und die Logik des Unendlichen," **DMVA 41**, 85–88.
1935 "Grundlagen einer allgemeinen Theorie der mathematischen Satzsysteme," **FM 25**, 136–146.

Zlot, William L.
1957 *The Role of the Axiom of Choice in the Development of the Abstract Theory of Sets* (Ph.D. dissertation: Columbia University).
1959 "Some Comments on the Role of the Axiom of Choice in the Development of Abstract Set Theory," **MG 32**, 115–122.
1960 "The Principle of Choice in Pre-Axiomatic Set Theory," **SCM 25**, 105–123.

Zoretti, Ludovic
1909 "La notion de ligne," **ENS (3) 26**, 485–497.

Zorn, Max
1935 "A Remark on Method in Transfinite Algebra," **BAMS 41**, 667–670.
1944 "Idempotency of Infinite Cardinals," **UCM 2**, 9–12.

Zuckerman, Martin M.
1969 "Some Theorems on the Axioms of Choice for Finite Sets," **ZML 15**, 385–399.
1969a "On Choosing Subsets of n-Element Sets," **FM 64**, 163–179.
1971 "Multiple Choice Axioms," in Scott 1971, 447–466.

Index of Numbered Propositions

This index includes all references to a numbered proposition *by its number.* If such a proposition is sometimes cited by a name, then a cross-reference is given as well. The page where a numbered proposition is first stated appears below in italics.

General Index

This index of persons and subjects is arranged word by word, not letter by letter; consequently, "logic, algebraic" and "logic, propositional" both precede "logicist." A page number occurs in italics when the indexed term was introduced or defined on that page. Expressions in non-Roman alphabets are alphabetized as if spelled out in English; for example, "γ-set" is treated as if it were spelled "gamma-set," and "T_1" as if it were "T-one." If a term consists of more than one word (*e.g.*, absolutely infinite multitude), then the gentle reader should be prepared to search the index under each word in the term in order to find the relevant entry.

Studies in the History of Mathematics and Physical Sciences

Volume 1
A History of Ancient Mathematical Astronomy
By **O. Neugebauer**
ISBN 0-387-**06995**-X

Volume 2
A History of Numerical Analysis from the 16th through the 19th Century
By **H. H. Goldstine**
ISBN 0-387-**90277**-5

Volume 3
I. J. Bienaymé: Statistical Theory Anticipated
By **C. C. Heyde** and **E. Seneta**
ISBN 0-387-**90261**-9

Volume 4
The Tragicomical History of Thermodynamics, 1822-1854
By **C. Truesdell**
ISBN 0-387-**90403**-4

Volume 5
A History of the Calculus of Variations from the 17th through the 19th Century
By **H. H. Goldstine**
ISBN 0-387-**90521**-9

Volume 6
The Evolution of Dynamics: Vibration Theory from 1687 to 1742
By **J. Cannon** and **S. Dostrovsky**
ISBN 0-387-**90626**-6

Volume 7
The Prehistory of the Theory of Distributions
By **J. Lützen**
ISBN 0-387-**90647**-9

Volume 8
Zermelo's Axiom of Choice: Its Origins, Development, and Influence
By **G. H. Moore**
ISBN 0-387-**90670**-3